Biocomputation and Biomedical Informatics:
Case Studies and Applications

Athina Lazakidou
University of Peloponnese, Greece

MEDICAL INFORMATION SCIENCE REFERENCE

Hershey · New York

Director of Editorial Content: Kristin Klinger
Senior Managing Editor: Jamie Snavely
Assistant Managing Editor: Michael Brehm
Publishing Assistant: Sean Woznicki
Typesetter: Sean Woznicki, Mike Killian
Cover Design: Lisa Tosheff
Printed at: Yurchak Printing Inc.

Published in the United States of America by
    Medical Information Science Reference (an imprint of IGI Global)
    701 E. Chocolate Avenue
    Hershey PA 17033
    Tel: 717-533-8845
    Fax: 717-533-8661
    E-mail: cust@igi-global.com
    Web site: http://www.igi-global.com/reference

Library of Congress Cataloging-in-Publication Data

Biocomputation and biomedical informatics : case studies and applications / Athina Lazakidou, editor.
    p. cm.

  Includes bibliographical references and index.
  Summary: "This book provides a compendium of terms, definitions, and explanations of concepts, processes, and acronyms"--Provided by publisher.

  ISBN 978-1-60566-768-3 (hardcover) -- ISBN 978-1-60566-769-0 (ebook) 1.
Bioinformatics. 2. Medical informatics. 3. Computational biology. I.
Lazakidou, Athina A., 1975-
  QH324.2.B56 2010
  570.285--dc22
                        2009017384

British Cataloguing in Publication Data
A Cataloguing in Publication record for this book is available from the British Library.

All work contributed to this book is new, previously-unpublished material. The views expressed in this book are those of the authors, but not necessarily of the publisher.

Table of Contents

Detailed Table of Contents

    Kostas Bethanis, Agricultural University of Athens, Greece
    Petros Giastas, Hellenic Pasteur Institute, Greece
    Trias Thireou, Agricultural University of Athens, Greece
    Vassilis Atlamazoglou, Agricultural University of Athens, Greece

Structural genomics or structural proteomics can be defined as the quest to obtain the three-dimensional structures of all proteins. Single-crystal X-ray crystallography provides the most direct, accurate and in most of the cases the only way of forming images of macromolecules. Using crystallography, three-dimensional images have been made of thousands of macromolecules, especially proteins and nucleic acids. These give detailed information about their activity, their mechanism for recognizing and binding substrates and effectors, and the conformational changes which they may undergo. This chapter presents the basic crystallographic procedure steps and a thorough survey of the computational software used most frequently by protein X-ray crystallographers. The determination of the structure of 2[4Fe-4S] ferredoxin from *Escherichia coli.* is examined as a case study of implementation of these steps and programs. Finally, some of the perspectives of the field of computational X-ray crystallography are noted showing the future developments in the ceaseless evolution of new methods and proliferation of new programs.

    Trias Thireou, Agricultural University of Athens, Greece; National Technical University
        of Athens, Greece
    Kostas Bethanis, Agricultural University of Athens, Greece
    Vassilis Atlamazoglou, Agricultural University of Athens, Greece; National Technical
        University of Athens, Greece

This chapter presents a particular cascade of computational steps in order to build a workflow for an *in silico* protein engineering approach. In this respect, all available information, in order to choose and computationally implement mutations, is described, employed and monitored. Some of the prerequisites of *in silico* protein engineering are access to various sequence and structure molecular biology databases, software tools for three dimensional molecular visualization and manipulation, sequence and structure alignment and comparison, molecular modelling and molecular docking. The implementation of these steps is demonstrated in the context of performing mutations of particular residues on the ligand pocket of a lipocalin protein family member, to derive the desired ligand binding properties. The example chosen for inclusion introduces the reader to all of the essentials of computational protein engineering experiments. More importantly, it provides insight into understanding and properly interpreting the data produced by these methods.

Chapter 3

    Dimitri Perrin, Dublin City University, Ireland
    Heather J. Ruskin, Dublin City University, Ireland
    Martin Crane, Dublin City University, Ireland

Biological systems are typically complex and adaptive, involving large numbers of entities, or organisms, and many-layered interactions between these. System behaviour evolves over time, and typically benefits from previous experience by retaining memory of previous events. Given the dynamic nature of these phenomena, it is non-trivial to provide a comprehensive description of complex adaptive systems and, in particular, to define the importance and contribution of low-level unsupervised interactions to the overall evolution process. In this Chapter, the authors focus on the application of the agent-based paradigm in the context of the immune response to HIV. Explicit implementation of lymph nodes and the associated lymph network, including lymphatic chain structure, is a key objective, and requires parallelisation of the model. Steps taken towards an optimal communication strategy are detailed.

Chapter 4

    Mohammad Haghpanahi, Iran University of Science and Technology, Iran
    Mohammad Nikkhoo, Iran University of Science and Technology, Iran
    Habib Allah Peirovi, Shaheed Beheshti University of Medical Science and Health Services, Iran

Computer aided tissue engineering integrates advances of multidisciplinary fields of biology, biomedical engineering, and modern design and manufacturing. It enables the application of advanced computer aided technologies and biomechanical engineering principles to derive systematic solutions for complex tissue engineering problems. After an introduction to tissue engineering, this chapter presents the recent development on computer aided tissue engineering, including computer aided tissue modeling, computer aided tissue scaffold informatics and biomimetic design, and computer aided biomanufacturing.

Chapter 5

    Isabel de la Torre Díez, University of Valladolid, Spain
    Roberto Hornero Sánchez, University of Valladolid, Spain
    Miguel López Coronado, University of Valladolid, Spain
    María Isabel López Gálvez, University of Valladolid, Spain

Electronic health record (EHR) refers to the complete set of information that resides in electronic form and is related to the past, present and future health status. EHR standardization is a key characteristic to exchange healthcare information. Health Level Seven (HL7) and Digital Imaging and Communications in Medicine (DICOM) are intensively influencing this process. This chapter describes the development and experience of a web-based application, TeleOftalWeb 3.2, to store and exchange EHRs in ophthalmology. We apply HL7 Clinical Document Architecture (CDA) and DICOM standards. The application has been built on Java Servlet and Java Server Pages (JSP) technologies. EHRs are stored in the database Oracle 10g. Its architecture is triple-layered. Physicians can view, modify and store all type of medical images. For security, all data transmissions were carried over encrypted Internet connections such as Secure Sockets Layer (SSL) and HyperText Transfer Protocol over SSL (HTTPS). The application verifies the standards related to privacy and confidentiality. TeleOftalWeb 3.2 has been tested by ophthalmologists from the University Institute of Applied Ophthalmobiology (IOBA), Spain. Nowadays, more than thousand health records have been introduced.

Chapter 6

A constantly increasing number of applications from various scientific fields are finding their way towards adopting Grid technologies in order to take advantage of their capabilities: the advent of Grid environments made feasible the solution of computational intensive problems in a reliable and cost-effective way. This book chapter focuses on presenting and describing how high performance computing in general and specifically Grids can be applied in biomedicine. The latter poses a number of requirements, both computational and sharing / networking ones. In this context, we will describe in detail how Grid environments can fulfill the aforementioned requirements. Furthermore, this book chapter includes a set of cases and scenarios of biomedical applications in Grids, in order to highlight the added-value of the distributed computing in the specific domain.

Chapter 7

With an increasingly mobile society and the worldwide deployment of mobile and wireless networks, the wireless infrastructure can support many current and emerging healthcare applications. This could fulfill the vision of "Pervasive Healthcare" or healthcare to anyone, anytime, and anywhere by removing locational, time and other restraints while increasing both the coverage and the quality. In this chapter the authors present applications and requirements of pervasive healthcare, wireless networking solutions and several important research problems. The pervasive healthcare applications include pervasive health monitoring, intelligent emergency management system, pervasive healthcare data access, and ubiquitous mobile telemedicine. On top of the valuable benefits new technologies enable the memory loss patients for independent living and also reduce the cost of family care-giving for memory loss and elder patients.

Chapter 8

Stelios Zimeras, University of Aegean, Greece

Anastasia N. Kastania, Athens University of Economics & Business, Greece

In recent years, biological research has been witness of a sea change mainly spearheaded by the advent of novel high throughput technologies that can provide unprecedented amounts of valuable data. This has given rise to novel field sharing the popular suffix 'omics'. Genomics/transcriptomics, proteomics, metabolomics, interactomics/regulomics and numerous other terms have been coined to categorize this ever increasing number of new fields. Biomarkers comprise the most critical tools for the early detection, diagnosis, prognosis and prediction of diseases providing key clues for drug development processes. A significant challenge is to define appropriate levels of specificity and sensitivity of new biomarkers in detecting complex diseases. The establishment of new biomarkers is not only an issue of optimizing wet lab experiments but also of designing appropriate and robust data analysis methods. Various approaches, like multivariate analysis methods as well as standard statistical tests have been applied to search for the important features in 'omics' data. Likewise, several methods, e.g. FDA, SVM, CART, nonparametric kernels, kNN, boosted decision stump and genetic algorithms, have been reported. However, it still remains an unsolved challenge to analyze and interpret the enormous volumes of 'omics' data.

Chapter 9

Computational Analysis and Characterization of Marfan Syndrome Associated

K. Sivakumar, Sri Chandrasekharendra Saraswathi Viswa Maha Vidyalaya University, India

Novel computational procedures and methods have been used to analyze, characterize and to provide more detailed definition of some Marfan syndrome associated human Fibrillin 1 proteins retrieved from NCBI Entrez protein database. Primary structure analysis reveals that the Marfan syndrome associated proteins are rich in cysteine and glycine residues. Extinction Coefficients of Marfan syndrome associated proteins at 280nm is ranging from 1490 to 259165 M^{-1} cm^{-1}. Expasy's ProtParam classifies most of the Marfan syndrome associated human Fibrillin 1 proteins as unstable on the basis of Instability index (II>40) and few proteins (AAB25244.1, 1EMO_A, Q504W9) as stable (II<40) proteins in the room temperature. The aliphatic index infers that the Fibrillin 1 proteins may become unstable at high temperature. GRAVY index of all the proteins indicates that all these proteins may interact equally and easily with water. The number of basic and acidic amino acids in each Marfan syndrome associated human Fibrillin 1 proteins correlates well with the corresponding pI computed. Secondary structure analysis shows that human Fibrillin 1 proteins are found to be with mixed secondary structural content. The average molecular weight of Marfan syndrome associated proteins calculated is 134086 Da. Scan-prosite server identified EGF-like domain, TGF-beta binding domain and extracellular sushi domain profiles in Marfan syndrome associated proteins.

In this chapter, an automatic determination algorithm for nuclear magnetic resonance (NMR) spectra of the metabolites in the living body by magnetic resonance spectroscopy (MRS) without human intervention or complicated calculations is presented. In such method, the problem of NMR spectrum determination is transformed into the determination of the parameters of a mathematical model of the NMR signal. To calculate these parameters efficiently, a new model called modified Hopfield neural network is designed. The main achievement of this chapter over the work in literature (Morita, N. and Konishi, O., 2004) is that the speed of the modified Hopfield neural network is accelerated. This is done by applying cross correlation in the frequency domain between the input values and the input weights. The modified Hopfield neural network can accomplish complex dignals perfectly with out any additinal computation steps. This is a valuable advantage as NMR signals are complex-valued. In addition, a technique called "modified sequential extension of section (MSES)" that takes into account the damping rate of the NMR signal is developed to be faster than that presented in (Morita, N. and Konishi, O., 2004). Simulation results show that the calculation precision of the spectrum improves when MSES is used along with the neural network. Furthermore, MSES is found to reduce the local minimum problem in Hopfield neural networks. Moreover, the performance of the proposed method is evaluated and there is no effect on the performance of calculations when using the modified Hopfield neural networks.

Chapter 11

The array of tools available to the medical practitioner for diagnosis of disease has experienced extremely rapid expansion over the past decades. Traditional "blood chemistries" and hematological testing have been augmented with immunoassays for serological testing and PCR-based assays for genomic screening. Rapid, inexpensive point-of-care assays with enhanced sensitivity and specificity have the potential for altering the manner in which medicine is practiced; pharmacogenomics and the advent of "personalized medicine" permit the tailoring of therapeutic pharmacologic regimens to the genetic makeup of an individual. Facilitating this are novel biosensing approaches for *in vitro* diagnostics, developed at the interface of engineering, physics, chemistry and biology. New discoveries promise to sustain the high rate of growth of this important field of research and development. This chapter examines recent advances in techniques for biosensing and *in vitro* biomedical diagnostics, building on progress in materials science, nanotechnology, semiconductor devices, and biotechnology. The importance of this topic is motivated through the presentation of case studies of biosensing applications within various medical specialties.

Chapter 12

Medical infrared imaging captures the temperature distribution of the human skin and is employed in various medical applications. Unfortunately, many of the conventional and commercial suites for image processing provide only very basic tools for the processing of medical thermal images which represent a challenging combination of both functional and morpho-structural imaging. In this chapter we discuss several more advanced approaches which in turn provide tremendous help to the clinician. As an example, it is often useful to cross-reference thermograms with visual images of the patient, either to see which part of the anatomy is affected by a certain disease or to judge the efficacy of the treatment. We show that image registration techniques can be effectively used to generate an overlay of visual and thermal images to provide a useful diagnostic visualisation. Image registration can also be performed based on two thermograms and we present a warping-based method for this. Segmenting the background from the foreground (i.e., the patient) is a crucial task and we highlight how this can be accomplished. Finally we show how descriptors, extracted from medical infrared images, can be usefully employed to search through a large database of cases as well as to aid in diagnosis.

Chapter 13

Diabetic retinopathy is recognised as one of the most common causes of blindness. Early diagnosis is important and is based on detection of features such as exudates during eye fundus image screening. In this chapter we show how areas corresponding to exudates can be automatically detected using a neural network that, following contrast enhancement and vessel and optic disc extraction steps, classifies each image pixel as exudate or non-exudate. Experimental results on an image set with known ground truth verify the usefulness of the presented approach.

Chapter 14

In this chapter the authors report about their experiences in designing, implementing, prototyping and evaluating a system for computer aided risk estimation of breast cancer. The strategy and architecture of "Hippocrates-mst" along with its functionalities are going to be presented. Also, the evaluation results in the clinical practice concerning the performance of "Hippocrates-mst" in the "Ippokrateio" University Hospital of Athens will be presented. The feedback from medical experts along with the new features of the system that are under development will be discussed.

Chapter 15

Vasileios G. Stamatopoulos, Biomedical Research Foundation of the Academy of Athens, Greece
George E. Karagiannis, Royal Brompton and Harefield NHS Trust, UK
Michael A. Gatzoulis, Royal Brompton and Harefield NHS Trust, UK
Anastasia N. Kastania, Biomedical Research Foundation of the Academy of Athens, Greece
Sophia Kossida, Biomedical Research Foundation of the Academy of Athens, Greece

This chapter presents the feasibility study of a virtual platform for medical related technology transfer, continuing medical education and e-conference. The concept extends the idea of live events (e.g. conferences, open day events) in one physical location. It exploits the creation of a virtual platform where the research world in the area of biomedicine, can showcase their success, interact and co-operate with the business community and collaborate on potentially valuable outcomes and learn without time or place restrictions. The main objective of the project was to offer a pilot service that can showcase the e-OpenDay market potential and technical feasibility. By developing a prototype and through user feedback and evaluation processes, a set of services was identified, developed and validated. The e-OpenDay project made clear that health information services are facing rapid development and expansion to wider markets and user groups. Based on the project results, a business plan was developed that showcased potential in commercial exploitation.

Chapter 16

Dolores A. Steinman, University of Toronto, Canada
David A. Steinman, University of Toronto, Canada

In the following chapter we will discuss the development of medical imaging and, through specific case studies, its application in elucidating the role of fluid mechanical forces in cardiovascular disease development and therapy (namely the connection between flow patterns and circulatory system disease - atherosclerosis and aneurysms) by means of computational fluid dynamics (CFD). The research carried in the Biomedical Simulation Laboratory can be described as a multi-step process through which, from the reality of the human body through the generation of a mathematical model that is then translated into a visual representation, a refined visual representation easily understandable and used in the clinic is generated. Thus, our daily research generates virtual representations of blood flow that can serve two purposes: a) that of a model for a phenomenon or disease or b) that of a model for an experiment (non-invasive way of determining the best treatment option).

Chapter 17

A. Maffezzoli, Università degli Studi di Milano-Bicocca, Italy
E. Wanke, Università degli Studi di Milano-Bicocca, Italy

In the present chapter authors want to expose new insights in the field of Computational Neuroscience at regard to the study of neuronal networks grown in vitro. Such kind of analyses can exploit the availability of a huge amount of data thanks to the use of Multi Electrode Arrays (MEA), a multi-channel technology which allows capturing the activity of several different neuronal cells for long time recordings. Given the possibility of simultaneous targeting of various sites, neuroscientists are so applying such recent technology for various researches. The chapter begins by giving a brief presentation of MEA technology and of the data produced in output, punctuating some of the pros and cons of MEA recordings. Then we present an overview of the analytical techniques applied in order to extrapolate the hidden information from available data. Then we shall explain the approach we developed and applied on MEAs prepared in our cell culture laboratory, consisting of statistical methods capturing the main features of the spiking, in particular bursting, activity of various neuron, and performing data dimensionality reduction and clustering, in order to classify neurons according to their spiking properties having showed correlated features. Finally the chapter wants to furnish to neuroscientists an overview about the quantitative analysis of in-vitro spiking activity data recorded via MEA technology and to give an example of explorative analysis applied on MEA data. Such study is based on methods from Statistics and Machine Learning or Computer Science but at the same time strictly related to neurophysiological interpretations of the putative pharmacological manipulation of synaptic connections and mode of firing, with the final aim to extract new information and knowledge about neuronal networks behavior and organization.

Chapter 18

Theodor Panagiotakopoulos, University of Patras, Greece
Maria-Anna Fengou, University of Patras, Greece
Dimitrios Lymberopoulos, University of Patras, Greece
Eduard Babulak, University of the South Pacific, Fiji

We have already moved away from traditional desktop-based computer technologies towards ubiquitous computing environments that progressively exist in our daily activity. This chapter introduces the concept of ubiquitous computing in the domain of healthcare as well as the prevalent technology its implementation depends on. This technology, named context-awareness, and a generic system for its realization are comprehensively described. Furthermore, we outline the main services that a context-aware system can provide and concluding we discuss the impact of ubiquitous computing in the healthcare domain. We aim at providing an overview of the technological proceedings in this area and through this understanding assist researchers to their brainstorming.

Chapter 19

Konstantinos Siassiakos, University of Piraeus, Greece
Athina Lazakidou, University of Peloponnese, Greece

Privacy includes the right of individuals and organizations to determine for themselves when, how and to what extent information about them is communicated to others. The growing need of managing large amounts of medical data raises important legal and ethical challenges. E-Health systems must be capable of adhering to clearly defined security policies based upon legal requirements, regulations and standards while catering for dynamic healthcare and professional needs. Such security policies, incorporating enterprise level principles of privacy, integrity and availability, coupled with appropriate audit and control processes, must be able to be clearly defined by enterprise management with the understanding that such policy will be reliably and continuously enforced. This chapter addresses the issue of identifying and fulfilling security requirements for critical applications in the e-health domain. In this chapter the authors describe the main privacy and security measures that may be taken by the implementation of e-health projects.

Preface

Biomedical Informatics is the scientific field that deals with the storage, retrieval, sharing, and optimal use of biomedical information, data, and knowledge for problem solving and decision making. It touches on all basic and applied fields in biomedical science and is closely tied to modern information technologies, notably in the areas of computing and communication. Biomedical informatics applications may be used for decision support, quality assurance, assistance in research studies, and resource allocation.

Biomedical Imaging Informatics defines the role of medical imaging and related technologies within the context of medical informatics decision support and improvement of patient care and outcome. The scope of biomedical imaging covers data acquisition, image reconstruction, and image analysis, involving theories, methods, systems, and applications. While tomographic and post-processing techniques become increasingly sophisticated, traditional and emerging modalities play more and more critical roles in anatomical, functional, cellular, and molecular imaging. The overall goal of "Biocomputation and Biomedical Informatics: Case Studies and Applications" is to promote research and development of biomedical imaging by publishing high-quality research articles and reviews in this rapidly growing interdisciplinary field.

Private and public research efforts worldwide are developing nanoproducts aimed at improving health care and advancing medical research. Some of these products have entered the marketplace, more are on the verge of doing so, and others remain more a vision that a reality. The potential for these innovations is enormous, but questions remain about their long-term safety and the risk–benefit characteristics of their usage.

Comparative analysis of genomes allows the rich source of biological genome sequence data to be most efficiently exploited. However, the rate at which microbial genomes are being sequenced is increasing rapidly. Soon the volume of data will put comparative analyses beyond the capability of the computing resources of most individual laboratories. The advent of Grid technology promises to provide resources for computation, data integration and collaboration in a way that is not addressed in current distributed computing technologies. Grid computing has therefore been identified as having major potential benefits for bioinformatics, particularly in the area of genome analysis and comparative genomics.

Improvements in healthcare delivery in recent years are rooted in the continued industry-wide investment in information technology and the expanding role of medical informatics. The main goal of this new publication is to provide innovative and creative ideas for improving communication environments in health and to explore all new technologies in medical informatics and health care delivery systems.

"Biocomputation and Biomedical Informatics: Case Studies and Applications" provides a compendium of terms, definitions and explanations of concepts, processes and acronyms. Additionally, this volume features short papers authored by leading experts offering an in-depth description of key terms and concepts related to different areas, issues and trends in various areas of **Biocomputation, Bioinformatics, and Biomedical Technologies**.

The topics of this book cover useful areas of general knowledge including Information and Communication Technologies related to Health, New Developments in Distributed Applications and Interoperable Systems, Applications and Services, **Biocomputation, Bioinformatics, and Biomedical Technologies**, Software environments for Biocomputation, Bioinformatics, and Biomedical Applications, Bioimaging and Biosensing Applications, Biocomputation and Knowledge Management in Drug Discovery and Developmens, Key Aspects, Components and Applications of Systems Biology, Biocomputation and Nanotechnology for Personalized Medicine, Data and Knowledge Mining in Biomedical Research, Modelling and Simulation in Biomedical Research, Computer Applications in Biomedicine, Health Care and Medicine.

This book is an excellent source of comprehensive knowledge and literature on the topic of Distributed Health, Biocomputation and Biomedical Informatics.

All of us who worked on the book hope that readers will find it useful.

Athina A. Lazakidou, Ph.D
Editor

Chapter 1
Macromolecular Crystallographic Computing

Kostas Bethanis
Agricultural University of Athens, Greece

Petros Giastas
Hellenic Pasteur Institute, Greece

Trias Thireou
Agricultural University of Athens, Greece

Vassilis Atlamazoglou
Agricultural University of Athens, Greece

ABSTRACT

Structural genomics or structural proteomics can be defined as the quest to obtain the three-dimensional structures of all proteins. Single-crystal X-ray crystallography provides the most direct, accurate and in most of the cases the only way of forming images of macromolecules. Using crystallography, three-dimensional images have been made of thousands of macromolecules, especially proteins and nucleic acids. These give detailed information about their activity, their mechanism for recognizing and binding substrates and effectors, and the conformational changes which they may undergo. This chapter presents the basic crystallographic procedure steps and a thorough survey of the computational software used most frequently by protein X-ray crystallographers. The determination of the structure of 2[4Fe-4S] ferredoxin from Escherichia coli. is examined as a case study of implementation of these steps and programs. Finally, some of the perspectives of the field of computational X-ray crystallography are noted showing the future developments in the ceaseless evolution of new methods and proliferation of new programs.

INTRODUCTION

Macromolecules are the principal non-aqueous components of living cells. Among the macromolecules (proteins, nucleic acids, and carbohydrates), proteins are the largest group. Enzymes are the most diverse class of proteins because nearly every chemical reaction in a cell requires a specific enzyme. To understand cellular processes, knowledge of

DOI: 10.4018/978-1-60566-768-3.ch001

the three-dimensional structure of enzymes and other macromolecules is vital. Two techniques are widely used for the structural determination of macromolecules at atomic resolution: X-ray diffraction of crystals and nuclear magnetic resonance (NMR). While NMR does not require crystals and provides more detailed information on the dynamics of the model in question, it can be used only for biopolymers with a molecular weight less than 20,000. X-ray crystallography can be applied to compounds with molecular weight up to at least 10^6. For many proteins, the difference is decisive in favour of X-ray diffraction (Drenth J., 1994). The pioneering work by Perutz and Kendrew on the structure of hemoglobin and myoglobin in the 1950's led to a slow but steady increase in the number of proteins whose structure was determining using X-ray diffraction. The introduction of sophisticated computer hardware and software dramatically reduced the time required to determine a structure while increasing the accuracy of the results. In recent years, recombinant DNA technology has further stimulated interest in protein structure determination. A protein that was difficult to isolate in sufficient quantities from its natural source can often be produced in arbitrarily large amounts using expression of its cloned gene in a microorganism. Also, a protein modified by site-directed mutagenesis of its gene can be created for scientific investigation and industrial application. Here, X-ray diffraction plays a crucial role in guiding the molecular biologist to the best amino acid positions for modification. Moreover, it is often important to learn what effect a change in a protein's sequence will have on its three-dimensional structure. Chemical and pharmaceutical companies have become very active in the field of protein structure determination because of their interest in protein and drug design.

As of January 2008, the Protein Data Bank (PDB) (Berman H. M., *et al.*, 2000), the world's largest repository of macromolecular models obtained from experimental data (called *experimental* models), contains more than 40,000 protein and nucleic-acid models determined by X-ray crystallography (Berman H. M., 2008). However, it should be noted that because many proteins appear in multiple forms -for example, wild types and mutants, or solo and also as part of protein-ligand or multiprotein complexes- the number of unique proteins represented in the PDB is only a fraction of the total number of models. In addition, the PDB holds roughly 4500 models, mostly proteins of fewer than 200 residues, that have been solved by NMR spectroscopy, which provides a model of molecule in solution, rather than in crystalline state. Finally, there are *theoretical* models, either built by analogy with the structure of known proteins having similar sequence, or based on simulations of protein folding (see chapter 'Describing methodology and applications of an *in silico* protein engineering approach'). Theoretical models are also available from databases other than PDB. All methods of obtaining models have their strengths and weaknesses, and they coexist happily as complementary methods (Rhodes G., 2006).

SINGLE CRYSTAL X-RAY CRYSTALLOGRAPHY: THE BASIC STEPS

Crystallography provides the most direct way of forming three-dimensional images of molecules. The most common experimental means of obtaining a detailed model of a large molecule, allowing the resolution of individual atoms, is to interpret the diffraction of X-rays from many identical molecules in an ordered array like a crystal. This method is called *single-crystal X-ray crystallography*. The three-dimensional structures of macromolecules, especially proteins and nucleic acids, give detailed information about their activity, their mechanism for recognizing and binding substrates and effectors, and the conformational changes which they may undergo. They show graphically the evolutionary relationships between molecules from widely separated systems and they give a wide view of the resemblances between different proteins, showing strong links in three-dimensional structure where the relationship between amino-acid sequences has dwindled to insignificance (Blow D., 2002).

The determination of a novel protein structure by single-crystal X-ray crystallographic analysis involves the following basic steps:

Cloning - Expression - Purification

Converting sequence information to biological reagents - choosing proper expression constructs, which includes purifying proteins rapidly and obtaining excellent structural samples - remains a significant problem for the production of proteins that feed the crystallographers and NMR spectroscopists. Thousands of proteins are expected to be produced in the next few years. The purified proteins will be valuable reagents for the entire research community (Edwards *et al.*, 2000).

Crystallization

Crystallization is often the rate limiting step in protein crystallography. There are many methods to crystallize biological macromolecules all of which aim at bringing the solution of macromolecules to a supersaturation state (for reviews see Ducruix, A., and Giece, R., 1999). Several methods of crystallization are now well established but application of these methods is still very much trial and error. Crystallization of a newly isolated protein can take weeks, months or years.

Data Collection

Once suitable protein crystals become available, X-ray diffraction data are collected. One of several procedures may be adopted for data collection. Most single crystal diffraction data are measured using monochromatic X-rays from either a rotating anode generator or from a synchrotron source.

Data Analysis

Data Reduction

X-ray data can be collected with 0, 1 and 2-dimensional detectors, 0-d (single counter) being the simplest and 2-d the most efficient in terms of measuring diffracted X-rays in all directions. Two-dimensional detectors have been used from the very beginning of X-ray diffraction studies, the year 1912. Initially the 2-d detector was made of X-ray sensitive film. At present, electronic and IP (phospholuminescent, best known by the trade name Image Plate) detectors dominate. To analyze single-crystal diffraction data collected with these detectors several computer programs have been developed. The 2-d detectors and related software are now used predominantly to measure and integrate diffraction from single crystals of biological macromolecules.

The goal of data collection is a set of consistently measured, indexed intensities for as many reflections as possible. After data collection, the raw intensities must be processed to improve their consistency and to maximize the number of measurements that are sufficiently accurate to be used.

In order to process the data, a crystallographer must first index the reflections within the multiple images recorded. This means identifying the dimensions of the unit cell and which image peak corresponds to which position in reciprocal space. *Indexing* is generally accomplished using an autoindexing routine. A byproduct of indexing is to determine the symmetry of the crystal, i.e., its space group. Having assigned symmetry, the data is then integrated. This converts the hundreds of images containing the thousands of reflections into a single file, consisting of records of the Miller index of each reflection, and the intensity for each reflection with the corresponding error estimation. Peaks that appear in two or more images have to be *merged* and finally *scaled* so that they have a consistent intensity scale. *Optimizing* the intensity scale is critical because the relative intensity of the peaks is the key information from which the structure is determined. Scaling and post-refinement are the final stages of such a procedure.

Initial Phasing

In a diffraction experiment one can only measure the intensities and diffraction angles of the diffracted beams. All information about the phases of the diffracted X-rays is lost. This phase information along with the amplitudes of the diffracted X-rays is essential for the solution of crystal structures and must be recovered. This is known as the phase problem (Taylor, G., 2003).

There are four approaches which may be taken in recovering phase information:

Ab Initio Phasing

Ab initio methods for solving the crystallographic phase problem rely on diffraction amplitudes alone and do not require prior knowledge of any atomic positions. Small-molecule structures of up to 100 unique atoms (not included hydrogen) are routinely solved *ab initio* using the so-called 'direct methods'. These direct methods are based on the positivity and atomicity of electron density which through a sophisticated probability theory and the assumption of approximately equal, resolved atoms lead to phase estimations from the measured intensities (Hauptman H. A, 1991). In keeping with recent practice, we will use the term 'direct methods' to refer to methods for solving the phase problem using probability theory and '*ab initio*' for methods that employ native data only, without the use of phase information from isomorphous derivatives or from anomalous scattering. Recent advances in *ab intio* direct methods (Hauptman H. A, 1997; Uson I. & Sheldrick G.M., 1999; Burla M. C., *et al.*, 2007) have enabled the solution of crystal structures of small proteins provided that the data are available to atomic resolution (<1.4 Å, for proteins this corresponds to very high resolution data).

Direct Methods are also proving to be useful for locating the heavy atoms of prepared isomorphous derivatives and the selenium atoms or other anomalous scatterers in the multiple wavelength anomalous diffraction phasing of large proteins at lower resolution (see the paragraphs 'Single or Multiple Isomorphous Replacement' and 'Anomalous X-ray scattering' below)

Molecular Replacement

When a homology model is available, it can be used as a search model in molecular replacement to determine the orientation and position of the molecules within the unit cell using methods first described by M. Rossmann and D. Blow (1962). As a rule of thumb a sequence identity > 25% is normally required and an r.m.s deviation of <2.0Å between the αC atoms of the model and the final new structure, although there are exceptions to this. Patterson methods are usually used to obtain first the orientation of the model in the new unit cell and then the translation of the correctly oriented model relative to the origin of the unit cell.

Single or Multiple Isomorphous Replacement (SIR or MIR)

If electron-dense metal atoms can be introduced into the crystal, direct methods or Patterson-space methods can be used to determine their location and to obtain initial phases. Typically, a crystallographer can introduce such heavy atoms either by soaking the crystal in a heavy atom-containing solution, or by co-crystallization (growing the crystals in the presence of a heavy atom). The created isomorphous heavy-atom derivatives (same unit cell, same orientation of protein in cell) give rise to measurable intensity changes which can be used to deduce the positions of the heavy atoms.

Isomorphous replacement has several problems: non-isomorphism between crystals (unit-cell changes, reorientation of the protein, conformational changes, changes in salt and solvent ions), problems in locating all the heavy atoms positions, occupancies and thermal parameters and errors in intensity measurements. Although it is the original method by which protein crystal structures were solved, it has largely been superseded by MAD phasing which is described in the following paragraph.

Anomalous X-Ray Scattering (Multi-Wavelength or Single-Wavelength Anomalous Diffraction - MAD or SAD Phasing)

The X-ray wavelength may be scanned past an absorption edge of an atom, which changes the scattering in a known way. In *MAD* case, by recording full sets of reflections at three different wavelengths (far below, far above and in the middle of the absorption edge) one can solve for the substructure of the anomalously diffracting atoms and thence the structure of the whole molecule. As in MIR phasing, the changes in the scattering amplitudes can be interpreted to yield the phases. The most popular method of incorporating anomalous scattering atoms into proteins is to express the protein in a methionine auxotroph (a host incapable of synthesising methionine) in a media rich in Seleno-methionine, which contains Selenium atoms (Ealick, S.E., 2000). A *MAD* experiment can then be conducted around the absorption edge, which should then yield the position of any methionine residues within the protein, providing initial phases.

It is becoming increasingly possible to collect data at just a single wavelength, typically at the absorption peak, and use density-modifcation protocols to break the phase ambiguity and provide interpretable maps. This is the *SAD* method described by Dodson E. (2003).

Model Building and Phase Refinement

Density Modification

It is rare that experimentally determined phases are sufficiently accurate to give a completely interpretable electron-density map. Experimental phases are often only the starting point for phase improvement using a variety of methods of density modification, which are also based on some prior knowledge of structure. Solvent flattening, histogram matching and non-crystallographic averaging are the main techniques used to modify electron density and improve phases.

Solvent flattening is a powerful technique that removes negative electron density and sets the value of electron density in the solvent regions to a typical value of 0.33 e $Å^{-3}$, in contrast to a typical protein electron density of 0.43 e $Å^{-3}$. Automatic methods are used to define the protein-solvent boundary, first developed by Wang (1985) and then extended into reciprocal space by Leslie (1988).

Histogram matching alters the values of electron-density points to concur with an expected distribution of electron-density values.

Non-crystallographic symmetry averaging imposes equivalence on electron-density values when more than one copy of a molecule is present in the asymmetric unit.

Density modification is often a cyclic procedure, involving back-transformation of the modified electron-density map to give modified phases, recombination of these phases with the experimental phases (so as not to throw away experimental reality) and calculation of a new map which is then modified and so the cycle continues until convergence. Density-modification techniques will not turn a bad map into a good one, but they will certainly improve promising maps that show some interpretable features.

Structure Refinement

The interpretation of the map produces an atomic model of the unit-cell contents. Given a model of some atomic positions, these positions and their respective Debye-Waller B-factors (accounting for the thermal motion of the atom) can be refined to optimize its agreement with the observed diffraction data, ideally yielding a better set of phases.

A new model can then be fit to the new electron density map and a further round of refinement is carried out. This continues until the correlation between the diffraction data and the model is maximized.

Validation of Protein-Structure Coordinates

As in all scientific measurements, the parameters that result from a macromolecular structure determination by X-ray crystallography (*e.g.* atomic coordinates and *B* factors) will have associated uncertainties. These arise not only from systematic and random errors in the experimental data but also in the interpretation of those data. Currently, the uncertainties cannot easily be estimated for macromolecular structures due to the computer- and memory intensive nature of the calculations required (Tickle I. J. *et al.*, 1998). Thus, more indirect methods are necessary to assess the reliability of different parts of the model, as well as the reliability of the model as a whole. Among these methods are those which rely on checking only the stereochemical and geometrical properties of the model itself, without reference to the experimental data (MacArthur M. W. *et al.*, 1994; Laskowski R. A. *et al.*, 1998).

Possibly the most telling and useful of the 'quality' indicators for a protein model is the Ramachandran plot of residue φ–ψ torsion angles. This can often detect gross errors in the structure (Kleywegt G. J. & Jones T. A., 1996*a,b*). In the original Ramachandran plot (Ramachandran G. N. *et al.*, 1963; Ramakrishnan C. & Ramachandran G. N, 1965), the 'allowed' regions were defined on the basis of simulations of dipeptides. Another parameter that seems to be a particularly sensitive measure of quality is the standard uncertainty (s.u.) of the χ torsion angles. Morris *et al.* (1992) found that the average values of a protein's χ_1 and χ_2 torsion angles are well correlated with the resolution at which the protein structure is solved.

Visualization

Visualization of the atomic coordinate data obtained from a crystallographic study is a necessary step in the analysis and interpretation of the structure. The scientist may use visualization for different purposes, such as obtaining an overview of the structure as a whole, or studying particular spatial relationships in detail. Different levels of graphical abstraction are therefore required. In some cases, the atomic details need to be visualized, while in other cases, high-level structural features must be displayed. In the study of protein 3D structures in particular, there is an obvious need to visualize structural features at a level higher than atomic. A common graphical 'symbolic language' has evolved to represent schematically hydrogen-bonded repetitive structures (secondary structure) in proteins. Cylinders or helical ribbons are used for α-helices, while arrows or ribbons show strands in β-sheets.

Computer programs for molecular modeling provide an interactive, visual environment for displaying and exploring models. The fundamental operation of computer programs for studying molecules is producing vivid and understandable displays – convincing images of models. Although the details of programming for graphics displays vary from one program to another, they all produce an image according to the same geometric principles. Functions such as side-chain placement, loop, ligand and fragment fitting, structure comparison, analysis and validation are routinely performed using molecular graphics. Lower resolution (d_{min} worse than 2.5 Å) data in particular need interactive fitting.

Deposition of the Structure

Once the model of a molecule's structure has been finalized (analysed and verified), it is often deposited in a crystallographic database such as the Protein Data Bank (PDB; for protein structures) or the Cambridge Structural Database (CSD; for small molecules).

The deposition is in the form of lists of atomic coordinates, which can be used to display and study the molecule with molecular graphics programs. The great majority of models are available through the PDB. The PDB structure files, which are called *atomic coordinate entries*, can be read within editor or word-processor programs and contain in addition to the coordinate list, a *header* or opening section with information about published papers on the protein, details of experimental work that produced the structure and other useful information. More on PDB file contents and other available information, as well as tutorials on using the PDB, can be found at their web site http://www.rcsb.org/pdb.

The analysis of the vast wealth of structural and functional macromolecular information contained in Protein Data Bank has been the subject of considerable work in order to advance knowledge beyond the collection of molecular coordinates (T.J. Oldfield, 2002; see also section 'Methodology of computational protein modification' in chapter 'Describing methodology and applications of an *in silico* protein engineering approach')

If the molecule under study plays an important biological or medical function, an intensely interested audience awaits the crystallographer's final molecular model. The audience includes researchers studying the same molecule by other methods, such as spectroscopy or kinetics, or studying metabolic pathways or diseases in which the molecule is involved. The model may serve as a basis for understanding the properties of the protein and its behaviour in biological systems. It may also serve as a guide to the design of inhibitors or to engineering efforts to modify its function by methods of molecular biology.

A SURVEY OF MACROMOLECULAR CRYSTALLOGRAPHY PROGRAMS IN WIDE USE

This section presents a survey of the computational software used most frequently by protein X-ray crystallographers in the structure determination of proteins and nucleic acids. Brief annotations on some of the most popular or frequently used programs in the crystallographic community are provided. Program descriptions have been liberally copied from those provided by the program authors where possible. For a more complete or comprehensive information the reader is referred to Part 25: '*Macromolecular crystallography programs*' of volume F of International Tables for Crystallography (Editors Rossmann, M. G., and Arnold. E, 2001). The reader is also referred to http://www.iucr.org/sincristop/logiciel/, which contains a compilation of a broad range of pro grams and software systems in crystallography, structural biology and molecular biology.

The program summaries are grouped according to the steps and methods described in the previous section 'Single-crystal x-ray crystallography: The basic steps' into the following categories and subcategories:

- Data Acquisition
- Phase determination and structure solution
  - Packages of crystallographic programs
  - Ab initio phasing
  - Molecular replacement
  - Isomorphous replacement (SIR-MIR), Anomalous X-Ray Scattering (MAD-SAD)
- Model building and refinement
  - Density-map modification
  - Structure refinement
  - Graphics and Model building
- Structure analysis and verification

Data Acquisition

HKL2000

The *HKL* suite (Otwinowski, Z. and Minor, W., 1997) is a package of programs intended for the analysis of X-ray diffraction data collected from single crystals, and consists of three programs: *XdisplayF* for visualization of the diffraction pattern, *Denzo* for data reduction and integration, and Scalepack for merging and scaling of the intensities obtained by Denzo or other programs.

The four most important recent developments in the data analysis of macromolecular diffraction measurements are autoindexing, profile fitting, transformation of data to a reciprocal-space coordinate system, and the demonstration that a single oscillation image contains all of the information necessary to derive the diffraction intensities from that image. The analysis and reduction of single crystal diffraction data consists of seven major steps. These are:

- **Denzo & XdisplayF**
  1. Visualization and preliminary analysis of the original, unprocessed, diffraction pattern.
  2. Indexing of the diffraction pattern.
  3. Refinement of the crystal and detector parameters.
  4. Integration of the diffraction maxima.
- **Scalepack**
  5. Finding the relative scale factors between measurements.
  6. Precise refinement of crystal parameters using the entire data set.
  7. Merging and statistical analysis of the measurements related by space group symmetry.

(Location: http://www.hkl-xray.com/. Operating systems: SGI, DEC Alpha, SUN and HP-UX. Type: binary. Distribution: commercial.)

MOSFLM

The *MOSFLM* suite of programs is designed to facilitate processing of rotation data collected on either image plate or film. The suite originates from the MOSCO system developed in Cambridge by Nyborg and Wonacott (Nyborg & Wonacott, 1977) but it has been extensively developed since that early version, primarily at Imperial College by A.J. Wonacott, P. Brick and A.G.W. Leslie, and more recently at LMB (Leslie, 1992). All necessary processing steps are now performed by a single program which incorporates routines for indexing, refinement, integration and display results. The basic procedure for data processing is independent of the type of detector (film, image plate or CCD) although there are a number of useful features which are only available for image plate data (particularly automatic updating of cell parameters and crystal orientation).

MOSFLM performs the actual integration of the reflection intensities. It generates the reflection list, reads the digitised image, integrates the spots and writes the intensities and standard deviations into the general file and mtz file. The image plate version has the additional capability of partially recorded reflections in the same manner as the *POSTCHK* program. The program can be run interactively making use of the graphical output options which is the most useful when first characterising a new crystal or when dealing with pathological cases. It has two main applications: (1) determination of crystal orientation, cell parameters and possible space group; and (2) autoindexing of images, generation of reflection lists and integration of diffraction spots.

The following steps are repeated until all diffraction images have been processed:

i. Initial refinement of detector parameters
ii. Location of reflections in outer region of the detector and further refinement
iii. Extracting the measurement boxes from the digitised image

iv. Post-refinement of crystal orientation, cell parameters and beam parameters

v. Formation of the standard profiles

vi. Reflection integration

vii. Orientation refinement by pattern matching

MOSFLM is distributed as part of the *CCP4* suite and runs on multiple platforms. Location: ftp://ftp.mrc-lmb.cam.ac.uk/. Operating systems: UNIX, VAX/VMS and WINDOWS. Type: source code and binary. Distribution: free academic.

XDS

The program package *XDS* (X-ray Detector Software) (Kabsch, 1988*a,b*, 1993) has been developed for the reduction of single-crystal diffraction data recorded on a planar detector by the rotation method using monochromatic X-rays. It includes a set of five programs:

1. *XDS* accepts a sequence of adjacent, non-overlapping rotation images from a variety of imaging plate, CCD and multiwire area detectors and produces a list of corrected integrated intensities of the reflections occurring in the images. The program assumes that each image covers the same positive amount of crystal rotation and that rotation axis, incident beam and crystal intersect at one point, but otherwise imposes no limitations on detector position, or directions of rotation axis and incident beam, or on the oscillation range covered by each image.

2. *XPLAN* provides information for identifying the optimal rotation range for collecting data. Based on detector position and unit-cell orientation obtained from evaluating one or a few rotation images using *XDS*, it reports the expected completeness of the data by simulating measurements at various rotation ranges specified by the user, thereby taking into account already-measured reflections.

3. *XSCALE* places several data sets on a common scale, optionally merges them into one or several sets of unique reflections, and reports their completeness and quality of integrated intensities.

4. *VIEW* displays rotation-data images as well as control images produced by *XDS*. It is used for checking the correctness of data processing and for deriving suitable values for some of the input parameters required by *XDS*. This program was coded in the computer language C by Werner Gebhard at the Max-Planck- Institut für medizinische Forschung in Heidelberg. The other programs are written in Fortran77, with the exception of a few C subroutines provided by Abrahams (1993) for handling compressed images.

5. *XDSCONV* converts reflection data files as obtained from *XDS* or *XSCALE* into various formats required by software packages for crystal structure determination. Test reflections previously selected for monitoring the progress of structure refinement may be inherited by the new output file, which simplifies the use of new data or switching between different structure determination packages.

XDS is not an interactive program. It communicates with the input file XDS.INP and during the run accepts only a change in specification of the last image to be included in the data set – a useful option when processing overlaps with data collection. Experience has shown that the most frequent obstacle in using the package is the indexing and accurate prediction of the reflections occurring in the images. Usually, the problems arise from incorrect specifications of rotation axis, beam direction or detector position and orientation, oscillation range, or wavelength. The occurrence of gross errors can be reduced by using file templates of XDS.INP specifically tailored to the actual experimental set-up which requires only small adjustments to the geometrical parameters. However, even small errors in the specification of the incident beam direction or the detector position may lead to indices which are all offset by one reciprocal-lattice point, particularly if the initial list of diffraction spots was obtained from a few images covering a small range of crystal rotation. It is recommended that all program steps are run on a few images to establish whether the indexing is correct and also to find reasonable values describing crystal mosaicity and spot size. Incorrect indexing may be apparent from large values of symmetry *R* factors or from comparison with a reference data set.

Location: http://xds.mpimf-heidelberg.mpg.de/. Operating systems: SGI, DEC Alpha, LINUX, Mac OS X. Type: source code and binary. Distribution: free academic.

BEST

BEST is a program for optimal planning of X-ray data collection from protein crystals (Popov A.N. and Bourenkov, G. P. 2003; Bourenkov G. P. and Popov A. N., 2006). The method employed in the program is based on the modelling of statistical characteristics of the data yet to be collected using the information derived from a few initial images. Diffraction anisotropy, crystal radiation damage, anomalous, geometrical restrictions (e.g. spot overlapping) and hardware limitations is taken into account. The functions of the program may be summarised as follows:

- To determine optimal plan of data collection that provides requested data statistics in an outer resolution shell either in the shortest total data collection time or with the minimum total radiation dose. The plan defines the total rotation range, scan speed, rotation range per frame and detector distance.
- To estimate the data statistics, relative number of overloaded reflections and level

of radiation damage for the data that will be collected according to a given set of data collection parameters.

- To estimate the total data collection time and dose required in order achieving given <I/SigI> ratio as a function of resolution.

BEST is part of the *CCP*4 suite. Location: http://www.embl-hamburg.de/BEST/ Operating systems: UNIX. Type: source code and binary. Distribution: free academic.

SCALA

The *SCALA* program (Evans P. R., 1993, 1997) scales together multiple observations of reflections, and (optionally) merges multiple observations into an averaged intensity. Various scaling models are implemented. The scale factor is a function of the primary beam direction, either as a smooth function of φ (the rotation angle), or expressed as batch (image) number. In addition, the scale may be a function of the secondary beam direction derived from the spatial coordinates of the measured spot on the detector. In this case, the scaling is an interpolated three-dimensional function similar to that described by Kabsch (1988a, b). The merging algorithm analyses the data for outliers and gives detailed analyses. It generates weighted means of the observations of the same reflection, after rejecting the outliers.

Location: *SCALA* is part of the *CCP*4 suite. Operating systems: UNIX and VAX/VMS. Type: source code and binary. Distribution: free academic.

DETWIN

DETWIN is a program that tests for merohedral twinning and detwins data. Twinned data is measured when two or more copies of the reciprocal lattice overlap (Yeates, T.O.,1997). Hence we need to deconvolute the two twinned components to obtain useable data. The detwinning is only possible if the twin fraction is not exactly 0.5. As it approaches this value the variances become extremely large. The occurrence of twinning can often be recognised from the intensity statistics of the data set. It is important to first check whether the crystal has a pseudo translation relating two or more molecules in the asymmetric unit. Such a translation will result in some reflection classes being very weak.

DETWIN carries out several tests for twinning as a function of both twinning fraction, and resolution. It reads either intensities or amplitudes from an MTZ file. This would normally be the output of *SCALA* (see the previous paragraph '*SCALA*') containing the measured intensities, but structure factor amplitudes can also be used as input. It can also detwin merohedrally twinned data for a given twinning fraction and write either corrected intensities or amplitudes for a given twinning fraction. The twin operator is required and for MTZ output, the chosen twin fraction. The output MTZ file contains corrected Is or Fs for a single twin fraction. If the file contains IMEAN, it can be run through *TRUNCATE* (see the next paragraph '*TRUNCATE*'), which tests whether the data has been successfully detwinned. The CCP4 GUI suggests possible twinning operators derived from the space group, or it can be entered by the user.

Location: *DETWIN* is part of the *CCP*4 suite. Operating systems: UNIX and VAX/VMS. Type: source code and binary. Distribution: free academic.

TRUNCATE

TRUNCATE (French G.S. and Wilson K.S., 1978) is a program that obtains structure factor amplitudes using Truncate procedure and/or generate useful intensity statistics. The standard use of the program is to read a file of averaged intensities (output from *SCALA*) and write a file containing mean amplitudes and the original intensities. If anomalous data is present then F(+), F(-), with the anomalous difference, plus I(+) and I(-) are also written out. The amplitudes are put on an approximate absolute scale using the scale factor taken from a Wilson plot. There are two ways in *TRUNCATE* to calculate the amplitudes from the intensities. The simplest is just to take the square root of the intensities, setting any negative ones to zero (keyword TRUNCATE NO). Alternatively, the "truncate" procedure (keyword TRUNCATE YES, the default) calculates a best estimate of F from I, sd(I), and the distribution of intensities in resolution shells. This has the effect of forcing all negative observations to be positive, and inflating the weakest reflections, because an observation significantly smaller than the average intensity is likely to be underestimated.

This program can be used even if the "truncate" procedure is not desired, since it produces some useful statistics on intensity distributions. These can indicate problems with the data, for instance if the data is extremely anisotropic (see the FALLOFF keyword) or if it is likely to be twinned. See the cumulative intensity distribution plot, which for a perfect twinning becomes sigmoidal, and the moments of E (or Z) which are different for twinned data than for untwinned. If there are indications of twinning they will be noted as warnings in the log file. The scale factor estimated from the Wilson plot is applied to the data and allows the data to be put on a (very approximate) absolute scale. This at least gives amplitudes of a sensible magnitude for further calculations. The program does not, however, apply any temperature factor.

Location: *TRUNCATE* is part of the *CCP*4 suite. Operating systems: UNIX and VAX/VMS. Type: source code and binary. Distribution: free academic.

Phase Determination and Structure Solution

Packages of Crystallographic Programs

CCP4

The *CCP4* program suite (Collaborative Computational Project, Number 4, 1994) is the program package most widely used by X-ray crystallographers in structure determination and analysis of macromolecules. The *CCP4* suite is an integrated set of programs for protein crystallography developed by close collaboration of crystallographers under an initiative by the UK Biotechnology and Biological Sciences Research Council (formerly the SERC). Some software developed elsewhere is also included. The *CCP4* suite contains programs for all aspects of protein crystallography, including data processing, data scaling, Patterson search and refinement, isomorphous and molecular replacement, structure refinement, phase improvement and density modification, and presentation of results. Individual program documentation is available, together with a PostScript version of the *CCP4* manual with content distinct from the program documentation. Runnable example files are also distributed with the suite.

Locations: http://www.dl.ac.uk/CCP/CCP4/main.html; http://www.sdsc.edu/Xtal/Xtal.html; ftp://ccp4a.dl.ac.uk/pub/ccp4/; ftp:// ftp.sdsc.edu/pub/sdsc/xtal/CCP4/ and ftp://ftp2.protein.osakau.ac.jp/mirror/ccp4/ccp4. Operating systems: UNIX, VAX/VMS, LINUX, Mac OS X and MS-WINDOWS. Type: source code and binary. Languages: Fortran and C. Distribution: free academic.

CNS

The *Crystallography & NMR System* (CNS) is an advanced software system for crystallographic NMR structure determination (Brünger A. T. *et al.*, 1998). The goals of CNS are: (1) to create a flexible computational framework for exploration of new approaches to structure determination; (2)

to provide tools for structure solution of difficult or large structures; (3) to develop models for analysing structural and dynamical properties of macromolecules; and (4) to integrate all sources of information into all stages of the structure-determination process. To meet these goals, algorithms were moved from the source code into a symbolic structure-determination language which represents a new concept in computational crystallography. The high-level CNS computing language allows definition of symbolic target functions, data structures, procedures and modules. The CNS program acts as an interpreter for the high-level CNS language and includes hard-wired functions for efficient processing of computing- intensive tasks. Methods and algorithms are therefore more clearly defined and easier to adapt to new and challenging problems. The result is a multi-level system which provides maximum flexibility to the user. The CNS language provides a common framework for nearly all computational procedures of structure determination. A comprehensive set of crystallographic procedures for phasing, density modification and refinement has been implemented in this language. Task-oriented input files written in the CNS language, which can also be accessed through an HTML graphical interface (Graham I. S., 1995), are available to carry out these procedures.

The multilayer architecture of CNS allows use of the system with different levels of expertise. The HTML interface allows the novice to perform standard tasks. The interface provides a convenient means of editing complicated task files, even for the expert. This graphical interface makes it less likely that an important parameter will be overlooked when editing the file. In addition, the graphical interface can be used with any task file, not just the standard distributed ones. HTML-based documentation and graphical output is planned in the future.

Most operations within a crystallographic algorithm are defined through modules and task files. This allows for the development of new algorithms and for existing algorithms to be precisely defined and easily modified without the need for source-code modifications. The hierarchical structure of CNS allows extensive testing at each level. For example, once the source code and CNS basic commands have been tested, testing of the modules and task files is performed. A test suite consisting of more than a hundred test cases is frequently evaluated during CNS development in order to detect and correct programming errors. Furthermore, this suite is run on several hardware platforms in order to detect any machine-specific errors. This testing scheme makes CNS highly reliable. Algorithms can be readily understood by inspecting the modules or task files. This self-documenting feature of the modules provides a powerful teaching tool. Users can easily interpret an algorithm and compare it with published methods in the literature.

Location: http://cns.csb.yale.edu/v1.0/. Operating systems: UNIX, SGI, SUN, DEC Alpha, HP, LINUX and Windows-NT. Type: source code. Languages: Fortran77 and C. Distribution: free academic.

SOLVE/RESOLVE

SOLVE/RESOLVE is a complete program package designed for automated crystallographic structure solution for MIR and MAD.

SOLVE (Terwilliger T. C. & Berendzen J., 1999) is a program that can carry out all the steps of macromolecular structure determination from scaling data to calculation of an electron density map, automatically. *SOLVE* does everything crystallographers do to solve an MIR or MAD structure, but automatically. It scales data, solves Patterson functions, calculates difference Fouriers, looks at a native Fourier to see if there are distinct solvent and protein regions, and can score partial MAD and MIR solutions to build up a complete solution. *SOLVE* has solved MIR and MAD structures with up to 66 heavy-atom sites.

RESOLVE (Terwilliger, T.C., 2000, 2003) is a program that improves electron density maps. It uses a statistical approach to combine experimental X-ray diffraction information with knowledge about the expected characteristics of an electron density map of a macromolecule. Most other approaches rely on phase recombination where the optimal statistical weighting of experimental and modified phases is not known. *RESOLVE* improves the electron density maps right after using *SOLVE* or another program to solve the structure. In addition it can find non-crystallographic symmetry in the heavy-atom sites and apply it automatically. Version 2.02 and higher can build a model as well. Versions 2.05 and higher can identify local patterns and use them to improve the phases. Versions 2.06 and higher can carry out iterative pattern id, fragment id, and model-building. RESOLVE versions 2.08 and higher can carry out automated ligand fitting as well.

Locations: http://www.solve.lanl.gov/, ftp://solve.lanl.gov/pub/ solve. Operating systems: SGI, SUN, HP, DEC and LINUX. Type: binary. Distribution: minor licence fee for academic users.

SHELX

SHELX (Sheldrick, 2008) is a set of programs for crystal structure determination from single-crystal diffraction data. Originally *SHELX* was intended only for small molecules. Since small-molecule direct methods and Patterson interpretation algorithms can be used to locate a small number of heavy atoms or anomalous scatterers the structure-solving program *SHELXS* has been used by macromolecular crystallographers for a number of years. More recently, improvements in cryo-crystallography, area detectors and synchrotron data collection have led to a rapid increase in the number of high-resolution (<2 Å) macromolecular data sets. The enormous increase in available computer power makes it feasible to refine these structures using algorithms incorporated in *SHELXL* (Sheldrick G.M. & Schneider T. R., 1997) that were initially designed for small molecules. These algorithms are generally slower but make fewer approximations (*e.g.* conventional structure-factor summation rather than fast Fourier transform (FFT)) and include features, such as anisotropic refinement, modelling of complicated disorder and twinning, estimation of standard uncertainties by inverting the normal matrix *etc.*, that are routine in small-moiety crystallography but, for reasons of efficiency, are not widely implemented in programs written for macromolecular structure refinement. This account will be restricted to features in *SHELX* of potential interest to macromolecular crystallographers. *SHELX* is written in a simple subset of Fortran77 that has proved to be extremely portable.

The following tasks for crystal structure determination and refinement are included in the corresponding *SHELX* programs:

- Heavy-atom location using *SHELXS* and *SHELXD*
- The Patterson map interpretation algorithm in *SHELXS*
- Integrated Patterson and direct methods: *SHELXD*
- Macromolecular refinement using *SHELXL* (Constraints and restraints - Least-squares refinement algebra -Full-matrix estimates of standard uncertainties - Refinement of anisotropic displacement parameters - Similar geometry and NCS restraints - Modelling disorder and solvent - Twinned crystals)
- *SHELXPRO* – protein interface to *SHELX*

Location: http://shelx.uni-ac.gwdg.de/SHELX/. Operating systems: UNIX, VMS, DOS and Windows. Type: binary. Language: Fortran77. Distribution: free academic.

PHENIX

The *PHENIX* (*Python-based Hierarchical ENvironment for Integrated Xtallography*) software suite (Adams P. D., 2002) is a highly automated system for macromolecular structure

determination that can rapidly arrive at an initial partial model of a structure without significant human intervention, given moderate resolution and good quality data. This achievement has been made possible by the development of new algorithms for structure determination, maximum-likelihood molecular replacement (*PHASER*), heavy-atom search (HySS), template and pattern-based automated model-building (*RESOLVE, TEXTAL*), automated macromolecular refinement (phenix.refine), and iterative model-building, density modification and refinement that can operate at moderate resolution (*RESOLVE*, AutoBuild). These algorithms are based on a highly integrated and comprehensive set of crystallographic libraries that have been built and made available to the community. The algorithms are tightly linked and made easily accessible to users through the *PHENIX* Wizards and the command line.

There are also a number of tools in *PHENIX* for handling ligands. Automated fitting of ligands into the electron density is facilitated via the LigandFit wizard. Besides being able to fit a known ligand into a difference map, the LigandFit wizard is capable to identify ligands on the basis of the difference density only. Stereo chemical dictionaries of ligands whose chemical description is not available in the supplied monomer library for the use in restrained macromolecular refinement can be generated with the electronic ligand builder and optimization workbench (*eLBOW*).

PHENIX builds upon Python, the Boost.Python Library, and C++ to provide an environment for automation and scientific computing. Many of the fundamental crystallographic building blocks, such as data objects and tools for their manipulation are provided by the Computational Crystallography Toolbox (cctbx). The computational tasks which perform complex crystallographic calculations are then built on top of this. Finally, there are a number of different user interfaces available in *PHENIX*. In order to facilitate automated operation there is the Project Data Storage

(PDS) that is used to store and track the results of calculations.

Location: http://www.phenix-online.org/. Operating systems: UNIX, LINUX and MAC OS-X . Type: source code and binary. Distribution: academic free.

Ab Initio Phasing: Direct Methods

SHAKE-AND-BAKE

Shake-and-Bake (*SnB*) (Weeks C. M. & Miller R., 1999) is a program that uses a dual-space direct-methods phasing algorithm based on the minimal principle to determine crystal structures of macromolecules. The program requires very high resolution data to 1.2 Å or better and $|E|$ values as input. It has been used in a routine fashion to solve difficult structures, containing as many as 1000 unique non-H atoms, that could not be solved by traditional reciprocal-space routines based on the tangent formula alone. Recently, SnB has also been used to determine the anomalously scattering substructures of selenomethionyl-substituted proteins containing as many as 160 Se sites. Generally, non-substructure applications require that diffraction data be measured to 1.1Å resolution or higher although some structures having several moderately heavy atoms (e.g. S or Cl) have been solved at 1.4Å. SAS or SIR substructure applications routinely use 3Å data and have been successful at a resolution as low as 5Å.

The current version, SnB v2.3, provides a graphical user interface for (i) computing normalized structure-factor magnitudes, (ii) the main Shake-and-Bake phasing algorithm, and (iii) visualization and molecule-editing facilities. SnB can conveniently be run in parallel on multiple processors for faster throughput. In addition, *SnB* is also available as part of the protein structure determination package, *BnP*. The *BnP* interface couples *SnB* to appropriate components of the *PHASES* package (see paragraph '*PHASES*')., thereby creating an automated pipeline from

intensity data to an unambiguous protein electron-density map (Weeks, C.M *et al.*, 2002).

Location: http://www.hwi.buffalo.edu/SnB/. Operating systems: UNIX, VMS and LINUX, MAC OS. Type: source code. Language: Fortran77. Distribution: free.

SIR2004

The *SIR* (Semi-Invariants-Representation) package has been developed for solving crystal structures by Direct Methods. The Representation theory, proposed by Giacovazzo (1977, 1980) allowed the derivation of powerful methods for estimating structure invariants (s.i.) and structure seminvariants (s.s.). The mathematical approach makes full use of the space group symmetry. SIR uses symmetry in a quite general way allowing the estimation and use of s.i. and s.s. in all the space groups.

SIR2004 (Burla M. C. *et al.*, 2005), is designed to solve *ab initio* structures of different size and complexity, up to proteins, provided that data resolution is no lower than 1.4-1.5Å. Data can be collected with X-Ray or electron sources. There is no limit to the number of reflections and to the number of atoms in the asymmetric unit. The maximum value allowed for |h|, |k|, |l| is 512. The maximum number of different atomic species is 8. Sir2004 includes several new features with respect to the previous version, Sir2002. New tools are represented by the use of procedures based on Patterson Methods as alternative to the application of the Tangent Formula in order to compute the starting phase set; the introduction of suitable figures of merit (FOM's) in order to recognize the correct trial solution and the application of new algorithms for solving *ab initio* protein structures (also for quasi-atomic resolution data), i.e. the use of the molecular envelope mask.

The range of options available to experienced crystallographers for choosing their own way of solving crystal structures is rather wide. However, scientists untrained in Direct Methods or people trustful in the SIR default mode often can solve crystal structures without personal intervention. The program has been designed to require a minimal information in input, work automatically reduce the user intervention and facilitate the interaction by means of a friendly graphic interface.

Location: http://www.ic.cnr.it/sir2004.php. Operating systems: Microsoft Windows and Unix (SGI, Compaq, Linux and others). Type: binary. Distribution: free academic.

MULTAN88

*MULTAN*88 (Main P. *et al.*, 1980) is a program that uses direct methods to determine crystal structures from single-crystal diffraction data. It can be used for very high resolution structure refinement and determination of heavy-atom positions.

Location: http://www.msc.com/ Operating systems: UNIX and VAX/VMS. Type: binary. Distribution: commercial.

Molecular Replacement

AMoRe

AMoRe (Navaza J., 1994) is a program package that carries out structure determination using molecular replacement. It reformats the data from the new crystal form, generates structure factors from the model, calculates rotation and translation functions, and applies rigid-body refinement to the solutions. *AMoRe* is part of the *CCP*4 suite.

Location: http://www.dl.ac.uk/CCP/CCP4/ dist/html/INDEX.html. Operating systems: UNIX, VAM/VMS and LINUX. Type: source code and binary. Distribution: free academic.

MOLREP

MOLREP (Vagin A. A. and Teplyakov A., 1997) is an automated program for molecular replacement which utilizes effective new approaches in data processing and rotational and translational searching. These include an automatic choice of all parameters, scaling by Patterson origin peaks and soft resolution cut-off. One of the cornerstones of the program is an original full-symmetry translation function combined with a packing function. Information from the model already placed in the cell is incorporated in both translation and packing functions.

MOLREP includes the following features:

- a full-symmetry Translation and Packing function.
- an automated choice of search parameters.
- scaling by Patterson origin peaks.
- soft low resolution cut-off.
- anisotropic correction of data.
- rigid body refinement.
- allows input of *a priori* knowledge of similarity and of the model.
- can use second fixed model.
- can check several peaks of Rotation function by computing Translation function and select result by correlation.
- if the number of monomers is known, MOLREP can position the input number of monomers in a simple run.
- can check and manage pseudo-translation.
- can improve the model before use.
- can use MTZ file.
- can compute only Cross Rotation or only Translation function.
- can compute Self Rotation function with PostScript plots.
- Spherically Averaged Phased Translation function.
- Phased Rotation and Phased Translation functions.
- fitting two models.

MOLREP is distributed as part of the *CCP*4 suite. Location: http://www.ysbl.york.ac.uk/~alexei/molrep.html. Operating systems: UNIX, LINUX, MAC OS and MS-WINDOWS. Type: source code and binary. Language: Fortran77. Distribution: free academic.

PHASER

PHASER (McCoy, A. J., 2007) is a program for phasing macromolecular crystal structures with maximum likelihood methods. Automated Molecular Replacement combines the anisotropy correction, likelihood enhanced fast rotation function, likelihood enhanced fast translation function, packing and refinement modes for multiple search models and a set of possible spacegroups. *PHASER* must be given the models that it will use for molecular replacement. A model in *PHASER* is referred to as an "ensemble", even when it is described by a single file. This is because it is possible to provide a set of aligned homologous structures as an ensemble, from which a statistically-weighted averaged model is calculated. A molecular replacement model is provided either as one or more aligned pdb files, or as an electron density map, entered as structure factors in an mtz file. Each ensemble is treated as a separate type of rigid body to be placed in the molecular replacement solution. An ensemble should only be defined once, even if there are several copies of the molecule in the asymmetric unit.

Location: http://www-structmed.cimr.cam.ac.uk/phaser/. Source installation of Phaser-2.1 is available through *CCP4* with the imminent of *CCP4-6.1*. Operating systems: UNIX, LINUX, MAC OS and MS-WINDOWS. Type: source code and binary. Distribution: free academic.

Isomorphous Replacement (SIR- MIR), Anomalous X-Ray Scattering (MAD-SAD)

As it has already been noted (see paragraph '*ab initio*-direct methods') direct methods are also proving to be useful for locating the heavy atoms of prepared isomorphous derivatives and the anomalous scatterers in the multiple wavelength anomalous diffraction phasing. Programs implementing algorithms based on direct methods like Shake an Bake, SIR2004, SHARP and SHELXD (contained in *SHELX* package) are used routinely to find heavy-atom substructure and/or anomalous scatterers.

MADSYS

MADSYS (Hendrickson W.A., 1991) is a software package developed over the years in Dr Wayne Hendrickson's laboratory for determining experimental phases of macromolecular structures by multi-wavelength anomalous diffraction (MAD). The package consists of a set of programs that carry out MAD data handling, determination of anomalous-scatterer sites, refinement of MAD sites, MAD phases calculation and structure refinement.

Location: http://convex.hhmi.columbia.edu/hendw/madsys/madsys.html. Operating system: UNIX. Type: binary. Distribution: free academic.

PHASES

The *PHASES* package (Furey W. and Swaminathan, S., 1997) was designed to deal with the major problem in macromolecular structure determination, *i.e.*, phasing the diffraction data. It focuses on the initial phasing of diffraction data from macromolecules by heavy-atom- and anomalous scattering- based methods. Also included are programs and procedures for phase improvement by non crystallographic symmetry averaging, solvent flattening, phase extension and partial structure phase combination. The programs and additional procedure scripts allow one to start with unique structure-factor amplitudes for native and/or derivative data sets and generate electron-density maps and skeletons that can be utilized in popular graphics programs for chain tracing and model building.

Location: http://www.imsb.au.dk/~mok/phases/phases.html. Operating systems: SGI, Sun, IBM R6000, ESV and DEC Alpha. Languages: Fortran77 and C. Distribution: free.

SHARP

SHARP (Statistical Heavy-Atom Refinement and Phasing) (La Fortelle E. de & Bricogne G., 1997) operates on reduced, merged and scaled data from SIR(AS), MIR(AS) and MAD experiments, refines the heavy-atom model, helps detect minor or disordered sites using likelihood-based residual maps, and calculates phase probability distributions for all reflections in the data set.

Location: http://Lagrange.mrc-lmb.cam.ac.uk/ sharp/SharpHome.phtml. Operating systems: IRIX and OSF1. Type: binary. Distribution: free academic.

MLPHARE

MLPHARE is a program for maximum-likelihood heavy-atom refinement and phase calculation. This program refines heavy-atom parameters and error estimates, then uses these refined parameters to generate phase information. The maximum number of heavy atoms that may be refined is 130 over a maximum of 20 derivatives. The program was originally written for MIR, but may also be used for phasing from MAD data, where the different wavelengths are interpreted as different 'derivatives'. *MLPHARE* is part of the *CCP*4 suite.

Location: http://www.dl.ac.uk/CCP/CCP4/ dist/html/mlphare.html. Operating systems: UNIX, VAX/VMS and LINUX. Type: source code and binary. Distribution: free academic.

Model Building and Refinement

Several programs that are used for density map modification and structure refinement are contained in the program packages described in Section 'Packages of crystallographic programs'. These include *CCP4, CNS, PHASES, SOLVE/ RESOLVE* and *SHELX*.

Density-Map Modification

DM/DMMULTI

DM (Cowtan K., 1994) is an automated procedure for phase improvement by iterated density modification. It is used to obtain a set of improved phases and figures of merit, using as a starting point the observed diffraction amplitudes and some initial poor estimates for the phases and figures of merit. *DM* improves the phases through an alternate application of two processes: real-space electron-density modification and reciprocal-space phase combination. *DM* can perform solvent flattening, histogram matching, multi-resolution modification, averaging, skeletonization and Sayre refinement, as well as conventional or reflection-omit phase combination. Solvent and averaging masks may be input by the user or calculated automatically. Averaging operators may be refined within the program. Multiple averaging domains may be averaged using different operators.

DM is part of the *CCP*4 suite. Operating systems: UNIX, VAX/VMS and LINUX. Type: source code and binary. Distribution: free academic.

SOLOMON

SOLOMON (Abrahams J. P. & Leslie A. G. W., 1996) is a program that modifies electron-density maps by averaging, solvent flattening and protein truncation. It can also remove overlapped parts of a mask between itself and its symmetry equivalents. *SOLOMON* is part of the *CCP*4 suite.

Location: http://www.dl.ac.uk/CCP/CCP4/ dist/html/solomon.html. Operating systems: UNIX, VAX/VMS and LINUX. Type: source code and binary. Distribution: free academic.

Structure Refinement

REFMAC

REFMAC (Murshudov G. N. *et al.*, 1997, 1999) is a macromolecular refinement program which has been integrated into the *CCP*4 suite. *REFMAC* can carry out rigid-body, restrained or unrestrained refinement against X-ray data, or idealization of a macromolecular structure. It minimizes the coordinate parameters to satisfy either a maximum-likelihood or least-squares residual. There are options to use different minimization methods. If the user wishes to invoke geometric restraints, the program *PROTIN*, which analyses the protein geometry and produces an output file containing restraints information, must be run prior to running *REFMAC*. *REFMAC* also produces an MTZ output file containing weighted coefficients for *SIGMAA*-weighted mFo-DFcalc and 2mFo-DFcalc maps, where 'missing data' have been restored.

Location: http://www.dl.ac.uk/CCP/CCP4/dist/html/refmac.html. Operating systems: UNIX, SGI, SUN, DEC and LINUX. Type: source code and binary. Distribution: free.

ARP/wARP

The Automated Refinement Procedure, *ARP/wARP* (Lamzin V. S. & Wilson K.S, 1993, 1997; Perrakis A. *et al.*, 1999), is a program package for automated model building and refinement of protein structures. It combines, in an iterative manner, reciprocal-space structure-factor refinement with updating of the model in real space to construct and improve protein models. The *real-space manipulation* of the model is achieved by mimicking user intervention *in silica*. Adding and/or deleting atoms (*model update*) and complete re-evaluation of the model to create a new one that better describes the electron density (*model reconstruction*) can achieve this aim.

The basic *ARP/wARP* applications comprise:

- Model building from initial phases
- Refinement of molecular-replacement solutions
- Density modification via averaging of multiple refinements
- Ab initio solution of metalloproteins
- Solvent building

Density-based atom selection for the whole structure is only possible if the X-ray data extend to a resolution where atomic positions can be estimated from the Fourier syntheses with sufficient accuracy for them to refine to the correct position. If the structural model is of reasonable quality, at 2.5 Å or better, at least a part of the

solvent structure or a small missing or badly placed part of the protein can be located. This provides indirect improvement of the whole structure. For automated model rebuilding, or for refining poor molecular-replacement solutions, higher resolution is essential. The general requirement is that the number of X-ray reflections should be at least six to eight times higher than the number of atoms in the model, which roughly corresponds to a resolution of 2.3 Å for a crystal with 50% solvent. However, the method can work at lower resolution or fail with a higher one, depending less on the quality of the initial phases and more on the internal quality of the data and on the inherent disorder of the molecule.

The X-ray data should be complete. If strong low-resolution data (*e.g.* 4 to 10 Å) are systematically missing, *e.g.* due to detector saturation, the electron density even for good models is often discontinuous. Because *ARP* involves updating on the basis of density maps, such discontinuity will lead to incorrect interpretation of the density and slow convergence or even uninterpretable output.

ARP/wARP is distributed as part of the *CCP*4 suite. Location: http://www.embl-hamburg.de/ARP/. Operating systems: UNIX, HPUX, IRIX and LINUX. Type: source code and binary. Language: Fortran77. Distribution: free academic.

BUSTER

BUSTER (Bricogne G., 1997*a*,*b*) is a program for recovering missing phase information by Bayesian inference. *BUSTER* has applications in maximum-likelihood refinement of partial structures, maximum-entropy structure completion for missing or ambiguous parts of a structure, and accurate electron-density reconstruction based on high resolution X-ray diffraction data. *BUSTER* is related to *SHARP*.

Location: http://lagrange.mrc-lmb.cam.ac.uk/buster/Buster Home.phtml. Operating systems: IRIX and OSF1. Type: binary. Distribution: free academic.

CRYSTWIV

CRYSTWIV (Thireou T. *et al.*, 2007) is a program for phase extension and refinement in macromolecular crystallography when initial phase estimates are available from isomorphous replacement or anomalous scattering. The program implements the twin variables (TwiV) method (Hountas A. & Tsoucaris G., 1995; Bethanis K. *et al*, 2000). The twin variables concept consists of the use of a set of auxiliary complex variables Ψ which are related to the normalised structure factors by a regression equation of standard probability theory derived from maximun determinant rule of Direct Methods (Tsoucaris G., 1970a,b). The (*TwiV*) algorithm is based on alternately transferring the phase information between the twin variable sets of E and Ψ values and it operates entirely in the reciprocal space. The phase extension and refinement is evaluated with the crystallographic symmetry test by deliberately sacrificing the spacegroup symmetry in the starting set, then using its re-appearance as a criterion for correctness (Tzamalis P. *et al.*, 2003).

(CrysTwiV) runs on the web (freely available at: http://btweb.aua.gr/ crystwiv/)

Graphics and Model Building

O

O (Jones T. A. *et al.*, 1991) is a general-purpose macromolecular modelling package. The program is aimed at scientists with a need to model, build and display macromolecules. Unlike other molecular modelling programs, *O* is a graphical display program built on top of a versatile database system. All molecular data are kept in this database, in a predefined data structure. However, any data can be stored in the database. Data produced by associated stand-alone programs can be stored very easily in the database and used by the program, for example for colouring of atoms. The powerful macro facility of *O* enables the user to customize the use of the program to satisfy his or her specific needs. The current version of *O* is mainly aimed

at the field of protein crystallography, bringing into use several new tools which ease the building of models into electron density, allowing it to be done faster and more correctly. Notably, some new auto-build options greatly enhance the speed of building and rebuilding molecular models.

Locations: http://kaktus/imsb.au.dk/_mok/o/; ftp://xray. bmc.uu.se/. Operating systems: IRIX, UNIX, ESV and MS WINDOWS. Type: binary. Distribution: free academic.

Turbo FRODO

FRODO (JonesT. A., 1978) is a general-purpose molecular-modelling program which can be used to model *de novo* macromolecules, polypeptides and nucleic acids from experimental 3D data obtained from X-ray crystallography and NMR, and to display the resulting models using various representations including van der Waals and Connolly molecular dot surfaces, as well as spline surfaces. *Turbo FRODO* is designed for ligand fitting and protein stacking. The user can interactively mutate a protein or chemically modify it, and evaluate the resulting conformational changes.

Location: http://afmb.cnrs-mrs.fr/TURBO_ FRODO/turbo.html. Operating systems: HPUX, IRIX and LINUX. Type: binary. Distribution: commercial.

QUANTA

QUANTA is an extensive library of crystallographic software programs that streamline and accelerate protein structure solution. *QUANTA* provides a powerful and comprehensive modelling environment for 2D and 3D modelling, simulation and analysis of macromolecules and small organic compounds.

Location: http://www.msi.com/life/products/ quanta/index.html. Operating system: SGI. Type: binary. Distribution: commercial

COOT

CCP4mg (Emsley P. & Cowtan K., 2004; Potterton L. *et al.*, 2004) is an initiative by *CCP4* to provide libraries and a molecular graphics application that is a popular system for representation, modelling, structure determination, analysis and validation. The aim is to provide a system that is easy to use and a platform for developers who wish to integrate macromolecular computation with a molecular-graphics interface. There are several modules to such graphical functionality; the protein model-building/map-fitting tools described here are only a part. These tools are available as a stand-alone software package, *COOT*. A map-fitting program has to provide certain functionality, which is not required by a molecular-display program. These functions include symmetry coordinates, electron-density map contouring and the ability to move the coordinates in various ways, such as model idealization or according to side-chain rotamer probabilities.

Coot attempts to provide more transparency, ease of use, better extendibility, (semi-)automated model-building methods and convenient integration with programs of the CCP4 suite. The various functions of Coot are split into 'standalone' classes in the sense that an attempt has been made to minimize the dependence of the classes on anything other than the above libraries. With portability in mind, special effort was made not to introduce GUI dependences into the interface to Coot's library of tools. Coot is event-driven; functions are only run as a result of user action (typically moving or clicking the mouse).

COOT is distributed as part of the *CCP*4 suite. Location: http://www.ysbl.york.ac.uk/~emsley/coot/. Operating systems: UNIX, LINUX, IRIX and MS-WINDOWS. Type: source code and binary. Distribution: free academic.

PYMOL

PyMOL is an open-source, user-sponsored, molecular visualization system. It is well suited to producing high quality 3D images of small molecules and biological macromolecules such as proteins. PyMOL is one of few open source visualization tools available for use in structural

biology. The Py portion of the software's name refers to the fact that it extends, and is extensible by the Python programming language. Almost a quarter of all published images of 3D protein structures in the scientific literature were made using PyMOL.

Location: http://www.pymol.org/ and http://pymol.sourceforge.net/ Operating systems: UNIX, LINUX, Mac OS and MS-WINDOWS. Type: source code and binary. Distribution: A controlled-access download system has been adopted for precompiled PyMOL builds (including betas). Access to these executables is now limited to paying customers but is free for students and teachers. However, current source code continues to be available at no cost, as are older precompiled builds. While the build systems for other platforms are open, the win32 build system is not. Non-students/teachers can either compile an executable from the source code or pay for a subscription to the support services to obtain access to pre-compiled executables.

MOLSCRIPT

MOLSCRIPT (Kraulis P. J., 1991) is a program for creating schematic or detailed molecular-graphics images in the form of PostScript plot files from molecular 3D coordinates, usually, but not exclusively, of protein structures.

The implementation of *MolScript* is based on two design principles: First, the program must allow both schematic and detailed graphical representations to be used in the same image. Second, the user must be able to control the precise visual appearance of the various graphics objects in as much detail as possible. Possible representations are simple wire models, CPK spheres, ball-and-stick models, text labels and Jane Richardson-type schematic drawings of proteins, based on atomic coordinates in various formats. Colour, greyscale, shading and depth cueing can be applied to the various graphical objects.

Location: http://www.avatar.se/molscript/. Operating system: UNIX. Type: source code and binary. Language: C. Distribution: free academic.

ORTEP

The Oak Ridge Thermal Ellipsoid Plot (ORTEP, version III) program (Burnett M. N. & Johnson C. K., 1996) is a computer program for drawing crystal-structure illustrations. Ball-and-stick type illustrations of a quality suitable for publication are generated with either spheres or thermal-motion probability ellipsoids, derived from anisotropic temperature-factor parameters, on the atomic sites. The program also produces stereoscopic pairs of illustrations that aid in the visualization of complex arrangements of atoms and their correlated thermal-motion patterns.

Location: http://www.ornl.gov/ortep/ortep. html. Operating systems: UNIX, LINUX, DOS, MacOS and Windows. Type: source code and binary. Language: Fortran77. Distribution: free.

RasMol

RasMol is a molecular-graphics program intended for the visualization of proteins, nucleic acids and small molecules. The program is aimed at display, teaching and generating high-quality images for publication. It is easy to use and produces beautiful space- filling three-dimensional colour images. *RasMol* reads in molecular coordinate files in a number of formats and interactively displays the molecule on the screen in a variety of colour schemes and representations. The X Windows version of *RasMol* provides optional support for a hardware dials box and accelerated shared memory rendering (*via* the XInput and MIT-SHM extensions) if available.

Location: http://www.umass.edu/microbio/ rasmol/. Operating systems: UNIX, VAX/VMS, Windows and MacOS. Type: source code. Distribution: free.

Raster3D

Raster3D (Bacon B. J. & Anderson W. F., 1988; Merritt E. A. & Murphy M. E. P., 1994; Merritt E. A.

& Bacon D. J., 1997) is a set of tools for generating high-quality raster images of proteins or other molecules. The core program renders spheres, triangles and cylinders with special highlighting, Phong shading and shadowing. It uses an efficient software Z-buffer algorithm that is independent of any graphics hardware. Ancillary programs process atomic coordinates from PDB files into rendering descriptions for pictures composed of ribbons, space-filling atoms, bonds, ball-and-stick *etc. Raster3D* can also be used to render pictures composed in Per Kraulis' program *MOLSCRIPT* in glorious 3D with highlights, shadowing *etc.* Output is pixel image files with 24 bits of colour information per pixel.

Location: http://www.bmsc.washington.edu/ raster3d/raster3d.html. Operating systems: DEC, SGI, ESV, SUN, IBM, HP and LINUX. Type: source code and binary. Distribution: free.

VMD

VMD (Visual Molecular Dynamics) is designed for the visualization and analysis of biological systems such as proteins, nucleic acids, lipid bilayer assemblies *etc.* It may be used to view more general molecules, as *VMD* can read standard PDB files and display the structure contained in them. *VMD* provides a wide variety of methods for rendering and colouring a molecule: simple points and lines, CPK spheres and cylinders, licorice bonds, backbone tubes and ribbons, cartoon drawings, and others. *VMD* can be used to animate and analyse the trajectory of a molecular-dynamics (MD) simulation. In particular, *VMD* can act as a graphical front end for an external MD program by displaying and animating a molecule undergoing simulation on a remote computer.

Location: http://www.ks.uiuc.edu/Research/ vmd/allversions. Operating systems: SGI, SUN, DEC Alpha, IBM AIX, HP-UX and LINUX. Type: binary. Distribution: free.

Structure Analysis and Verification

NACCESS

NACCESS is a stand-alone program that calculates the accessible area of a molecule from a PDB format file. It can calculate the atomic and residue accessibility for both proteins and nucleic acids. The program uses the Lee B. & Richards F. M. (1971) method, whereby a probe of given radius is rolled around the surface of the molecule, and the path traced out by its centre is the accessible surface. Typically, the probe has the same radius as water (1.4Å) and hence the surface described is often referred to as the solvent accessible surface. The calculation makes successive thin slices through the 3D molecular volume to calculate the accessible surface of individual atoms.

Location: http://sjh.bi.umist.ac.uk/naccess. html. Operating systems: UNIX, SGI, Sun, HP, DEC and LINUX. Type: source code. Language: Fortran77. Distribution: free academic.

PROCHECK

PROCHECK (Laskowski R. A. *et al.*, 1993), is a widely used program for checking the stereochemical quality of a protein structure. The aim of *PROCHECK* is to assess how normal, or, conversely, how unusual, the geometry of the residues in a given protein structure is, as compared with stereochemical parameters derived from well refined high-resolution structures. It computes a number of stereochemical parameters for the given protein model and compares them with 'ideal' values obtained from a database of well refined high resolution protein structures in the Protein Data Bank. The results of these checks are output in easy-to-understand coloured plots in PostScript format. Significant deviations from the derived standards of normality are highlighted as being 'unusual'. The program's primary use is during the refinement of a protein structure; the highlighted regions can direct the crystallographer to parts of the structure that may have problems and which may need attention.

It should be noted that outliers may just be outliers; they are not necessarily errors. Unusual features may have a reasonable explanation, such as distortions due to ligand binding in the protein's active site. However, if there are many oddities throughout the model, this could signify that there is something wrong with it as a whole. Conversely, if a model has good stereochemistry, this alone is not proof that it is a good model of the protein structure. Because the program requires only the 3D atomic coordinates of the structure, it can check the overall 'quality' of any model structure: whether derived experimentally by crystallography or NMR, or built by homology modelling.

PROCHECK is in fact a suite of separate Fortran and C programs which are run successively *via* a shell script. The programs first 'clean up' the input PDB file, re-labelling certain side-chain atoms according to the IUPAC naming conventions, then calculate all the protein's stereochemical parameters to compare them against the norms, and finally generate the PostScript output and a detailed residue-by-residue listing. Hydrogen and atoms with zero occupancy are omitted from the analyses and, where atoms are found in alternate conformations, only the highest-occupancy conformation is retained.

Location: http://ww.biochem.ucl. ac.uk/~roman/procheck.html. It has also been incorporated into the *CCP*4 suite of programs. Operating systems: UNIX, VAX/VMS and Windows. Type: source code. Distribution: free

CRYSTALLOGRAPHIC STUDY OF THE 2[4FE-4S] FERREDOXIN FROM *ESCHERICHIA COLI*

The sequential procedure of the X-ray crystallographic structure determination of the 2[4Fe-4S] ferredoxin from *Escherichia coli* is described as a case study of implementation of the basic crystallographic steps and programs presented in the previous sections. A typical crystallographic work comprise the crystallization of the protein, data collection, phasing (structure solution), model building, reciprocal and real space refinement.

Function – Sequence and Structural Features of 2[4Fe-4S] Ferredoxins

The 2[4Fe-4S] ferredoxin (Fds) from the bacterium *Escherichia coli* (EcFd) belongs to the *Allochromatium vinosum* (AlvinFd) subfamily and is believed that it is solely involved in electron transfer processes (Holm, R. H. *et al.*, 1996). However, the nature and the complete list of redox partners for the 2[4Fe-4S] Fds have not been clearly identified in most of the cases. Its structure determination made possible a reliable comparison with other available structures of [4Fe-4S] containing Fds, in an effort to rationalize the unusual electrochemical properties of the particular subfamily. On the basis of their aminoacid sequences, the Alvin-like Fds are fairly homologues, as the two iron-sulfur clusters are bound to the polypeptide by conserved motifs containing the eight necessary cysteine residues. The main secondary structure elements of EcFd are a well defined 3.5 turn C-terminus α-helix, a cluster-joining one-turn α-helix (Cys49-Cys53), a 3_{10} helix (Cys14-Cys18), two antiparallel β-sheets, one between Ile23-Met25 and Tyr30-Ile32 and a very short one between residues Leu2-Leu3 and Val60-Lys61 and four prominent loops. Conserved water molecules, either on the surface of the molecule or buried, which is a peculiar feature in Alvin-like Fds, are also present in the EcFd structure.

Protein Crystallization: Data Collection

The crystallisation set up was incorporated into a sealed glove bag filled with nitrogen using the hanging drop method. EcFd crystals were grown by mixing 1.0 μL of protein solution at a concentration of 18 mg/ml in 20 mM Tris buffer at pH 7.0, 50 mM NaCl, with 1.0 μL of reservoir solution.

The macroscopically best crystals of EcFd (Figure 1) were hexagonal rods and grew from reservoir solutions containing 0.5 M $CaCl_2$, 100 mM Tris-HCl pH 9.0, 20% w/v PEG 4000, after seeding of the drops with micro crystals grown in earlier attempts. Cryoprotection was achieved by soaking of the crystal for a few seconds in a solution consisting of 80% original well solution plus 20% (v/v) glycerol. Then crystals were flash-cooled in liquid nitrogen and later placed in a 100 K cryostream at the beamline X13 with fixed-wavelength (0.81 Å) at EMBL/DESY, Hamburg (DORIS positron storage ring), equipped with the Mar 165-mm CCD detector and Oxford Cryosystems cryostream.

After manual inspection of the initial diffraction images, for assessing the suitability of the mounted crystal for data collection, program BEST was used, in order to determine the optimal data collection strategy. It performs simple but crucial calculations to optimize parameters, such as crystal to detector distance, exposure time (or dose) to the X-ray beam, starting φ angle and angular displacement per frame. Its major achievement is

Figure 1. Crystals of EcFd, ~150μm long, developed in a hanging drop, diffracting at 1.65Å

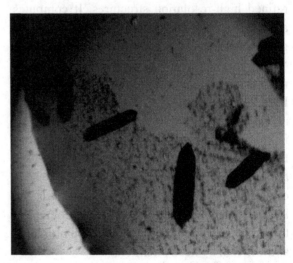

the recording of data sets characterized by completeness, high redundancy, with good intensity statistics in the minimum time, avoiding in parallel undesirable side effects due to overexposure, spot overlapping etc.

The processing of the diffraction patterns (Figure 2) was performed using the programs *DENZO* (data reduction and integration) and XDisplayF (visualization of the images), while the final data set was extracted by the program Scalepack (merging and scaling of the intensities), all parts of the HKL package.

Space Group Determination: Structure Solution

The determination of the real space group proved a difficult task, due to the occurrence of merohedral twinning. The lattices of individual crystals overlapped producing a single diffraction pattern, which appeared with higher diffraction symmetry (point group 622) than the real (point group 6) one. The solution of the structure could not be achieved before detwinning of the data (program DETWIN) even after the determination of the correct space group. Therefore the twinning was analysed first

Figure 2. Diffraction pattern of the EcFd crystal, visualized by the XDisplayF program

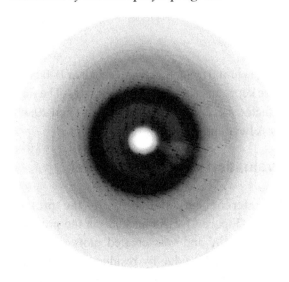

by the program TRUNCATE. The structure was first solved in the space group $P3_2$ by molecular replacement using the program PHASER and the coordinates of the *Allochromatium vinosum* Fd (PDB code 1BLU) (Dauter Z. *et al.*, 1997) after the two (4Fe-4S) clusters and the water molecules had been removed.

The model produced by the molecular replacement was incomplete, concerning labile side chains and the C-terminus. The initial R was 49% and this high value was attributed to the missing clusters, whose density was revealed in the difference Fourier map (Figure 3). However, the six independent molecules of EcFd/asymmetric unit were related in pairs by the same 2-fold axis, which was coinciding with the 3_2 axis. The latter revealed the existence of a 6_2 axis, thus the data were reprocessed and detwinned in space group $P6_2$ in which the structure was solved again by the program *PHASER*. The initial R was 48% and a subsequent cycle of rigid body refinement gave the positions of the cluster atoms more accurately than in space group $P3_2$.

Structure Refinement

Insertion of the clusters' atoms improved the structure (R=31%) and the refinement proceeded using *SHELXL*, because it uses the unmodified twinned Fo which are refined against Fc that take into account both parts of the twinned crystal. The relevant equation is: $F_c^2 = \alpha F_{c1}^2 + (1-\alpha) F_{c2}^2$, where, α is the twin fraction and the two parts c1 and c2 are related by the twin law k,h,-l. This twin law describes a real space rotation about the diagonal **a+b.** The twin fraction was given an initial value of 0.37 as indicated by the Britton plot and the cumulative distribution of *H* (Figure 4).

The protein model was completed with the graphics programs *O* and *COOT* after the initial rigid body refinement. The isotropic restrained refinement, which was subsequently implemented, was interrupted by regular inspection of the electron density and refitting of the model (real space

Figure 3. Difference map Fo-Fc contoured at 3σ of the area of cluster I (a). The cluster model built, based on electron density map (b). Carbon atoms are shown in green, nitrogen in blue, oxygen in red, sulphur in yellow and iron in deep red colour. The figure has been drawn with the graphics program PYMOL

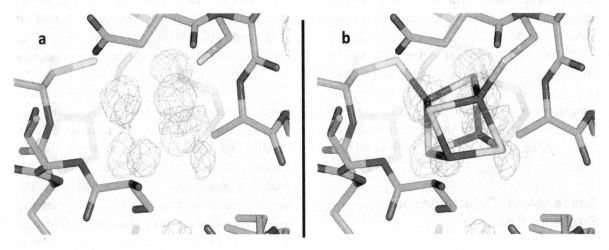

Figure 4. Plots using data from TRUNCATE that indicate that the crystal is merohedrally twinned. Estimation of the twin fraction by plotting the cumulative fractional intensity difference of acentric twin-related intensities, H= |I1 – I2|/(I1 + I2), as a function of H (a). Estimation of α by Britton plot. The number of negative intensities after detwinning is plotted as a function of the assumed value of α (b)

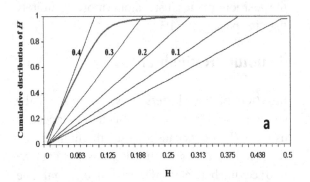 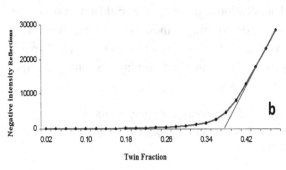

refinement) with the graphics program Coot, where necessary. In order to eliminate bias coming from the initial model, omit maps (Figure 5) of all protein regions were generated and minor corrections of the backbone and the side chains were performed, through the real space refinement procedure. The water molecules were determined automatically using the *SHELXWAT*) contained in package *SHELX*) algorithm and their positions were confirmed with the program Coot. Their occupancies were constrained to 1, 0.75 or 0.5

after an initial refinement in which their B-factors were kept constant at reasonable values. The final R factor was 21.1% for all reflections and 20.4% for reflections with F>4σ(F) (Table 1).

Evaluation of the Final Model

The stereochemical quality of the protein model (Figure 6) was checked via the program *PROCHECK* and was created plots analyzing its overall and residue-by-residue geometry. The

Figure 5. Omit map of the area Thr45-Thr47, calculated using the final model and omitting its part 42-49, contoured at 2.0σ. Even though the structure suffered from twinning the electron density map was fairly interpretable

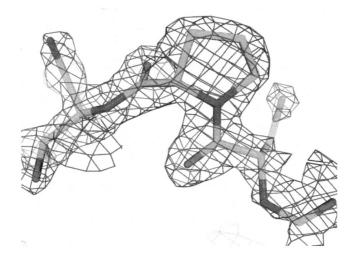

assessment was based on known structures determined with the same accuracy (high resolution limit). The effect of the merohedral twinning on the quality of the structure was apparent in all quality plots.

FUTURE DEVELOPMENTS - PERSPECTIVES

The field of computational methods for X-ray crystallographic structure solution is currently in a state of rapid flux, arising both from internal development of the methods employed, and from external pressures of new applications of the technology. These may be characterised as follows (Cowtan K., 2003):

New applications: High throughput crystallography. The success of genome projects has given rise to new demands for structural studies, with an emphasis on speed and automation. Large numbers of targets for structural studies may be generated automatically from the genome sequences, and to determine structures for a significant number of these targets will require systems which can be run with either high rapidity or parallelism with-

out the intervention of an operator. This raises a number of issues. To start the process, data must be captured automatically from the experimental apparatus, from the crystallisation through to the data collection. Every stage of the structure solution must be reasonably fast or easily parallelisable. Finally, the methods must be robust against most common difficulties, in order to run without human intervention in the majority of cases.

New methods are being used in the structure solution calculation. Bayesian approaches such as those proposed by Bricogne (1988) are increasingly being adopted throughout the structure solution process. Statistical representation of information places different types of information on a similar, objective scale, so that information from many sources can be combined. However, if different types of data from different sources are to be combined, data organisation and representation become significant problems. All the necessary data must be provided to the application in a consistent form, with sufficient organisational information (or 'metadata') that the application will know how to use each datum. Bayesian methods, employing statistical representations of information, are now standard in phasing and

Table 1. Crystallographic data and refinement details

Space group	P6$_2$
Cell dimensions (Å)	65.536 65.536 132.338
Temperature (K)	100
Wavelength (Å)	0.8131 Å
Resolution (Å)	30.0 - 1.65 Å (1.67-1.65 Å)*
Unique reflections	36216
Redundancy	11.5
R$_{sym}$ (%)	6.6 (40.9)*
Completeness (%)	99.2 (100)*
I/σ(I) overall	18.9 (2.0)*
Refinement program	SHELX97
wR2 (%)	46.12
R$_{cryst}$ (R1), all reflections (%)	21.10
R$_{cryst}$ (R1), F > 4σ(F) (%)	20.37 R$_{free}$ (%)=26.43
RMSD bonds (Å)	0.009
RMSD angles (°)	2.2
Mean B$_{eq}$ of main chain atoms (Å^2)	17.1
Residues in most favoured regions	201
Residues in additional allowed regions Residues in generously allowed regions	25 9

* values in parentheses are for the outer shells

$R_{sym} = \Sigma |I_h - <I_h>| / \Sigma I_h$

$R_{cryst} = \Sigma (|F_{obs}| - |F_{calc}|) / \Sigma |F_{obs}|$

$wR2 = (\Sigma[w(F_o^2 - F_c^2)^2] / \Sigma[w F_o^2)^2])^{1/2}$

refinement (La Fortelle E. de and Bricogne G., 1997; Murshudov G. N. *et al.*, 1997), and are becoming standard in the area of phase improvement (Terwilliger T. C., 1999). This trend will continue over the next few years.

Moreover, newly rising fields like Ultra-high-resolution X-ray crystallography of macromolecules promise to provide new insights into the structure–function relationships of biomacromolecules. The picture emerging from macromolecular structures at this resolution is far more complex than previously understood, requiring for its study improved tools for structure refinement, analysis and annotation. Such a goal is not yet attained on a regular basis for macromolecules of biological interest, but a number of efforts are being made

in this direction. The use of Quantum Mechanical (QM) methods in X-ray crystallography opens new possibilities in structure analysis. We can further the analysis of the molecular properties by exploring the actual reactivity parameters of a given atom. In order to do so we must first fit a QM wavefunction to the experimental density (Bethanis *et al.*, 2002, 2008; Jayatilaka D. and Grimwood D. J., 2001). This level of detail would allow the chemical characterization of the interaction between a potential drug and a pharmaceutical target (well beyond the purely geometrical characterization of the interaction), and the identification of the sources of potency and selectivity of lead compounds (e.g. atomic charges) embedded in the finer details of the electron density.

Figure 6. The backbone of the three crystallographic independent molecules of the asymmetric unit of EcFd. View perpendicular the bc plane. Sulfur and Iron atoms are shown in yellow and red respectively

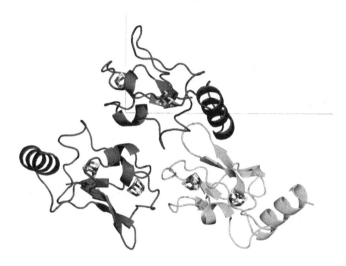

New programming languages greatly facilitate the development of new software. Object oriented programming, either using compiled languages such as C++ or scripting languages such as Python, provides a basis for much greater re-use of software components, increasing productivity for the developer. New object-oriented software frameworks for crystallography are under development. An additional influence is access to new developers and development tools. Much existing crystallographic software is written in Fortran, and yet few universities are now training Fortran programmers. Fortran development tools are increasingly lagging behind their more modern counterparts, and most libraries for modern numerical techniques such as genetic algorithms, neural networks, and automatic differentiation are more easily accessed from other languages, in particular C.

Currently there is a wide range of crystallographic software under development, including larger community projects such as Clipper/CCP4, CCTBX/Phenix, DANSE, and Age Concern. These do not only develop new methods, but also harvest the knowledge of previous generations.

REFERENCES

Abrahams, J. P. (1993). *Compression of X-ray images.* Jt CCP4 ESF–EACBM Newsl. *Protein Crystallogr.*, *28*, 3–4.

Abrahams, J. P., & Leslie, A. G. W. (1996). Methods used in the structure determination of bovine mitochondrial F_1 ATPase. *Acta Crystallographica. Section D, Biological Crystallography*, *52*, 30–42. doi:10.1107/S0907444995008754

Adams, P. D., Grosse-Kunstleve, R. W., Hung, L. W., Ioerger, T. R., McCoy, A. J., & Moriarty, N. W. (2002). PHENIX: building new software for automated crystallographic structure determination. *Acta Crystallographica. Section D, Biological Crystallography*, *58*, 1948–1954. doi:10.1107/S0907444902016657

Bacon, D. J., & Anderson, W. F. (1988). A fast algorithm for rendering space-filling molecule pictures. *Journal of Molecular Graphics*, *6*, 219–220. doi:10.1016/S0263-7855(98)80030-1

Berman, H. M. (2008). The Protein Data Bank: a historical perspective . *Acta Crystallographica. Section A, Foundations of Crystallography*, *64*, 88–95. doi:10.1107/S0108767307035623

Berman, H. M., Westbrook, J., Feng, Z., Gilliland, G., Bhat, T. N., & Weissig, H. (2000). The Protein Data Bank. *Nucleic Acids Research, 28*, 235–242. doi:10.1093/nar/28.1.235

Bethanis, K., Tzamalis, P., Hountas, A., Mishnev, A. F., & Tsoucaris, G. (2000). Upgrading the twin variables algorithm for large structures. *Acta Crystallographica. Section A, Foundations of Crystallography, 56*, 105–111. doi:10.1107/S01087767399013355

Bethanis, K., Tzamalis, P., Hountas, A., & Tsoucaris, G. (2002). Ab initio determination of a crystal structure by means of the Schrödinger equation . *Acta Crystallographica. Section A, Foundations of Crystallography, 58*, 265–269. doi:10.1107/S0108767302003781

Bethanis, K., Tzamalis, P., Hountas, A., & Tsoucaris, G. (2008). Convergence study of a Schrödinger-equation algorithm and structure-factor determination from the wavefunction . *Acta Crystallographica. Section A, Foundations of Crystallography, 64*, 450–458. doi:10.1107/S0108767308010416

Blow, D. (2002), *Outline of Crystallography for Biologists*, Oxford University Press

Bourenkov, G. P., & Popov, A. N. (2006). A quantitative approach to data-collection strategies . *Acta Crystallographica, D62*, 58–64.

Bricogne, G. (1988). *A Baysian statistical theory of the phase problem. i. a multichannel maximum entropy* formalism for constructing generalised joint probability distributions of structure factors. *Acta Crystallographica. Section A, Foundations of Crystallography, 44*, 517–545. doi:10.1107/S010876738800354X

Bricogne, G. (1997a). Ab initio macromolecular phasing: a blueprint for an expert system based on structure factor statistics with built in stereochemistry. *Methods in Enzymology, 277*, 14–19. doi:10.1016/S0076-6879(97)77004-6

Bricogne, G. (1997b). Efficient sampling methods for combinations of signs, phases, hyperphases, and molecular orientations. *Methods in Enzymology, 276*, 424–448. doi:10.1016/S0076-6879(97)76070-1

Brünger, A. T., Adams, P. D., Clore, G. M., DeLano, W. L., Gros, P., & Grosse-Kunstleve, R. W. (1998). Crystallography & NMR System (CNS): a new software suite for macromolecular structure determination. *Acta Crystallographica. Section D, Biological Crystallography, 54*, 905–921. doi:10.1107/S0907444998003254

Burla, M. C., Caliandro, R., Camalli, M., Carrozzini, B., Cascarano, G. L., & De Caro, C. (2007). IL MILIONE: a suite of computer programs for crystal structure solution of proteins . *Journal of Applied Crystallography, 40*, 609–613. doi:10.1107/S0021889807010941

Burla, M. C., Camalli, M., Carrozzini, B., Cascarano, G. L., De Caro, C., & Giacovazzo, C. (2005). SIR2004: an improved tool for crystal structure determination and refinement. *Journal of Applied Crystallography, 38*, 381–388. doi:10.1107/S002188980403225X

Burnett, M. N., & Johnson, C. K. (1996), *ORTEPIII: Oak Ridge thermal ellipsoid plot program for crystal structure illustrations*. Report ORNL-6895. Oak Ridge National Laboratory, Tennessee, USA.

Collaborative Computational Project Number 4 *Acta Cryst.* (1994), *D50*, pp. 760-763.

Cowtan, K. (1994). DM: an automated procedure for phase improvement by density modification. CCP4 ESF-EACBM Newsl. *Protein Crystallogr., 31*, 34–38.

Cowtan, K. (2003). An Overview of some developments in Crystallographic Computing Methods worldwide . *Crystallography Reviews, Vol., 9*(1), 73–80. doi:10.1080/0889311031000069326

Dauter, Z., Wilson, K. S., Sieker, L. C., Meyer, J., & Moulis, J. M. (1997). Atomic resolution (0.94Å) structure of Clostridium acidurici ferredoxin. Detailed geometry of [4Fe-4S] clusters in a protein. *Biochemistry*, *36*, 16065–16073. doi:10.1021/bi972155y

de La Fortelle, E., & Bricogne, G. (1997). Maximum-likelihood heavy-atom parameter refinement in the MIR and MAD methods. *Methods in Enzymology*, *276*, 472–494. doi:10.1016/S0076-6879(97)76073-7

Dodson, E. (2003). Is it Jolly SAD? *Acta Crystallographica. Section D, Biological Crystallography*, *59*, 1958–1965. doi:10.1107/S0907444903020936

Drenth, J. (1994), *Principles of Protein X-ray Crystallography*, Springer-Verlag

Ducruix, A., & Giece, R. (1999), *Crystallization of Nucleic Acids and Proteins*, Oxford University Press

Ealick, S. E. (2000). Advances in multiple wavelength anomalous diffraction crystallography. *Current Opinion in Chemical Biology*, *4*(5), 495–499. doi:10.1016/S1367-5931(00)00122-8

Edwards, A. M., Arrowsmith, C. H., Christendat, D., Dharamsi, A., Friesen, J. D., Greenblatt, J. F., & Vedadi, M. (2000), *Protein production: feeding the crystallographers and NMR spectroscopists*, Nature Structural Biology, structural genomics supplement, Nov., pp. 970 - 972.

Emsley, P., Cowtan, K. (2004), *Coot: model-building tools for molecular graphics* Acta Cryst. *D***60**, Part 12 Sp. Iss. 1, pp. 2126-2132

Evans, P. R. (1993), *Data reduction*. In *Proceedings of the CCP4 study weekend. Data collection and processing*, edited by L. Sawyer, N. W. Isaacs & S. Bailey, pp. 114–122.

Evans, P. R. (1997), *Scaling of MAD data*. In *Proceedings of the CCP4 study weekend. Recent advances in phasing*, edited by M. Winn, Vol. **33**, pp. 22–24.

French, G. S., & Wilson, K. S. (1978). On the treatment of negative intensity observations . *Acta Crystallographica. Section A, Crystal Physics, Diffraction, Theoretical and General Crystallography*, *34*, 517–525. doi:10.1107/S0567739478001114

Furey, W., & Swaminathan, S. (1997). PHASES-95: a program package for the processing and analysis of diffraction data from macromolecules. *Methods in Enzymology*, *277*, 590–620. doi:10.1016/S0076-6879(97)77033-2

Giacovazzo, C. (1977). A general approach to phase relationships: the method of representations . *Acta Crystallographica. Section A, Crystal Physics, Diffraction, Theoretical and General Crystallography*, *33*, 933–944. doi:10.1107/S0567739477002253

Giacovazzo, C. (1980). The method of representations of structure seminvariants. II. New theoretical and practical aspects . *Acta Crystallographica. Section A, Crystal Physics, Diffraction, Theoretical and General Crystallography*, *36*, 362–372. doi:10.1107/S0567739480000836

Graham, I. S. (1995), *The HTML sourcebook*. John Wiley and Sons.

Hauptman, H. A. (1991). The phase problem of x-ray crystallography . *Reports on Progress in Physics*, 1427–1454. doi:10.1088/0034-4885/54/11/002

Hauptman, H. A. (1997). Phasing methods for protein crystallography . *Current Opinion in Structural Biology*, *7*(5), 672–680. doi:10.1016/S0959-440X(97)80077-2

Hendrickson, W. A. (1991). Determination of macromolecular structures from anomalous diffraction of synchrotron radiation. *Science, 254,* 51–58. doi:10.1126/science.1925561

Holm, R. H., Kennepohl, P., & Solomon, E. I. (1996). Structural and Functional Aspects of Metal Sites in Biology . *Chemical Reviews, 96,* 2239–2314. doi:10.1021/cr9500390

Hountas, A., & Tsoucaris, G. (1995). Twin Variables and Determinants in Direct Methods. *Acta Crystallographica. Section A, Foundations of Crystallography, 51,* 754–763. doi:10.1107/S0108767395004661

Jayatilaka, D., & Grimwood, D. J. (2001). Wavefunctions derived from experiment. I. Motivation and theory . *Acta Crystallographica. Section A, Foundations of Crystallography, 57,* 76–86. doi:10.1107/S0108767300013155

Jones, T. A. (1978). A graphics model building and refinement system for macromolecules. *Journal of Applied Crystallography, 11,* 268–272. doi:10.1107/S0021889878013308

Jones, T. A., Zou, J.-Y., Cowan, S. W., & Kjeldgaard, M. (1991). Improved methods for building protein models in electron density maps and the location of errors in these models. *Acta Crystallographica. Section A, Foundations of Crystallography, 47,* 110–119. doi:10.1107/S0108767390010224

Kabsch, W. (1988a). Automatic indexing of rotation diffraction patterns. *Journal of Applied Crystallography, 21,* 67–72. doi:10.1107/S0021889887009737

Kabsch, W. (1988b). Evaluation of single-crystal X-ray diffraction data from a position-sensitive detector. *Journal of Applied Crystallography, 21,* 916–924. doi:10.1107/S0021889888007903

Kabsch, W. (1993). Automatic processing of rotation diffraction data from crystals of initially unknown symmetry and cell constants. *Journal of Applied Crystallography, 26,* 795–800. doi:10.1107/S0021889893005588

Kleywegt, G. J., & Jones, T. A. (1996a). Phi/psi-chology: Ramachandran revisited. *Structure (London, England), 4,* 1395–1400. doi:10.1016/S0969-2126(96)00147-5

Kleywegt, G. J., & Jones, T. A. (1996b). Efficient rebuilding of protein structures. *Acta Crystallographica. Section D, Biological Crystallography, 52,* 829–832. doi:10.1107/S0907444996001783

Kraulis, P. J. (1991). MOLSCRIPT: a program to produce both detailed and schematic plots of protein structures. *Journal of Applied Crystallography, 24,* 946–950. doi:10.1107/S0021889891004399

Lamzin, V. S., & Wilson, K. S. (1993). Automated refinement of protein models. *Acta Crystallographica. Section D, Biological Crystallography, 49,* 129–147. doi:10.1107/S0907444992008886

Lamzin, V. S., & Wilson, K. S. (1997). Automated refinement for protein crystallography. *Methods in Enzymology, 277,* 269–305. doi:10.1016/S0076-6879(97)77016-2

Laskowski, R. A., MacArthur, M. W., Moss, D. S., & Thornton, J. M. (1993). PROCHECK: a program to check the stereochemical quality of protein structures. *Journal of Applied Crystallography, 26,* 283–291. doi:10.1107/S0021889892009944

Laskowski, R. A., MacArthur, M. W., & Thornton, J. M. (1998). Validation of protein models derived from experiment. *Current Opinion in Structural Biology, 8,* 631–639. doi:10.1016/S0959-440X(98)80156-5

Lee, B., & Richards, F. M. (1971). The interpretation of protein structures: estimation of static accessibility. *Journal of Molecular Biology, 55,* 379–400. doi:10.1016/0022-2836(71)90324-X

Leslie, A. G. W. (1988), *Profile fitting.* In: J.R. Helliwell, P.A. Machin and M.Z. Papiz, Editors, Proceedings of the CCP4 Study Weekend, Daresbury Laboratory.

Leslie, A.G.W. (1992), *Recent changes to the MOSFLM package for processing film and image plate data,* Joint CCP4+ESF-EAMCB Newsletter on Protein Crystallography, No. 26.

MacArthur, M. W., Laskowski, R. A., & Thornton, J. M. (1994). Knowledge-based validation of protein structure coordinates derived by X-ray crystallography and NMR spectroscopy. *Current Opinion in Structural Biology, 4,* 731–737. doi:10.1016/S0959-440X(94)90172-4

Main, P., Fiske, S. J., Hull, S. E., Lessinger, L., Germain, G., Declercq, J.-P., & Woolfson, M. M. (1980), *MULTAN80. A system of computer programs for the automatic solution of crystal structures from X-ray diffraction data.* Universities of York, England, and Louvain, Belgium.

McCoy, A. J., Grosse-Kunstleve, R. W., Adams, P. D., Winn, M. D., Storoni, L. C., & Read, R. J. (2007). Phaser crystallographic software . *Journal of Applied Crystallography, 40,* 658–674. doi:10.1107/S0021889807021206

Merritt, E. A., & Bacon, D. J. (1997). Raster3D: photorealistic molecular graphics. *Methods in Enzymology, 277,* 505–524. doi:10.1016/S0076-6879(97)77028-9

Merritt, E. A., & Murphy, M. E. P. (1994). Raster3D version 2.0. A program for photorealistic molecular graphics. *Acta Crystallographica. Section D, Biological Crystallography, 50,* 869–873. doi:10.1107/S0907444994006396

Morris, A. L., MacArthur, M. W., Hutchinson, E. G., & Thornton, J. M. (1992). Stereochemical quality of protein structure coordinates. *Proteins, 12,* 345–364. doi:10.1002/prot.340120407

Murshudov, G. N., Vagin, A. A., & Dodson, E. J. (1997). Refinement of macromolecular structures by the maximum-likelihood method. *Acta Crystallographica. Section D, Biological Crystallography, 53,* 240–255. doi:10.1107/S0907444996012255

Murshudov, G. N., Vagin, A. A., Lebedev, A., Wilson, K. S., & Dodson, E. J. (1999). Efficient anisotropic refinement of macromolecular structures using FFT. *Acta Crystallographica. Section D, Biological Crystallography, 55,* 247–255. doi:10.1107/S090744499801405X

Navaza, J. (1994). AMoRe: an automated package for molecular replacement. *Acta Crystallographica. Section A, Foundations of Crystallography, 50,* 157–163. doi:10.1107/S0108767393007597

Nyborg, J. and A.J. Wonacott A. J. (1977), *The Rotation Method in Crystallography,* U.W. Arndt & A. J. Wonacott, eds, North Holland Publishing Co.

Oldfield, T. J. (2002). Data Mining the Protein Data Bank: Residue Interactions. *Proteins, 49,* 510–528. doi:10.1002/prot.10221

Otwinowski, Z., & Minor, W. (1997), *Processing of X-ray Diffraction Data Collected in Oscillation Mode,* Methods in Enzymology, Volume **276**: Macromolecular Crystallography, part A, pp.307-326.

Perrakis, A., Morris, R., & Lamzin, V. S. (1999). Automated protein model building combined with iterative structure refinement. *Nature Structural Biology, 6,* 458–463. doi:10.1038/8263

Popov, A. N., & Bourenkov, G. P. (2003). Choice of data-collection parameters based on statistic modeling . *Acta Crystallogr. D, 59,* 1145–1153. doi:10.1107/S0907444903008163

Potterton, L., McNicholas, S., Krissinel, E., Gruber, J., Cowtan, K., & Emsley, P. (2004). Developments in the CCP4 molecular-graphics project . *Acta Crystallographica. Section D, Biological Crystallography, 60,* 2288–2294. doi:10.1107/S0907444904023716

Ramachandran, G. N., Ramakrishnan, C., & Sasisekharan, V. (1963). Stereochemistry of polypeptide chain configurations. *Journal of Molecular Biology, 7,* 95–99. doi:10.1016/S0022-2836(63)80023-6

Ramakrishnan, C., & Ramachandran, G. N. (1965). Stereochemical criteria for polypeptide and protein chain conformations. II. Allowed conformations for a pair of peptide units. *Biophysical Journal, 5,* 909–933. doi:10.1016/S0006-3495(65)86759-5

Rhodes, G. (2006), *Crystallography Made Crystal Clear*, Elsevier.

Rossmann, M. G., & Arnold, E. (2001), Editors of International Tables for Crystallography, Volume **F**: Crystallography of biological macromolecules, Part 25: *Macromolecular Crystallography Programs*, pp. 685-743, International Union of Crystallography.

Rossmann, M. G., & Blow, D. M. (1962). The detection of sub-units within the crystallographic asymmetric unit . *Acta Crystallographica, 15,* 24–31. doi:10.1107/S0365110X62000067

Sheldrick, G. M. (2008). A short history of SHELX . *Acta Crystallographica. Section A, Foundations of Crystallography, 64,* 112–122. doi:10.1107/S0108767307043930

Sheldrick, G. M., & Schneider, T. R. (1997). SHELXL: high resolution refinement. *Methods in Enzymology, 277,* 319–343. doi:10.1016/S0076-6879(97)77018-6

Taylor, G. (2003). The phase problem . *Acta Crystallogr. D, 59,* 1881–1890. doi:10.1107/S0907444903017815

Terwilliger, T. C. (2000). Maximum likelihood density modification . *Acta Crystallographica. Section D, Biological Crystallography, 56,* 965–972. doi:10.1107/S0907444900005072

Terwilliger, T. C. (2003). Automated main-chain model building by template matching and iterative fragment extension . *Acta Crystallographica. Section D, Biological Crystallography, 59,* 38–44. doi:10.1107/S0907444902018036

Terwilliger, T. C., & Berendzen, J. (1999). Automated MAD and MIR structure solution . *Acta Crystallographica. Section D, Biological Crystallography, 55,* 849–861. doi:10.1107/S0907444999000839

Thireou, T., Altamazoglou, V., Levakis, M., Eliopoulos, E., Hountas, A., Tsoucaris, G., & Bethanis, K. (2007). CrystTwiv: a webserver for automated phase extension and refinement in X-ray crystallography . *Nucleic Acids Research, 35,* W718–W722. doi:10.1093/nar/gkm225

Tickle, I. J., Laskowski, R. A., & Moss, D. S. (1998). Error estimates of protein structure coordinates and deviations from standard geometry by full-matrix refinement of B- and B2-crystallin. *Acta Crystallographica. Section D, Biological Crystallography, 54,* 243–252. doi:10.1107/S090744499701041X

Tsoucaris, G. (1970a). A new method for phase determination. The 'maximum determinant rule' . *Acta Crystallographica. Section A, Crystal Physics, Diffraction, Theoretical and General Crystallography, 26,* 492–499. doi:10.1107/S0567739470001298

Tsoucaris, G. (1970b). The strengthening of direct methods of crystal structure determination by use of data from isomorphous compounds . *Acta Crystallographica. Section A, Crystal Physics, Diffraction, Theoretical and General Crystallography*, *26*, 499–501. doi:10.1107/S0567739470001304

Tzamalis, P., Bethanis, K., Hountas, A., & Tsoucaris, G. (2003). The crystallographic symmetry test for the correctness of a set of phases . *Acta Crystallographica. Section A, Foundations of Crystallography*, *59*, 28–33. doi:10.1107/S0108767302018810

Usón, I., & Sheldrick, G. M. (1999). Advances in direct methods for protein crystallography . *Current Opinion in Structural Biology*, *9*(5), 643–648. doi:10.1016/S0959-440X(99)00020-2

Vagin, A. A., & Teplyakov, A. (1997). MOLREP: an Automated Program for Molecular Replacement J. *Appl. Cryst.*, *30*, 1022–1025. doi:10.1107/S0021889897006766

Wang, B. C. (1985). Resolution of phase ambiguity in macromolecular crystallography . *Methods in Enzymology*, *115*, 90–112. doi:10.1016/0076-6879(85)15009-3

Weeks, C. M., Blessing, R. H., Miller, R., Mungee, R., Potter, S. A., & Rappleye, J. (2002). Towards automated protein structure determination: BnP, the SnB-PHASES interface . *Zeitschrift fur Kristallographie*, *217*, 686–693. doi:10.1524/zkri.217.12.686.20659

Weeks, C. M., & Miller, R. (1999). The design and implementation of SnB version 2.0. *Journal of Applied Crystallography*, *32*, 120–124. doi:10.1107/S0021889898010504

Yeates, T. O. (1997). Detecting and Overcoming Crystal Twinning . *Methods in Enzymology*, *276*, 344–358. doi:10.1016/S0076-6879(97)76068-3

KEY TERMS AND DEFINITIONS

2[4Fe-4S] ferredoxin (Fds): The 2[4Fe-4S] ferredoxin (Fds) from the bacterium *Escherichia coli* (EcFd) belongs to the *Allochromatium vinosum* (AlvinFd) subfamily and is believed that it is solely involved in electron transfer processes. However, the nature and the complete list of redox partners for the 2[4Fe-4S] Fds have not been clearly identified in most of the cases.

Crystal Structure Solution Methods: The methods of solving the phase problem. The standard methods are: *ab initio* - Direct Methods, Molecular Replacement, Isomorphous Replacement, Anomalous Scattering.

Crystallographic Computing: The field of computational methods for X-ray crystallographic structure solution

Data Reduction: The procedure of indexing, merging, scaling and optimizing the collected X-ray diffraction data in order to improve their consistency and to maximize the number of measurements that are sufficiently accurate to be used in the crystal structure determination.

High Throughput Crystallography: The success of genome projects has given rise to new demands for structural studies, with an emphasis on speed and automation. This high throughput structure determination requires systems which can be run with either high rapidity or parallelism without the intervention of an operator.

Model Building - Structure Refinement: The parameters of the initially obtained atomic model can be refined to optimize its agreement with the observed diffraction data, ideally yielding a better set of phases. Then, a new model can be fit to the new electron density map and a further round of refinement is carried out. This continues until the correlation between the diffraction data and the model is maximized

Molecular Graphics: Visualization of the atomic coordinate data obtained from a crystallographic study is a necessary step in the analysis and interpretation of the structure. Computer programs for molecular modeling provide an interactive, visual environment for displaying and exploring models.

Structural Genomics or Structural Proteomics: The quest to obtain the three-dimensional structures (relative atomic positions) of all proteins. The 3-D protein models give detailed information about proteins' activity, their mechanism for recognizing and binding substrates and effectors, and the conformational changes which they may undergo.

Structure Deposition: Once the model of a molecule's structure has been finalized (analysed and verified), it is often deposited in a crystallographic database such as the Protein Data Bank (PDB, for protein structures) or the Cambridge Structural Database (CSD, for small molecules). The deposition is in the form of lists of atomic coordinates, which can be used to display and study the molecule with molecular graphics programs.

The Crystallographic Phase Problem: In an X-ray diffraction experiment, the intensities of the reflections are measured but it is not normally practicable to measure their relative phases. The calculation of an electron density map, which provides an interpretable picture of a molecule, requires both intensities and phases. Thus, it is essential for the solution of crystal structures to recover the lost phase information. This is known as the *'crystallographic phase problem'*.

X-Ray Crystallography: The interpretation of the pattern obtained from the diffraction of X-rays from a crystal (an ordered array of identical molecules) in order to determine a detailed model of the molecule allowing the resolution of individual atoms.

Chapter 2
Describing Methodology and Applications of an In Silico Protein Engineering Approach

Trias Thireou
Agricultural University of Athens, Greece; National Technical University of Athens, Greece

Kostas Bethanis
Agricultural University of Athens, Greece

Vassilis Atlamazoglou
Agricultural University of Athens, Greece; National Technical University of Athens, Greece

ABSTRACT

This chapter presents a particular cascade of computational steps in order to build a workflow for an in silico protein engineering approach. In this respect, all available information, in order to choose and computationally implement mutations, is described, employed and monitored. Some of the prerequisites of in silico protein engineering are access to various sequence and structure molecular biology databases, software tools for three dimensional molecular visualization and manipulation, sequence and structure alignment and comparison, molecular modelling and molecular docking. The implementation of these steps is demonstrated in the context of performing mutations of particular residues on the ligand pocket of a lipocalin protein family member, to derive the desired ligand binding properties. The example chosen for inclusion introduces the reader to all of the essentials of computational protein engineering experiments. More importantly, it provides insight into understanding and properly interpreting the data produced by these methods.

ASPECTS OF PROTEIN DESIGN & ENGINEERING

Due to the complex nature of the protein-folding problem, the numerous attempts of *de novo* protein design have not lead to a major success. Therefore, more 'realistic' approaches have become accepted towards protein engineering, in order to acquire novel functions (DeGrado W.F., 1997; Desjarlais J.R. and Mayo S. L, 2002)

Indeed there is no need to design a protein sequence from scratch to do protein engineering. A

DOI: 10.4018/978-1-60566-768-3.ch002

general strategy is rational design, in which the scientist uses detailed knowledge of the structure and function of the protein to make desired changes. In this sense and in the context of this chapter, we will focus on the concept of scaffolds that can be equipped with artificial binding sites, an approach that has gained recent interest. The term 'scaffold' is being used to describe some kind of natural protein architecture onto which unrelated structural elements can be incorporated and thus new biochemical activities created (Skerra A., 2000).

A few types of protein folds have been examined for this purpose. For example single immunoglobulin (Ig) domains and helical bundles, have been found to be useful for the generation of biomolecules with the ability of binding to other proteins. These scaffolds, however, are hardly capable of complexing small ligands (Skerra A., 2000).

In this respect, the lipocalin protein family appears to be a promising model system. Lipocalins constitute a family of small, robust proteins that typically transport or store biological compounds, which are either of low solubility or are chemically sensitive, including vitamins, steroid hormones, odorants and various secondary metabolites. The artificial lipocalins recognizing specific ligands, termed "anticalins", could provide an alternative to recombinant antibody fragments, with interesting applications in biotechnology and medicine.

Computational approaches offer significant potential for engineering protein structure and function, and can be combined with experimental testing to gain new insights into the fundamental properties of proteins, in rational structure based design. These methods offer the potential to drive structure and function manipulations for implementing *in silico* protein engineering experiments. In this respect, the number of experiments necessary to better understand a protein's function can be significantly reduced.

In this chapter, we present and discuss a particular cascade of computational steps in order to build a workflow for *in silico* protein engineering. In this respect, all available information, in order to choose and implement *in silico* mutations, is described, employed and monitored. Some of the prerequisites of *in silico* protein engineering are access to various sequence and structure molecular biology databases, software tools for three dimensional molecular visualization and manipulation, sequence and structure alignment and comparison, molecular modelling and molecular docking.

An in-depth description of key terms and concepts related to the above mentioned procedure is being described. Additionally the implementation of these steps is demonstrated in the context of performing mutations of particular residues on the ligand pocket of a lipocalin protein family member, to derive the desired ligand binding properties. The example chosen for inclusion introduces the reader to all of the essentials of computational protein engineering experiments. More importantly, it provides insight into understanding and properly interpreting the data produced by these methods.

PROTEIN SCAFFOLDS

As mentioned above, rather than attempting to design from scratch, one can utilize a number of nature provided templates to improve on. At the very least, these templates already fulfill the criteria for stably folding proteins.

Therefore, the idea is to utilize structurally well-defined polypeptide frameworks for the introduction of novel functions, by locally reshaping a part of the protein surface that is thought to be less important for the protein folding process or its stability (Skerra A., 2000).

Hence an ideal protein scaffold should provide a rigid folding unit, which spatially brings together several exposed loops (Ku J, Schultz PG. 1995). Since protein folding is a highly cooperative event, a critical issue is how it can be assessed whether a certain number of amino acid mutations can be tolerated in the chosen loop region, without affecting ability to adopt a defined structure or without becoming less stable. The answer is that a protein scaffold should exhibit the particular feature of having structurally separated the stability of its structural conformation and the local shape and molecular recognition function of its active site.

Three general criteria that seem to characterize an applicable scaffold are: (i) the protein family should possess a well-defined hydrophobic core, which is structurally superimposable among its individual members and can provide a major contribution to the free energy of folding; (ii) it should possess a solvent-accessible active site or binding pocket, which is spatially well separated from the core, and which is ideally involved in the recognition of clearly different targets; (iii) it should have a rather low sequence homology among the members of the family and with various biochemical functions (Skerra A., 2000).

Once such a promising scaffold has been selected, a whole series of currently available biochemical methods may be applied in order to modify it to derive the desired ligand binding function.

Antibodies as a Nature's Paradigm for Protein Engineering

Antibodies (also known as immunoglobulins) are gamma globulin proteins that are found in blood or other bodily fluids of vertebrates. They are typically made of basic structural units - each with two large heavy chains and two small light chains - to form, for example, monomers with one unit, dimers with two units or pentamers with five units. Antibodies comprise a natural type of biomolecular scaffold which is utilized by the immune system of higher organisms against pathogenic invaders or their toxins.

Although the general structure of all antibodies is very similar, a small region at the tip of the protein is extremely variable, allowing a very large number of antibodies with slightly different tip structures to exist. This region is known as the hypervariable region. Each of these variants can bind to a different target, known as an antigen.

The unique part of the antigen recognized by an antibody is called an epitope. These epitopes bind with their antibody in a highly specific interaction, called induced fit, which allows antibodies to identify and bind only their unique antigen.

Antibodies owe their unique recognition capability to a modular type of structure, which has been well identified (Padlan, 1994; Bork et al., 1994). On the first level, they consist of two types of domains which are either of constant or of variable character, although both are based on the structurally well-conserved so-called Ig fold.

Altogether six hypervariable loops, also called complementarity-determining regions (CDRs), three within each variable domain, form the combining site. This structural principle explains the success of this protein class as an almost universal molecular tool used in the biological sciences for the complexation of a vast number of molecules.

The scaffold-like properties of the Ig fold have been recognized from the structural point of view. The application of antibody fragments as scaffolds has been promoted by the development of molecular library techniques, especially with the help of bacterial phage display.

This technology has proven to be practically useful for the preparation of artificial binding proteins on the basis of Ig fragments. However, there are some intrinsic disadvantages of the immunoglobulin architecture. Antibodies possess considerable size. In terms of economy of production and several technical applications, smaller entities are desired. Second, antibodies are composed of two different polypeptide chains, which can lead to unstable association and necessitates complicated cloning steps. Third, there are six hypervariable loops in the combining site, which are difficult to manipulate simultaneously and all of which are probably not needed. Therefore, smaller proteins are certainly advantageous for biotechnological purposes.

Other Systems

The evaluation of potential protein scaffolds is based on the structural plasticity of the binding site that may be created on the basis of a given scaffold, as well as the affinities and specificities that have been experimentally achieved for specific ligands.

In the first respect, scaffolds, which only provide a single loop for variation, i.e. most protease inhibitors and the engineered enzymes with loop insertions in their active sites, are probably of a less broad use. In the second respect, there is discrimination between macromolecular 'antigens' and low molecular weight 'haptens' as target ligands.

Many of the scaffolds tested so far are probably capable of complexing other proteins, especially because these macromolecular targets usually provide grooves into which exposed loop segments or convex surface regions can intimately bind.

In the case of small haptens the situation is even more critical. Although several scaffolds (including single Ig domains, helix bundles, etc) have been investigated in this regard, the affinities were usually low, and dependence in terms of the macromolecular hapten carrier could be observed.

The choice of a suitable polypeptide scaffold opens the way to many applications of protein design. Some other examples are briefly mentioned.

Affibodies

The first non-immunoglobulin protein scaffold was described in 1995. These artificial a-helical receptor proteins are called affibodies. Their sequence can be modified in order to obtain the appropriate ligand binding properties. Affibodies have a wide range of applications as selective biotechnology reagents.

Streptavidin

Streptavidin system has also been used for its strong (optimal binding in the order of ~pM) but often irreversible binding of small vitamins. It is a good diagnostic but not transporting system.

Protease Inhibitors

Protease inhibitors are widely known small and very stable protein systems. In the majority of cases, their active site is composed of a small number of peptides, with a varying amino acid sequence in a loop form. Therefore these systems are used as structural frameworks for structurally confined peptide loops.

Lipocalins

The lipocalin protein family consists of small extracellular proteins, that are mainly responsible for the storage and transport of chemically sensitive or poorly soluble compounds, such as vitamins, steroids and metabolic products (Flower D.R, 1995; Flower D.R, 1996; Schlehuber S. and Skerra A., 2000; Bishop R.E., 2000).

Lipocalins display unusually low levels of overall sequence conservation, with pairwise sequence identity often falling below 20%. However, the kernel lipocalins share three characteristic conserved sequence motifs, while the outlier lipocalins, a group of more divergent family members, share only one. This motif can be used as a diagnostic of family membership (Flower D.R, 1996).

Despite their weak sequence homology, lipocalin crystal structures are highly conserved and comprise a single eight-stranded antiparallel beta-barrel, which encloses an internal ligand-binding site. Six of the seven loops linking the beta-strands of the barrel are short b-hairpins, whereas the first one is a large X loop forming a lid folded back to partially close the internal binding site. The two ends of the b-barrel are topologically distinct. The

Open end of the molecule has four b-hairpins that form the entrance to the ligand-binding pocket. The N-terminal polypeptide chain crosses the other end via a conserved 3_{10} helix affecting closure of the barrel (Closed end). Lipocalins that reveal extraordinary similarity with each other, in terms of their secondary structure composition, are called "prototypic". The root mean square deviation (RMSD) value for the structural superposition of their beta-barrel is smaller than 1.3Å. Other members of the family may possess additional secondary structure elements, be intertwined dimers or carry inverted strands in the beta-barrel (Flower D.R, 1996).

Lipocalins are characterized by three molecular-recognition properties: ability to bind small hydrophobic molecules, ability to bind to specific cell-surface receptors and formation of complexes with soluble macromolecules (Flower D.R, 1996).

The lipocalin family affords several benefits for protein engineering applications in biotechnology and medicine.

Biochemical and structural findings demonstrate that the beta-barrel of prototypic lipocalins represents a stable folding unit that can support loops with highly divergent length, amino acid composition and conformation at its open end (Schlehuber S. and Skerra A., 2002).

The lipocalins binding site can adopt extremely different shapes ranging from a wide, funnel-like opening to the solvent, to a closed cavity that fully encapsulates the ligand. Therefore, it can be effectively reshaped in order to complex various ligands with high affinity and specificity.

Lipocalins usually consist of a single small thermostable polypeptide chain. Since they are typically secretory proteins, they have a varying number of disulfide bonds, which appear to be generally dispensable for formation of the lipocalin fold. Moreover, few members of the family are glycosylated and their four hypervariable loops can be more easily manipulated at the genetic level. Thus lipocalins can be easily produced as recombinant proteins using bacterial expression systems (Skerra A., 2000; Muller H.N. and Skerra A., 1993).

Lipocalins' Function

The lipocalins are generally classified as extracellular transport proteins. The human plasma retinol-binding protein (RBP), the first lipocalin for which a 3D structure was resolved, is complexed to transthyretin, another plasma protein, and transports the poorly soluble and oxidation-prone vitamin A from the liver to peripheral tissues. Although some family members are not well characterized functionally, it is clear that the lipocalins have significant roles in several biological functions, including regulation of cell homoeostasis and modulation of the immune response, pheromone activity, olfaction, cryptic coloration, enzymatic synthesis and clearance of endogenous and exogenous compounds (Flower D.R, 1996).

Contrary to antibodies, there is no physiological mechanism for functional variation of the lipocalins, which were apparently optimized during evolution to serve specialized tasks. For example, the human lipocalins exhibit distinct binding functions and have similar orthologues in other mammals or vertebrates (e.g. the human RBP and the porcine protein differ by only 12 amino acid residues located far from the retinol-binding site).

Lipocalins research and the expected applications in biotechnology and medicine, resulted in a broader interest in this protein family, which is reflected in the corresponding special issue of Biochimica Biophysica Acta (Akerstrom B. et al., 2000).

Anticalins

Engineered lipocalins created for the complexation of prescribed target molecules at their reshaped ligand pocket are called anticalins and have potential applications in biotechnology, medical diagnosis and therapy (Skerra A., 2001).

Initial studies were based on Bilin-Binding Protein (BBP) from Pieris brassicae, which normally complexes biliverdin IX (Blv) at its rather wide and shallow ligand binding pocket (Huber R. et al., 1987). Sixteen residues distributed across the four loops of the open end of the ligand barrel were identified by molecular modeling and subjected to random mutagenesis. A mutant library of 3.7×10^8 independent transformants was generated and different variants were selected via bacterial phagemid display. Those variants were able to complex small ligands, namely fluorescein, digoxigenin, phthalic acid esters and doxorubicin, with high affinity and specificity, exhibiting dissociation constants in the order of nM (Schlehuber S. and Skerra A., 2005; Schlehuber S. et al., 2000; Beste G. et al., 1999).

Structural studies revealed that despite random mutagenesis, the beta-barrel architecture remained intact, while the loop region exhibited considerable backbone plasticity. Conformational changes of the backbone and rearrangement of aromatic side chains within the binding site created shape complementarity with the ligand. Specific interactions are formed by hydrogen bonds, which are partially mediated by water molecules (Korndorfer I.P. et al., 2003).

Anticalins appear to have high physico-chemical stability and they can be massively produced in E. coli (Beste et al., 1999). Lipocalins are also well suited for the construction of functional fusion proteins, which could serve as valuable reagents for bioanalytical purposes (e.g. fusion with alkaline phosphatase) (Schmidt T.G.M. and Skerra A., 1994; Schlehuber S. et al., 2000).

Moreover, anticalins that are based on human lipocalins and recognize macromolecular protein targets can be generated (Schlehuber S. et al., 2000). Since large macromolecules cannot penetrate into the ligand-binding site, side chains at more exposed positions, close to the tips of the four loops at the open end of the beta-barrel, should be subjected to random mutagenesis. Using this approach, anticalins with specificities for disease-related cell surface receptors such as cytotoxic T lymphocyte-associated antigen and affinities in the nM range were generated.

Ligand's Binding Modification

A basic reason for designing a modified protein is the implementation of stereochemical changes that will induce functional alterations, necessary for the binding of a specific ligand.

In many cases, protein function depends on the specific binding of certain ligands. Enzymes as catalysts of chemical reactions can be affected by other molecules. Inhibitors are molecules that decrease enzyme activity; activators are molecules that increase activity. Additionally receptors also show discrimination and specific binding to ligand molecules.

Binding modification is based on the selection of the appropriate residues that should change, using the available information from the experimentally resolved complexes of the specific ligand.

METHODOLOGY OF COMPUTATIONAL PROTEIN MODIFICATION

The methodology relies on the exploitation of all available information for selecting the appropriate mutations and creating an engineered protein with the desired properties (DeGrado W.F., 1997, Desjarlais J.R. and Mayo S. L., 2002). Some suggestive (definitely not the only) steps that may be applied are detailed below (Steipe, 1998).

- Find natural proteins that possess several of the desired properties and could be used in the computational protein engineering process.
- Find structural models for the protein, using the Protein Data Bank (PDB, http://

www.rcsb.org/pdb/home/home.do)
(Berman H.M. et al., 2000), or representative subsets of PDB, which have been generated using clustering algorithms based on sequence (http://bioinfo.tg.fh-giessen.de/pdbselect/) (Hobohm U. and Sander C., 1994)) or structure similarity (http://ekhidna.biocenter.helsinki.fi/dali_new/start) (Holm L. and Sander C., 1997).

- Perform a multiple alignment and search protein family databases to identify critical conserved residues. The HSSP database (http://swift.cmbi.kun.nl/swift/hssp/) contains proteins of known structure that were aligned with homologous sequences (Sander C. and Schneider R., 1991). The InterPro database (http://www.ebi.ac.uk/interpro/) of protein families, domains, repeats and sites provides information about identifiable features found in known proteins (Mulder N.J. et al., 2007).

- Structural alignment. Alignment and comparison of protein structures is a challenging task. While the efficiency of sequence alignment algorithms derives from their locality i.e., the score for an aligned residue pair is not influenced by the rest of the alignment, aligning protein structures may require the taking of contextual information into account (Holm L. and Sander C, 1994). One of the most widely used structural alignment programs is DALI (http://ekhidna.biocenter.helsinki.fi/dali_server/).

- The structural alignment of the proteins of interest are then inspected for interactions that may need to be conserved, side chain rotamers, or alternative loop sequences that may generate specific local backbone conformations.

Sequence and Structural Data

Protein sequences can be acquired from the curated protein sequence database SWISS-PROT (http://

www.expasy.org/sprot/), while atomic coordinates of the experimentally resolved 3D structures can be obtained from the structural database PDB (Protein Data Bank, http://www.rcsb.org/pdb).

Protein Sequence Multiple Alignment

A multiple sequence alignment facilitates identification of critical conserved residues and determination of frequencies of occurrence of amino acids in specific locations, indicating which residues may be optimal in these locations. This is very useful in designing experiments to test and modify the function of specific proteins, in predicting the function and structure of proteins and in identifying new members of protein families.

Most multiple sequence alignment programs use heuristic methods rather than global optimization, because identifying the optimal alignment between more than a few sequences of moderate length is prohibitively computationally expensive.

ClustalW is a widely used multiple sequence alignment computer program. It produces biologically meaningful multiple sequence alignments of divergent sequences. It calculates the best match for the selected sequences, and lines them up so that the identities, similarities and differences can be seen. Evolutionary relationships can be seen via inspecting Cladograms or Phylograms (Thompson J.D. et al, 1994).

Interpro

InterPro is a database of protein families, domains, motifs, repeats and sites in which identifiable features found in known proteins can be applied to new protein sequences. It combines a number of databases (referred to as member databases) that use different methodologies and a varying degree of biological information on well-characterized proteins to derive protein signatures (methods).

A sequence motif is a locally conserved region of a sequence, or a short sequence pattern shared by a set of sequences. The term "motif" most often refers to any sequence pattern that is predictive of a molecule's function, a structural feature, or family membership. The identified motifs are very useful to identify critical conserved residues that should not change during the *in silico* experiments.

Structural Alignment Software: DALI and DaliLite

Comparison of protein structures may reveal interesting biological similarities that may not be detectable using sequence comparison methods. Many measures are available, often based on intermolecular distances upon rigid body super-imposition (Holm L. and Sander C, 1994).

The use of intramolecular geometrical relationships, such as distances, to describe protein structures has the advantage of being independent of the coordinate frame.

The Dali algorithm uses a detailed description of internal geometry, in the form of Calpha-Calpha distance matrices and builds up an optimal alignment using Monte Carlo optimization to combine pairs of matching fragments (matching submatrices of the distance matrices) into larger consistent sets of pairs.

The Dali server is a network service for comparing protein structures in 3D. The coordinates of a query protein structure is compared, using the Dali algorithm, against those in the Protein Data Bank (PDB).

DaliLite is a standalone version of Dali for pairwise structure comparison and structure database searching. It provides a web interface to view the results, multiple alignments and 3D superimpositions of the structures and can be found at http://www.ebi.ac.uk/DaliLite/.

Exclusion Volume Maps

Depth perception is an important issue in three dimensional manipulation on a two dimensional display. Volumes represented by points, lines or a wire mesh are still difficult to comprehend, and comparison of two volume projections on a flat screen is entirely dependent on the relative orientation of the projections.

Exclusion volume maps can facilitate the study of closely interacting molecules, such as host and guest. They can be used as a tool for finding the position of guests and for molecular modeling studies, since they provide an outline for the design of the wire model of the guest (E Eliopoulos and IM Mavridis).

Analysis of Protein Ligand Interactions

LIGPLOT automatically generates schematic diagrams of protein-ligand interactions, including hydrogen bonds and hydrophobic contacts, from the 3D coordinates of a PDB file (Wallace A.C., 1995). It is available free to academic institutions by anonymous ftp from: ftp.biochem.ucl.ac.uk

Protein Structure Validation and Analysis

Stereochemical validation of protein model structures is an important aspect of molecular modeling. PROCHECK (www.biochem.ucl.ac.uk/~roman/procheck/procheck.html) is a suite of programs for assessing the "stereochemical quality" of a given protein structure producing a number of PostScript plots analysing its overall and residue-by-residue geometry (Laskowski R. A., 1993). Unusual regions highlighted by PROCHECK are not necessarily errors, but may be unusual features for which there is a reasonable explanation Nevertheless they are regions that should be checked carefully.

Contact Analysis Methods

All-atom contact analysis is an independent, complementary approach for assessing macromolecular structure quality. Evaluating atomic packing is especially useful in order to either study a mutation or a sidechain conformational change within a protein, or analyze the all-atom contacts between two molecules.

Probe generates "contact dots" at points on the van der Waals surface of atoms which are in close proximity to other atoms, whereas Reduce can be used to add hydrogens, since meaningful analysis of molecular contact surfaces requires that all atoms are considered. These programs, plus Prekin and Mage used for viewing and exploration, are available from the web site http://kinemage.biochem.duke.edu/

Molecular Visualization & Modeling

The ability to flexibly visualize various aspects of a molecular system is extremely important. Visual examination often provides key insights to properties such as intra- and inter-molecular interactions.

Swiss-PdbViewer (http://www.expasy.org/spdbv/) is a molecular graphics analysis program for displaying and manipulating 3D structures (Guex N. and Peitsch M.C.,1997). It includes a version of the GROMOS 43B1 force field, which allows evaluating the energy of a structure as well as repairing distorted geometries through energy minimization.

Energy Minimization

Following a residue mutation or a different side chain rotamer selection, energy minimization of the 'distorted' structure geometry could release internal constraints, by moving atoms.

A modified version of the Newton-Raphson method, Adopted Basis Newton-Raphson (ABNR) that maintains excellent convergence properties but in a much shorter time is a good choice. Each step of the ABNR method begins with a steepest descents stage. Then the bond lengths and angles that change the most are noted and only

these coordinates are used in a second stage of Newton-Raphson minimization.

In order to release local constraints and remove close contacts, a number of 100 ABNR energy minimization steps along with an energy based criterion (energy gradient tolerance = 0.01) can be used.

Molecular Docking Simulation

Molecular docking studies the structure of the intermolecular complex formed between two or more constituent molecules. It is a useful tool in drug discovery efforts. Especially, protein–ligand docking occupies a very special place in the general field of docking, because of its usage as a primary component in many drug discovery programs (Sousa SF et al., 2006).

As far as protein–ligand docking methods are concerned, the docking problem can be rationalized as the search for the specific ligand conformations within a given targeted protein, when the structure of the protein is known or can be estimated. The binding affinity prediction refers to how well the ligand binds to the protein (*scoring*).

Docking methods can be described as a combination of a *search* algorithm and a *scoring* function. The two critical elements in a search algorithm are *speed* and *effectiveness*. The scoring function should represent the thermodynamics of interaction of the protein–ligand system in an adequate manner, so as to distinguish the true binding modes and to rank them accordingly. Additionally, it should be fast enough to allow its application to a large number of explored binding modes.

One of the most popular docking programs is Autodock. It uses a Lamarckian genetic algorithm (LGA), but incorporates also a Monte Carlo simulated annealing and a traditional genetic algorithm. However, the last two are not as efficient and reliable as the LGA. LGA search method implements a methodology of global optimization with local search. The program uses a five-term force field-based function, comprising a Lennard-Jones 12-6 dispersion term, a directional 12-10 hydrogen bonding term, a coulombic electrostatic potential, an entropic term, and an intermolecular pairwise desolvation term. Autodock reports docked energies, which include the intermolecular and intramolecular interaction energies and are used during the dockings, and predicted free energies, which include the intermolecular energy and the torsional free energy (Morris GM et al., 1998).

The new version of AutoDock (version 4.0) also encompasses receptor side chain flexibility in addition to the set of functions already included in the previous version.

A CASE STUDY: SELECTING AND IMPLEMENTING MUTATIONS OF PARTICULAR RESIDUES OF THE HYDROPHOBIC LIGAND POCKET OF B-LACTOGLOBULIN

In this section, we present a particular cascade of computational steps that form a workflow for an *in silico* protein engineering approach. In this respect, all available information, in order to choose and implement *in silico* mutations, is described, employed and monitored.

We examine the features of this procedure in the context of selecting and implementing mutations of particular residues of the hydrophobic ligand pocket of β-lactoglobulin, in order to accommodate the hydrophilic cardiac steroid digoxigenin (DOG).

Practical Advantages of β-lactoglobulin

Beta-Lactoglobulin (Blg) is the major protein component of whey from the milk of many mammals that binds a wide range of small hydrophobic ligands. It is readily isolated from bovine milk in large quantities and has been well studied by a wide variety of physical and biochemical techniques. It is remarkably resistant to acid denaturation, and although some specific moiety (-ies) in the protein cannot return to the native conformation from a denatured state, such a conformational difference between renatured and native forms has no affect on the biological function of ligand binding. Moreover, the Blg gene and its transcription have been well studied and the production of transgenic mice and sheep is well established.

DigA16: An Anticalin that Binds Digoxigenin

DigA16 is an engineered lipocalin derived via reshaping of the natural ligand pocket of the Pieris brassicae bilin-binidng protein (BBP), in order to bind digoxigenin (DOG) (Beste G. et al., 1999; Schlehuber S. et al., 2000; Korndorfer I.P. et al., 2003).

DigA16 carries 21 amino acid mutations with respect to the BBP, 17 of which are distributed in the proximity of the entrance to the ligand pocket and the other 4 were introduced elsewhere to facilitate cloning and improve stability.

Despite substitution of 21 side-chains the tertiary structures of BBP and DigA16, either in the complexed or uncomplexed state, are very similar. The largest deviations are seen at the four loops forming the entrance to the binding site, while the superposition of just the 58 Ca-positions of the b-barrel results in a r.m.s.d. of 0.84Å. Especially for loop 1, a new seven amino acid a-helical segment with two turns appears, stabilized by the hydrogen bonds between the mutated residues Arg58 and Ser60 and the unmutated residue Tyr39, and the packing of the newly introduced side-chain of His35 between the side-chains of Tyr39 and of the mutated residue Leu127 (Korndorfer I.P. et al., 2003).

RESULTS OF THE DESCRIBED PROCEDURE

Multiple Sequence Alignment

A multiple sequence alignment facilitates identification of critical conserved residues and determination of frequencies of occurrence of amino acids in specific locations, indicating which residues may be optimal in those locations.

ClustalW and the default parameter values were used: gap open penalty - 10.0, protein gap extension penalty - 0.2 and the Gonnet substitution matrix.

Based on a multiple sequence alignment of representative members of the lipocalin protein superfamily, the residues of β-lactoglobulin that belong to the three conserved sequence motifs of the lipocalins were recognized. These residues (Gly17, Trp19, Tyr20, Thr97, Asp98, Tyr99, Tyr102 and Arg124) remained immutable in the process of designing / modifying the engineered lipocalin.

Studying the Experimentally Solved Complex

Digoxigenin is bound to digA16 in a central position at the open end of the b-barrel, mainly via van der Waals interactions with four aromatic side-chains (Tyr39, Tyr88, Phe114, and Trp129) forming the largest contacts. Tyr39 and Trp129 occur already in the BBP, whereas Tyr88 replaced Leu and Phe114 replaced Tyr. Moreover hydrogen bonds are formed with digoxigenin polar hydroxyl and oxo groups, which are partially mediated by water molecules. Despite the hydrophilic character of the ligand pocket, DOG is almost fully trapped within the binding site of DigA16, with a remaining accessible surface of 5%.

Structural Alignment of the β-lactoglobulin and the Complex of digA16 with Digoxigenin

The structural alignment of the experimentally resolved structures of β-lactoglobulin (PDB code: 1BEB) and the complex of digA16 with digoxigenin (PDB code: 1LKE) was produced using Dali-Lite (figure 1). The rms fit of the Ca's was 2.81 Å.

It appears that the binding site of digA16 is more wide and deep compared to β-lactoglobulin. Additionally if DOG was placed in β-lactoglobulin maintaining the same position as in digA16-DOG complex, it would clash with L5 turn (connecting strands E and F of β-lactoglobulin). Therefore a different position of digoxigenin in the binding site of β-lactoglobulin should be selected.

Finding Binding Modes of the Ligand

The technique of exclusion volume maps combined with available information was used to identify possible positions of the ligand in the protein binding site.

The orientation of digoxigenin in the binding site of β-lactoglobulin is determined by the sugar moiety attachment to the 3-OH group to form the cardiac glycoside digoxin.

Two possible positions – binding modes of digoxigenin in β-lactoglobulin were suggested as shown in figure 2.

One of these binding modes will be used for further study and analysis (*binding mode 1*). Using PROCHECK and Probe, the β-lactoglobulin residues clashing with digoxigenin were identified. AutoDock was used to estimate the binding affinity of digoxigenin to β-lactoglobulin.

The position determined by exclusion volume was used as the starting ligand position in the automated docking simulation, avoiding primary studies of "blind docking" and using a finer grid for the representation of the binding site. A total of 10 docking configurations were determined

Figure 1. Structural superposition of β-lactoglobulin (light gray) and digA16 complexed with digoxigenin

Figure 2. Exclusion Volume Maps along the ligand axis for suggested binding mode 1 (a) and 2 (b)

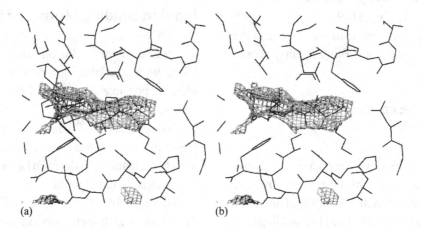

(a)   (b)

for each docking calculation. The choice of a "preferable" docking configuration was based on the values of the binding free energy, the number of configurations in a cluster and the rms distance of the resulting against the initial position of the ligand.

The positive value of the resulted energy term listed in table 1, indicates clashes – "bad contacts" between atoms, in agreement with PROCHECK and Probe results.

Selection and *in silico* Implementation of Mutations

Reshaping the natural binding site to increase protein-ligand affinity involved selecting appropriate mutations based on the following criteria:

- Optimisation of ligand – protein interactions
  - Selection of polar residues to improve binding site-ligand complementarity

Table 1. AutoDock results for native and mutated BLG-DOG binding

	Docked Energy (kcal/mol)	Free Energy of Binding (kcal/mol)	K_i (M)	rms (Å)
BLG *binding mode 1*	41.36	41.67	– *	0.87
BLG_MUT1	-10.32	-10.01	4.6×10^{-8}	0.97

* Dissociation constant is not calculated due to positive $\Delta G_{binding}$ value.

Table 2. Proposed BLG mutations for binding mode 1.

BLG	BLG_MUT1
Ile12	Asn12
Val15	Thr15
Val41	Ser41
Leu46	Asn46
Leu54	Asn54
Ile56	Leu56
Ile71	Phe71
Val92	Ala92
Met107	Tyr107

and to create hydrogen bonds, in order to increase binding affinity and specificity

○ Selection of aromatic residues for better "*packing*" of the ligand

• Elimination / minimisation of ligand-protein clashes

• Selection, where possible, of amino acid substitutions that exist in the corresponding locations of the multiple sequence alignment and are likely to be better tolerated.

• Conservation of residues of the hydrophobic core / of the characteristic protein family motifs, hence minimizing potential deleterious effects on the protein folding.

• Selection of a limited number of mutations and experimental confirmation of protein folding and ligand binding, before suggesting additional mutations.

During *in silico* mutagenesis, side chain rotamers of mutated residues were selected so as to form the intended protein - ligand interactions and minimise clashes with other protein residues.

Based on the above criteria, the mutations listed in table 2 were selected to improve binding of *digoxigenin* to β-lactoglobulin, and they were in silico implemented using Swiss-PdbViewer.

Energy Minimization and Structural Evaluation of the Engineered Protein

Close contacts between mutated and native residues revealed by contact analysis, created repulsion forces and increased system's dynamic energy. Therefore, elimination of possible "bumps" by energy minimization is carried out prior to protein structure validation.

Mutated residues causing "bad contacts" in β-lactoglobulin can be identified using PROCHECK and Probe.

Figure 3. Interactions between mutated BLG (set 1) and DOG. Dashed lines indicate hydrogen bonds between protein and ligand.

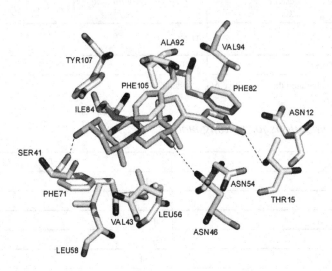

Since the engineered protein resulted by incorporating few residue mutations in the experimentally resolved structure of β-lactoglobulin, geometrical restrictions can be imposed, "fixing" ligand position, protein main chain and native side chains, during the energy minimization procedure.

Using PROCHECK, the Ramachandran plots of the engineered protein structure and the experimentally resolved structure of β-lactoglobulin were produced and compared. The number of residues found in the regions characterized as "disallowed" and "generously allowed" remained constant and all stereochemical, main chain and side chain parameters were within the allowable range of values, providing an indication of the quality of the proposed engineered structures.

Studying the *in silico* Engineered Complex

The ligand binding affinity to the computationally engineered protein was estimated using automated docking techniques. Protein residues interacting with the ligand were identified using LIGPLOT. AutoDock results are presented in table 1 (*BLG_MUT1*), while residues interacting with digoxigenin are shown in figure 3 and atoms forming hydrogen bonds are listed in table 3.

The vast majority of the resulted different docking configurations of digoxigenin to β-lactoglobulin were very similar in binding orientation and calculated binding energy and with small r.m.s against the initial position. The estimated negative ΔG values for the complexes of the proposed modified structures indicated improved

Table 3. Hydrogen Bonds between mutated BLG (set 1) and DOG (d: donor, a: acceptor)

BLG_MUT1 residue	distance (Å)	BLG_MUT1 atom	DOG atom
Thr15	3.08	OG1 (*d*)	*O23 (a)*
Ser41	3.33	OG (*d*)	*O32 (a)*
Asn46	3.12	ND2 (*d*)	*O12 (a)*

binding affinity of digoxigenin, resulting from the elimination of intermolecular close contacts and the improvement of protein-ligand interactions (hydrogen bonds formation / steric, electrostatic and hydrophobic complementarity).

Of course the experimental validation of the above mentioned computational results remains a critical necessity.

CONCLUSION

Structure-based computational procedures, combined with experimental testing, have emerged as a robust approach to gain new insights into the fundamental properties of proteins. These methods offer the potential to drive structure and function manipulations for implementing *in silico* protein engineering experiments. This approach is particularly useful for the estimation – prediction of the possible effect of mutations before they are actually implemented, and contributes considerably in understanding the mechanisms of interaction and reducing the required cost and time in the laboratory.

In this chapter, a particular cascade of computational steps in order to build a workflow for an *in silico* protein engineering approach is presented. These are some suggestive but definitely not the only steps of a protein engineering approach that exploits protein scaffolds. Some of the prerequisites are access to various sequence and structure molecular biology databases, software tools for three dimensional molecular visualization and manipulation, sequence and structure alignment and comparison, molecular modelling and molecular docking.

The implementation of these steps is demonstrated in the context of performing mutations of particular residues on the ligand pocket of a lipocalin protein family member, to derive the desired ligand binding properties.

Computational protein engineering is an iterative cycle of in silico design and evalua-tion, followed by production of the sequence by genetic engineering, expression, biophysical and structural characterization and ligand binding mode verification. Experimental confirmation is essential for further improvement of ligand binding characteristics.

REFERENCES

Akerstrom, B., Flower, D. R., & Salier, J. P. (2000). Lipocalins: unity in diversity. *Biochimica et Biophysica Acta, 1482*, 1–8.

Berman, H. M., Westbrook, J., Feng, Z., Gilliland, G., Bhat, T. N., & Weissig, H. (2000). The Protein Data Bank. *Nucleic Acids Research, 28*, 235–242. doi:10.1093/nar/28.1.235

Beste, G., Schmidt, F. S., Stibora, T., & Skerra, A. (1999). Small antibody-like proteins with prescribed ligand specificities derived from the lipocalin fold. *Proceedings of the National Academy of Sciences of the United States of America, 96*, 1898–1903. doi:10.1073/pnas.96.5.1898

Bishop, R. E. (2000). The bacterial lipocalins. *Biochimica et Biophysica Acta, 1482*, 73–83.

Bork, P., Holm, L., & Sander, C. (1994). The immunoglobulin fold structural classification, sequence patterns and common core. *Journal of Molecular Biology, 242*, 309–320.

DeGrado, W. F. (1997). PROTEIN DESIGN: Enhanced: Proteins from Scratch. *Science, 278*, 80–81. doi:10.1126/science.278.5335.80

Desjarlais, J. R., & Mayo, S. L. (2002). Computational protein design. *Current Opinion in Structural Biology, 12*, 429–430. doi:10.1016/S0959-440X(02)00343-3

Doig, A. J., (n.d.). *Protein Engineering - Introduction & Lecture Summaries*. 2PAB: Physical and Analytical Biochemistry Lectures, Department of Biomolecular Sciences, UMIST.

Eliopoulos, E., & Mavridis, I. M. (1996). Molecular graphics approaches in structure prediction and determination. In G. Tsoucaris et al. (eds.), *Crystallography of Supramolecular Compounds* (pp. 491-498), NATO ASI Series: Mathematical and Physical Sciences.

Flower, D. R. (1995). Multiple molecular recognition properties of the lipocalin protein family. *Journal of Molecular Recognition, 8*, 185–195. doi:10.1002/jmr.300080304

Flower, D. R. (1996). The lipocalin protein family: structure and function. *The Biochemical Journal, 318*, 1–14.

Guex, N., & Peitsch, M. C. (1997). SWISS-MODEL and the Swiss-Pdb Viewer: An environment for comparative protein modeling. *Electrophoresis, 18*, 2714–2723. doi:10.1002/elps.1150181505

Hobohm, U., & Sander, C. (1994). Enlarged representative set of protein structures. *Protein Science, 3*, 522–525.

Holm, L., & Sander, C. (1994). Searching protein structure databases has come of age. *Proteins, 19*, 165–173. doi:10.1002/prot.340190302

Holm, L., & Sander, C. (1997). Dali/FSSP classification of three-dimensional protein folds. *Nucleic Acids Research, 25*, 231–234. doi:10.1093/nar/25.1.231

Huber, R., Schneider, M., Mayr, I., Müller, R., Deutzmann, R., & Suter, F. (1987). Molecular structure of the bilin binding protein (BBP) from Pieris brassicae after refinement at 2.0 A resolution. *Journal of Molecular Biology, 198*, 499–513. doi:10.1016/0022-2836(87)90296-8

Korndorfer, I. P., Schlehuber, S., & Skerra, A. (2003). Structural mechanism of specific ligand recognition by a lipocalin tailored for the complexation of digoxigenin. *Journal of Molecular Biology, 330*, 385–396. doi:10.1016/S0022-2836(03)00573-4

Ku, J., & Schultz, P. G. (1995). Alternate protein frameworks for molecular recognition. *Proceedings of the National Academy of Sciences of the United States of America, 92*, 6552–6556. doi:10.1073/pnas.92.14.6552

Laskowski, R. A., MacArthur, M. W., Moss, D. S., & Thornton, J. M. (1993). PROCHECK: a program to check the stereochemical quality of protein structures. *Journal of Applied Crystallography, 26*, 283–291. doi:10.1107/S0021889892009944

Morris, G. M., Goodsell, D. S., Halliday, R. S., Huey, R., Hart, W. E., Belew, R. K., & Olson, A. J. (1998). Automated docking using a Lamarckian genetic algorithm and an empirical binding free energy function. *Journal of Computational Chemistry, 19*, 1639–1662. doi:10.1002/(SICI)1096-987X(19981115)19:14<1639::AID-JCC10>3.0.CO;2-B

Mulder, N. J., Apweiler, R., Attwood, T. K., Bairoch, A., Bateman, A., & Binns, D. (2007). New developments in the InterPro database. *Nucleic Acids Research, 35*, D224–D228. doi:10.1093/nar/gkl841

Muller, H. N., & Skerra, A. (1993). Functional expression of the uncomplexed serum retinol binding protein in Escherichia coli. Ligand binding and reversible unfolding characteristics. *Journal of Molecular Biology, 230*, 725–732. doi:10.1006/jmbi.1993.1194

Padlan, E. A. (1994). Anatomy of the antibody molecule. *Molecular Immunology, 31*, 169–217. doi:10.1016/0161-5890(94)90001-9

Sander, C., & Schneider, R. (1991). Database of homology derived protein structures and the structural meaning of sequence alignment. *Proteins, 9*, 56–68. doi:10.1002/prot.340090107

Schlehuber, S., Beste, G., & Skerra, A. (2000). A novel type of receptor protein, based on the lipocalin scaffold, with specificity for digoxigenin. *Journal of Molecular Biology, 297,* 1105–1120. doi:10.1006/jmbi.2000.3646

Schlehuber, S., & Skerra, A. (2001). Duocalins: engineered ligand-binding proteins with dual specificity derived from the lipocalin fold. *Biological Chemistry, 382,* 1335–1342. doi:10.1515/BC.2001.166

Schlehuber, S., & Skerra, A. (2002). Tuning ligand affinity, specificity, and folding stability of an engineered lipocalin variant a so-called anticalin using a molecular random approach. *Biophysical Chemistry, 96,* 213–228. doi:10.1016/S0301-4622(02)00026-1

Schlehuber, S., & Skerra, A. (2005). Lipocalins in drug discovery: from natural ligand-binding proteins to anticalins. *Drug Discovery Today, 10,* 23–33. doi:10.1016/S1359-6446(04)03294-5

Schmidt, T. G. M., & Skerra, A. (1994). One-step affinity purification of bacterially produced proteins by means of the "Strep tag" and immobilized recombinant core streptavidin. *Journal of Chromatography. A, 676,* 337–345. doi:10.1016/0021-9673(94)80434-6

Skerra, A. (2000). Engineered protein scaffolds for molecular recognition. *Journal of Molecular Recognition, 13,* 167–187. doi:10.1002/1099-1352(200007/08)13:4<167::AID-JMR502>3.0.CO;2-9

Skerra, A. (2000). Lipocalins as a scaffold. *Biochimica et Biophysica Acta, 1482,* 337–350.

Skerra, A. (2001). Anticalins: a new class of engineered ligand-binding proteins with antibody-like properties. *Journal of Biotechnology, 74,* 257–275.

Sousa, S. F., Fernandes, P. A., & Ramos, M. J. (2006). Protein–Ligand Docking: Current Status and Future Challenges. *PROTEINS: Structure, Function, and Bioinformatics, 65,* 15–26. doi:10.1002/prot.21082

Steipe, B. (1998). Protein Design Concepts. In P. v. R. Schleyer, et al. (Ed.), *The Encyclopedia of Computational Chemistry* (pp. 2168-2185). Chichester, UK: John Wiley & Sons.

Thompson, J. D., Higgins, D. G., & Gibson, T. J. (1994). ClustalW: improving the sensitivity of progressive multiple sequence alignment through sequence weighting, position specific gap penalties and weight matrix choice. *Nucleic Acids Research, 22,* 4673–4680. doi:10.1093/nar/22.22.4673

Wallace, A. C., Laskowski, R. A., & Thornton, J. M. (1995). LIGPLOT: a program to generate schematic diagrams of protein-ligand interactions. *Protein Engineering, 8,* 127–134. doi:10.1093/protein/8.2.127

KEY TERMS AND DEFINITIONS

Engineered Protein - Modified Protein: A protein, which was artificially changed as far as the order, the number or the type of its amino-acids are concerned, in order to alter the structure and consequently its function.

In Silico: It is an expression used to mean "performed on computer or via computer simulation." The phrase is coined in analogy to the Latin phrases *in vivo* and *in vitro* which are commonly used in biology.

Ligand: A ligand (latin *ligare* = to bind) is a substance that is able to bind to and form a complex with a biomolecule to serve a biological purpose. The tendency or strength of binding is called affinity. Ligands include *substrates*, *inhibitors*, *activators*, and *neurotransmitters*.

Protein Engineering: The design and construction of new proteins with novel or desired functions by modifying amino acid sequences.

Protein Family: A protein family is a group of evolutionarily related proteins. The use of this term is somewhat context dependent; it may indicate large groups of proteins with the lowest possible level of detectable sequence similarity, or very narrow groups of proteins with almost identical sequence, function, and three-dimensional structure, or any kind of group in-between. For this reason, additional terms such as protein class, protein group, and protein sub-family are in use.

Protein Scaffold: A structurally well-defined polypeptide framework for the introduction of novel functions by locally reshaping a part of the protein surface that is thought to be less important for the protein folding process or its stability.

Chapter 3
In Silico Biology:
Making the Most of Parallel Computing

Dimitri Perrin
Dublin City University, Ireland

Heather J. Ruskin
Dublin City University, Ireland

Martin Crane
Dublin City University, Ireland

ABSTRACT

Biological systems are typically complex and adaptive, involving large numbers of entities, or organisms, and many-layered interactions between these. System behaviour evolves over time, and typically benefits from previous experience by retaining memory of previous events. Given the dynamic nature of these phenomena, it is non-trivial to provide a comprehensive description of complex adaptive systems and, in particular, to define the importance and contribution of low-level unsupervised interactions to the overall evolution process. In this chapter, the authors focus on the application of the agent-based paradigm in the context of the immune response to HIV. Explicit implementation of lymph nodes and the associated lymph network, including lymphatic chain structure, is a key objective, and requires parallelisation of the model. Steps taken towards an optimal communication strategy are detailed.

INTRODUCTION

Biological systems are typically complex and adaptive. They are complex, involving large numbers of organs, cells, molecules, as well as their interactions. They are adaptive, because their behaviour evolves over time, and can change and learn from experience, e.g. through memory in the context of immune responses.

Key principles of complex adaptive systems, (e.g. RNA folding or immune response), are *emergence* and *self-organisation*. Emergence refers to patterns of system evolution arising from an abundance of simple, low-level, interactions, (see e.g. Corning (2002)). For the immune system, this is particularly relevant, as the response is obtained from multiple cell interactions throughout the body. Self-organisation refers to increased complexity obtained without intervention from outside sources, (e.g. De Wolf and Holvoet (2005)). Again, this is an obvious property of both the infection mechanisms of HIV and the immune response to those.

DOI: 10.4018/978-1-60566-768-3.ch003

Given these two properties, a full description of a complex adaptive and definition of the way in which low-level unsupervised interactions, (and their relative importance), lead to its overall evolution, are far from trivial.

In this chapter, we introduce concepts and approaches designed to gain insight into complex systems, focusing on immune models. Concepts of bottom-up and top-down programming are explained. Approaches include mathematical, shape-space models, cellular automata and agent-based models.

We, subsequently, further detail the agent-based paradigm, which is very suited to biological systems. Limitations of existing agent-based immune models are analysed, forming a basis for the model objectives as case study.

The model structure is detailed next, with particular emphasis on the importance of the balance between agent diversity and agent population size, and on the need for an explicit implementation of the lymph network.

Due to the computational requirements of such a model, a parallel implementation is necessary. Efforts towards an optimal parallel implementation are detailed, along with a presentation of MPI, (*de facto* standard for parallel programs).

Finally, we present some important model results, and reflect on the applicability of a similar development framework for other biological systems.

MODELLING COMPLEX SYSTEMS

Two categories of complex system modelling are discussed, *top-down* and *bottom-up* designs (Bohringer and Rutherford, 2008). The main concept of a top-down design is to break down a system into several components, expected to be easier to manipulate and understand. The overall system is formulated and specified, but without going into details of its parts. In an iterative process, each component is then defined in more detail and, if necessary, split into lower-level subsystems. This process, repeated until the entire specification is obtained for its base elements, involves use of black boxes which facilitate model development, but may also hinder model validation if these fail to elucidate elementary mechanisms of the system studied.

In a bottom-up approach, individual base components are detailed and designed, and then linked together. These form more complex systems, which are again linked, in an iterative process, and the top-level model increasingly emerges. This approach is, therefore, particularly suited to complex adaptive systems, which in their structure demonstrate both emergence and self-organisation. The remainder of this Section considers several examples of top-down/bottom-up design, grouped in three families: mathematical, shape-space, and agent-based models.

In the context of HIV research, mathematical models were first introduced to study the epidemiological aspect of infection, (i.e. spread in a human population). This early focus was motivated by the need to understand dynamics of the infection and population threat, but also by a lack of detailed biological information, which ruled out models of pathogenesis. Although data, even on spread of HIV, were sparse, models were developed, which focused on specific "at risk" groups, e.g. early work by Anderson (1988). More accurate medical information and improved computing resources have now led to considerably more advanced epidemiological models, e.g. Naresh et al. (2006). In the context of this research, the mathematics of pathogenesis are directly relevant. Models using differential equations (DE), to reproduce variations of cell counts and viral loads appeared in the late 1990s, (see e.g. Perelson and Nelson (1999)) and are typical of top-down designs. These have been refined for each count variation by detailing different DEs to account for viral production, and drug therapies such as RT inhibitors or protease inhibitors. These models, however, can not currently cover whole infection progression, and each focuses on either long-term, mid-term, or short-term variations.

Immunological models based on the shape-space paradigm were first introduced to account for dynamics of antibody-antigen bindings (Perelson and Oster, 1979). The main concept of this bottom-up approach is to represent each clonotype by N integers and, therefore, to consider clonotypes as points in an N-dimensional Euclidean space. In that shape space, (space of clonotypes), two cells sharing clonotype are located on the same point. Each cytotoxic lymphocyte clonotype '*c*' is surrounded by a sphere of radius '*r*'. To any antigen '*a*' within this sphere is applied a clearance pressure inversely proportional to the distance between '*a*' and '*c*' in the shape space. Interested readers may refer to Burns (2005, and references therein) for more details on this, (such as a discussion on adequate values for N).

Recent work includes considerations of real space, and formation of hybrid models (Burns and Ruskin, 2004; Ruskin and Burns, 2005; 2006). The focus is on emergent principles of CD8 cell clonotype repertoire and its distribution and differentiation, with emphasis on systemic self-organisation, (for which the shape space paradigm is particularly suited). Hybridisation of shape space is obtained through use of a stochastic model of the lymph system as stimulus to the network shape space model. Emergent topology obtained from this model resulted in introduction of a theoretical network architecture for the immune system. It includes α and β nodes: the latter correspond to CD8 cells that act only against that antigen, which stimulated its activation, while the former represent those which also effect clearance pressure on subsequent APCs. The argument outlined by the authors suggested that disruption (or suppression) of α nodes results in a significantly degraded pathogen clearance, compared to β node disruption. This was proposed as a possible cause of individual variations in the latency period.

The Cellular Automata, (CA), paradigm is another popular example of bottom-up model design. Even though the concept was introduced in the 1940s, (through the work of Ulam and von Neumann), the paradigm gained large popularity only in the 1970s, with the introduction of a two-state, two-dimensional cellular automaton, Conway's "Game of Life" (Gardner, 1970). On a 2D grid, cells have eight neighbours, and two possible states, living or dead. At each time step, cells are updating using simple defined rules, e.g. a live cell with fewer than two live neighbours dies, while a live cell with more than three live neighbours dies.

The popularity of this interpretation of the paradigm can be attributed to its obvious analogy with living systems, (e.g. rules which embody death by overcrowding or competition), but also to this CA being a perfect and simple illustration of concepts of emergence and self-organisation. Pattern evolution in this CA is well documented, with example entities such as blocks, gliders and pulsars (Berlekamp et al., 2004).

More realistically, when including more than two possible states, Cellular Automata provide a powerful modelling paradigm. Since the early efforts of e.g. Celada and Seiden (1992); Seiden and Celada (1992), CA have been widely used to investigate immune events, and several models, in particular, provide useful insights into some aspects of the immune response to HIV infection.

CA-based immune models outlined the importance of viral mutation on dynamics of immune cell population (Mannion et al., 2000; Pandey et al., 2000; Mannion et al., 2002; Ruskin et al., 2002). A threshold value was identified, under which steady-state density of immune cells is larger than that of HIV, (i.e. dominant to deficient phase transition). These authors also provided one of the rare attempts to account for cell mobility and, subsequently, variable viral load.

Another CA model focused on latency period and treatment solutions (Benyoussef et al., 2003). Combining a mean field approximation method and CA simulations, these authors reproduced the three-phase evolution of HIV infection and identified a threshold for treatment, (a combination of protease inhibitors and RT inhibitors), above which virus load decreases over time. They also indicated that such treatment would need to continue for years even if viral load falls under detectable limits.

AGENT-BASED MODELS

The agent-based paradigm can be considered, in some aspects, as an extension of the CA approach. In this approach, the key abstraction elements are agents with both spatial and temporal positioning. The generally-accepted properties for an intelligent agent, there being no unique definition, are given by Wooldridge and Jennings (1995):

- **Autonomy:** it can act without any intervention and has some control over its actions and its internal state.
- **Social behaviour:** it can interact with other agents through a specific language.
- **Reactivity:** it can scan part of its environment and change its behaviour to take advantage of it.
- **Proactivity:** it not only reacts to its environment but also acts and takes initiatives, to satisfy identified goals.

It is theoretically possible, though counterintuitive, to design an agent-based approach and implement it without explicit references to agent-like entities in the code. In practice, efficient realisation of the agent-based paradigm requires both.

Advantages of the approach include modelling efficiency, robustness, interoperability between existing systems, and reasonably intuitive solving of problems for which data, expertise and control are distributed, Jennings et al. (1998). The approach is thus particularly useful in the context of Natural Sciences, as it permits reciprocity between agents and biological entities, as well as between interactions of the real system and exchanges between agents.

Agent-based models implementing several agents are referred to as multi-agent systems. These systems provide a generic framework for model development and have been widely applied. Well-documented examples include air traffic scheduling, (Cammarata et al., 1983), intensive care unit management, (Hayes-Roth et al., 1989), vehicle monitoring, (Durfee, 1998), and control of robots, (Mizoguchi et al., 1999).

While these examples are developed *ab initio*, there are several development environments which can be used for agent-based models, such as Swarm (Minar et al., 1996), Cougaar (Helsinger et al., 2004) or JAMES (Uhrmacher et al., 2000). The latter, for instance, is a Java-based framework aimed at modelling and simulation, and permits creation of mobile agents. These development solutions, however, are not best suited to the context of this study, which demands a very large agent population, with extensive optimization requirements at every level of implementation.

From the few examples above, (the list is not intended to be exhaustive), it is evident that the agent-based paradigm is a versatile approach and has, naturally, also been considered to model the immune response to HIV.

An early example of such models, (Bernaschi and Castiglione, 2001; Castiglione et al., 2004; Baldazzi et al., 2006), is a refinement of the Celada-Seiden model, (IMMSIM), particularly in terms of performance and fidelity to the real immune system. Implemented entities include CD4 and CD8 T cells, B cells, macrophages and dendritic cells. These interact, based on location in a 2D or 3D lattice, and according to an affinity function that depends on values of bit strings representing their respective binding sites. Interactions lead to changes of internal state. This approach successfully reproduced the three-phase disease progression.

Another successful attempt at reproducing typical HIV progression was recently proposed (Zhang et al., 2005). Here, types of agents are limited to T cells and HIV virions, with the objectives of simulating large populations, and improving detail on aspects such as immune memory and HIV sequence representation. The former objective is achieved through introduction of a global memory repertory, which is empty when simulation starts and which subsequently stores all HIV genomes that have been recognized and were targeted by immune responses. The latter objective is achieved by introduction of finite-size binary arrays, which each represent a viral strain, a measure of similarity between these arrays, (based on Hamming distance), and a "recognition probability" as a function of this similarity.

The Need for a Refined, Large-Scale, Agent-Based Model of HIV Infection

Models of HIV progression have been developed for the last twenty years, with some successes reported, as detailed above. There are, however, a number of limitations that need to be addressed in order to improve realism with respect to the biological system modelled.

Inability to deal with subsystems small enough to be informative on the interactions involved, which is crucial in the context of HIV progression, is a major limitation of top-down models, especially mathematical ones. Immune response typically is bottom-up, and these models may not, therefore, be the most suited to describe it.

The shape-space paradigm is very elegant, and hybrid models based on this provide new insight into self-organisation of the immune response. It is, however, sometimes difficult to explain observed emergent features in biological terms (Burns, 2005). This may be attributed to the fact that such models, even though bottom-up, can rarely achieve sufficient refinement of the individual base components. Extensions proposed to current models may improve this situation (Burns, 2005).

CA and agent-based design provide a natural choice in terms of structure of the biological system being considered and significant biomedical studies, involving direct tracking of HIV viral genotypes in local microenvironments, provide further validation for these approaches, (e.g. infected cells releasing virions will only induce infection of local targets (Cheynier et al., 1994)). It seems evident that local interactions and cell densities are more important to disease progression and experience than overall cell counts in the body (see e.g. Grossman et al., 1998, and subsequently). As a bottom-up approach, the agent-based paradigm offers the best prospects for detailed local solutions, and hence for examination of variable disease factors.

Inherent to any agent-based approach is choice of agent types, where several permit improved tuning of the model to the observed system. It is particularly relevant in the context of immune response, as several lineages of cells are involved. However, there is a balance to be achieved between the range of agent types and size of their populations, due to the computational cost involved. This is important in the immune context, as overall behaviour emerges from a very large number of interactions, (involving many cells). It is not clear that sufficiently large cell populations have hitherto been modelled. The theoretical limit in Bernaschi and Castiglione (2001) is not explicitly specified, but appears to be of the order of two million, and it is debatable whether this is enough to account for HIV progression and complex evolution through the whole body.

Another essential aspect is that of spatial location of immune response elements. Most of the immune response to HIV takes place in the lymph nodes, as opposed to within circulating blood, but no clear modelling approach has been explicitly described for this. There are indications, e.g. in Baldazzi et al. (2006), that this overall structure is recognised as important, but no report of large-scale validation tests could be found at the time of writing. Any novel agent-based approach needs to explicitly account for lymph nodes and, specifically, the lymph network in order to gain insight on the importance of cell mobility.

In the context of a bottom-up design and need for detailed modelling of cell-level interactions, a refined temporal granularity is also required, to account for cells passing through the lymph network, (which occurs within minutes), even though this adds to the overhead on code in order to increase computational efficiency.

These two conditions, (explicit structure for the lymph network, and high granularity), permit accurate implementation of cell mobility, which is an essential aspect of the immune response, as activation involves not only affinity between epitopes, but also physical contact.

Memory of previous responses is also important, (see Zhang et al. (2005)), although the implementation here is slightly counter-intuitive, and features a centralised control. Memory at cell-level is potentially more useful in light of the role of localised interactions.

A common feature of the most advanced models is incorporation of viral mutation and loading. This is core to viral model success but consideration of variable properties, (e.g. to account for less stable strains, or for those with higher probabil-

ity of successful infection), is also important for sophisticated model development.

Finally, current implementations for antigen recognition are based on lock-and-key or on naive distances, and response is binary: there is complete recognition, or no recognition at all. This contradicts recent immunological understanding (Brehm et al., 2002) and structure proposed by Ruskin and Burns (2006), and requires explicit improvement in any newly-proposed model.

Including all these new features requires extensive code optimization, and parallel implementation, and the costs are, additionally, a longer development process and additional tests. The gains, however, are significant. The objective is a model which permits large-scale simulations and inclusion of layered, localised, effects, and which accounts for multiple facets of the biological immune response.

Overall Model Structure

As immune response to HIV is predominantly cell-mediated, it can be modelled using three types of agents, (based on agent-cell reciprocity): CD4, CD8 and APC. A fourth type of agent represents HIV virions. Agents evolve in lymph

Figure 1. Agent interactions

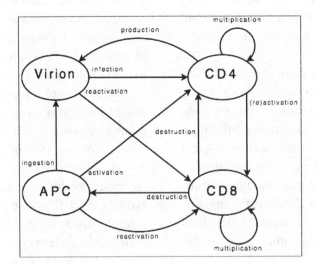

nodes modelled as 2D matrices in which each element is a 3D physical neighbourhood able to host several agents of all types.

Each agent type corresponds to a specific class in the C++ implementation, and inherits from a common, abstract base class. This class contains attributes and functions needed for management of ageing and of location within matrices. Being abstract, this class does not correspond to any agent during simulations. For immune (CD4, CD8, APC) and viral agents, the number in each matrix element is limited to 10 and 20, respectively, to ensure realistic neighbourhood size for interactions. An element can, for instance, contain 8 CD4 agents, 7 CD8, 5 APC and 17 virions. The reciprocity between agents and biological entities, (cells and virions), mirrors real-system interactions between the latter. These are summarised in Figure 1.

Each agent has complete knowledge of its parameters and, therefore, of its internal state. Knowledge on its environment is, however, limited and temporary. Information on the environment is, indeed, limited to presence of targets for interaction in the neighbourhood. Agents have no memory of the evolution of their neighbourhood, as is the case with biological entities of the real system.

All internal parameters are coded as integers, (or lists of). Age is involved in the internal state of all agents. Other parameters are type-specific, and are given in Table 1.

Direct implementation of the age in the base class would be ill-advised, as it would require updating it for all agents at each iteration, even when this information is not immediately required. This can be quite slow, especially as agent counts increase. A more suitable alternative is to save the iteration number at which the agent was created. No repetitive update is required, and the difference between the current iteration number and the "birth date" of the agent provides its age when needed.

Different viral strains mean different properties are needed for the associated viral agents: these are not recognised by the same set of immune agents, may have distinct mutation rates, etc. Explicitly implementing all these properties within each agent would make them too "big", in terms of memory usage, (to be avoided for large simulations). The solution here is to use a single integer, coding the viral strain. This identification can then be used to access strain-specific properties, which are stored in a large array, (representing tens of thousands of potential strains). An

Table 1. Agent parameters and initial internal state

Agent type	Parameter	Value stored	Initial value
Virion	Viral strain	Strain ID	Fixed value
CD4	Clonotype	Clonotype ID	Fixed value
	Activation	Responsible strain	0
	Multiplication	Expansion status	0
	Infection	Responsible strain	0
	Memory	Past activation status	Fixed value
CD8	Clonotype	Clonotype ID	Fixed value
	Activation	Responsible strain	0
	Multiplication	Expansion status	0
	Memory	Past activation status	Fixed value
APC	Antigen list	Strain IDs	Empty list

interesting property, here, is the identification of which immune clonotypes recognise each strain. As previously outlined, antigen recognition must be refined with respect to lock-and-key concepts. Proposed here is that two lists of clonotypes should be available. The first list corresponds to clonotypes for which recognition is certain, (i.e. $p=1$), and the second accounts for those for which recognition is not perfect, (i.e. $p<1$). An important characteristic is that when an agent from the second list recognises the viral strain, the associated clonotype moves from the second list into the first. This is critical to the realism of the model, since it allows us to introduce some adaptability.

Memory allocation has been carefully considered. A naive implementation of agents within the matrix representation of a lymph node would allow for arrays of agents within each element. Each array would store agents of a given type and localisation of agents would, therefore, be trivial, with agent movement between elements involving simple deletion from one array and addition to another.

Memory allocation for agents also offers an intuitive solution: memory could be dynamically allocated whenever an agent is created, and freed when one is destroyed.

Both these solutions would be suitable for a small-scale system, but would not be efficient for a large model such as the one studied here. In the first place, memory allocation is one of the slowest operations on a computer, and would be intensively solicited here. At each iteration, a large number of agents are created or destroyed. Successions of dynamic allocations and deallocations would be used and would, therefore, significantly hinder efficient computation. Moreover, as immune response is initiated, agent count will sharply increase, and may reach values close to theoretical limits of the model. The main advantage of dynamic allocation, (i.e. using only as much memory as is needed at any time point), would not apply in such a simulation.

Consequently, our implementation is based on static memory allocation. Large arrays of agents, (one for each agent type), are allocated when simulation starts, and store the maximum number of agents which can be present in the whole matrix at the same time. Each matrix element then only stores integers, used as offsets, to locate agents currently held in these arrays.

Intra-node mobility refers to agent movements within each matrix. Overall movement is based on flow, created by the fact that agents enter the node on one side, and exit on the other. The required distribution of agents and time spent in each node is stochastically governed. Using the implementation and memory allocation strategy detailed, updates of agent localisation are easily managed: (*i*) creation of an agent requires initialising its internal state and saving the offset, to access this state, in the matrix element; (*ii*) deletion of an agent involves deletion of this offset; a new agent can subsequently be created in this space; (*iii*) movement from one matrix element to the next requires transferring the offset value to the destination element.

Although implementing intra-node mobility is straight-forward, inter-node mobility, involving agents moving from one node to another, requires more attention. As key defense units, lymph nodes are distributed throughout the body. Humans have around 500 lymph nodes, and lymphocytes constantly circulate through these. In our model, agents leave a node when they reach its exit point. They are then added to a transfer list, which is communicated to other nodes on a regular basis. From the list, they are then added to the entry point of their new current node.

To guarantee efficient cell circulation through the body, connectivity is an essential property of the lymph network: a cell newly produced in the thymus must be able to reach any lymph, and efficient immune response implies interactions between nodes, (by means of cell exchanges).

The lymph network, however, is not equivalent to a complete graph: between any given pair of nodes, there is a path, but not necessarily a direct connection. In contrast, the lymph network is organised as a set of chains. These "clusters" of nodes can be found in the neck, chest, abdomen, underarm, etc.

The circulation of immune cells between nodes is not trivial. It is type-dependent, (Witherden et al., 1990), activation-dependent, (Mackay et al., 1990), and is also tissue-specific, (Kunkel and Butcher, 2002). As might be expected, these patterns are affected by immune response, (Cahill et al., 1976).

The lymph network structure implies that the final destination of an agent leaving a node can theoretically be any node, but its immediate destination is limited to a small subset of nodes. Selection of final destination is random, but not based on uniform distributions, to account for the patterns mentioned above. Selection of the immediate destination is based on the lymph network structure modelled. This structure is generated through an automated technique reproducing the lymphatic chain structures for any number of nodes, which guarantees a realistic topology irrespective of the simulation scale.

Inter-node mobility, of course, implies simulation of significant number of nodes, and this implies increased computing load. A parallel implementation is employed to address this final requirement.

OPTIMAL PARALLEL IMPLEMENTATION

The approximately 500 nodes representing those of the human body can not be handled on a single-processor computer, because of limitations in both available memory, (it is not possible to allocate a large number of nodes), and computing power, (it would take too long to simulate all nodes). The only solution is to consider a parallel implementation of the lymph network.

Local interactions within each node have been implemented by a regular algorithm. Since node size is constant, there is little variation in the agent counts of the various nodes, and we can expect local iterations to finish around the same time. Our problem is thus regular and loosely synchronous, and the model is suited for a parallel implementation.

Identification of possible "divisions" within the program is essential. In some programs, a set of instructions is repeated several times, with each iteration independent of the others. Monte-Carlo simulations are a typical example here. In such cases, several iterations can run at the same time, using a time parallelisation.

This technique can not be applied to the lymph network model however: each iteration i uses the final state of the agents after iteration $i-1$, so that two iterations can not run at the same time. Yet, the problem does have a parallel nature, in the sense that each node is largely independent of the others, apart from cell transfer.

We can, therefore, use a spatial parallelisation, based on a reciprocity between the lymph nodes and the computer nodes of a cluster. Each lymph node of the model is assigned to a computer node of the parallel architecture, and a communication network is designed to mimic cell mobility along the lymph network.

Spatial parallelisation has previously been investigated, e.g. in the context of Monte-Carlo simulations, for HIV infection (Hecquet et al., 2007), with the main disadvantage in that case being the communication overload. The parallelisation strategy, detailed above, guarantees that communication is kept to a minimum. Indeed communication between computer nodes is only used to represent actual cell exchanges; we do not add any communication overhead due to the parallelisation itself, apart from initial problem splitting and final results gathering.

Note: This is an important point, as communication is often the bottle-neck in parallel implementations. This is usually a consequence of hardware architecture, where physical data transfer is significantly slower than computing operations. The communication time is the sum of the actual data transfer time, (data size divided by bandwidth), and of a constant component called network latency, (corresponding to the time delay between communication initiation and actual start of data transfer). Current hardware configurations have a latency ranging from 1.5 to 100 μs. This may appear small, but can significantly reduce model efficiency if multiple communications are involved over time.

During iterations of the parallel model, lists of "migrating" agents are the only information that need be exchanged between cluster nodes. A communication strategy is, therefore, needed as the decision basis for the frequency and the method of list transfer.

A first consideration, to limit data transfer steps, is to use a single list for agent transfer, (rather than as many lists as agent types). A balance must also be found between minimal communication and need for realistic exchange of information. In what follows, we introduce the communications protocol used for implementation.

In brief, MPI, (or Message-Passing Interface) is a communications protocol used for parallel implementation of programs. MPI provides support for point-to-point and collective communications, enquiry routines to query the execution environment, as well as constants and data-types (Gropp et al., 1999a;b).

MPI is a low-level library, which provides an interface to C, C++ and Fortran 90. MPI is, therefore, a language-independent protocol. Portability was a priority during its development, and MPI is platform-independent. This is crucial, as the range of parallel platforms is heterogeneous.

This, along with high performance and scalability, made MPI de facto standard and most current distributed-memory computers offer MPI implementations. MPI is taught and used widely, which, together with the availability of open-source

implementations and the large body of programs that require MPI, (including both Research models and commercial products), guarantees long-term legacy of MPI and sustainability of our model. MPI supports two communication types:

- *Point-to-point communication.* Data is sent from one node to another. Default communications are blocking: the send call blocks until the send buffer can be reclaimed. This implies that after the send, the sender can safely over-write the contents of a variable used for communication. The situation is similar on the receiving end: the receive function blocks until the receive buffer actually contains the contents of the message. If needed, MPI also supports non-blocking communications, which allow possible overlap of message transmission with computation, or of multiple message transmissions.
- *Collective communication.* This is used when information located on one cluster node must be shared with all the others, or when information scattered over the nodes must be gathered on one of them. A single instruction, called by all involved nodes, replaces a loop of point-to-point transmissions.

An initial strategy is to use point-to-point communication only, and to exchange data after each iteration. From here on, we will refer to this strategy as "strategy 1", and improvements to this as "1.x".

Strategy 1 implies point-to-point communications between each pair of nodes. First, each node must create *n-1* sublists, (with n the number of nodes); each sublist contains the agents going from the current node to another specific node. Exchange of lists between the nodes involved can then take place, and requires *n(n-1)* list transfers.

For MPI, each list transfer requires two communications, with one parameter of MPI commu-

Table 2. Naive strategies tested on small configurations

Cluster configuration		4 nodes	10 nodes	16 nodes
Strategy 1	Local iterations (s)	69	71	68
	Communication (s)	57	206	324
Strategy 2	Local iterations (s)	67	65	63
	Communication (s)	62	377	1239

nication routines being the size of the transmission. It means that, to send a list containing m values, we must first send 'm' in a one-integer message, and then send the list. Strategy 1, therefore, leads to $2n(n-1)$ point-to-point communications.

Another strategy is to exclusively use collective communications. In the following, we refer to this as "strategy 2", and use the same naming convention as above for future improvements.

Here, each node will send the whole list to all the others. The first step is, therefore, the communication process, which represents n collective list transfers, i.e. $2n$ actual MPI collective communications. The next step is performed locally by each node: searching in all received lists for the agents which are entering the node.

Strategies 1 and 2 were tested on a local cluster, for small configurations, (i.e. less than 16 nodes), and short simulations. Results were discussed in Perrin et al. (2006c), and are here displayed in Table 2. The first remark is that communication is, indeed, a bottle- neck. Tested strategies are of course naive, and connectivity on this cluster is known not to be very efficient, but these first results still highlight the need for an efficient communication strategy.

It is also important to note that local iteration time is largely constant and independent of communication strategies. This is a confirmation of the parallel nature of the problem.

Strategy 2 performs very poorly. This is due to the fact that, with this approach, destination nodes receive more data than they actually need, since they receive information about all the agents which left the host node, irrespective of their destination. There is a reduction in the number

of communication steps, (n, compared to $n(n-1)$ with strategy 1), but this is obtained at too heavy a price: excessive data transfer eliminates strategy 2 as a viable solution.

Performances obtained from strategy 1, despite need for some improvement, are more encouraging, as communication overhead is more reasonable. One solution is to consider the frequency at which lists are shared between nodes. More efficient communication implies sending non-empty lists. Obviously, the longer the interval between list transfers, the bigger this list gets, and the likelihood of sending an empty list decreases. In Table 3, computation times are shown for 20,000 iterations, when communication is performed at the end of every iteration or every other time-step. The program appears slightly faster when we communicate data less often. However the gain is not significant for very low agent count: in this case, few agents are scattered in the lymph node and are less likely to reach the exit point, even over several time-steps. The improvement is highest for medium agent count: for a high count, it is likely that at least one agent will reach the exit point during each iteration, and iterations leading to an empty list are, therefore, less common, but do occur. We observe an improvement when sending only at every other iteration; this pattern is confirmed if we wait three, four, or five iterations before sending the lists.

There are, however, two disadvantages to this strategy. There is a memory concern, since an ever bigger list is resource-consuming but, more importantly, there are biological considerations involved. A time-step is equivalent to fifty seconds, and the number of iterations must therefore be

Table 3. Influence of the list transfer frequency on communication time

Communication frequency	Every iteration	Every other iteration
Low agent count	377 s	-1.48%
Medium agent count	982 s	-34.90%
High agent count	2187 s	-10.80%

kept close to the actual time estimated for a cell to commute from one node to another. Separating the communication phases by more than five iterations is thus less realistic and should be avoided. Further improvements on the communication protocol were, therefore, indicated.

In addition to eliminating solutions based on collective communication, early tests also provided a basis for improving strategy 1. In fact, on several occasions, direct transfer between every couple of nodes, due to MPI constraints detailed above, would lead to sending information about an empty list, thus slowing the program down. For this approach to be efficient, we need a node to act as the intermediary between these pairs of nodes, with all the nodes sending their list to this one. On this node, the agents are sorted according to their destination, and, to every node, a list is sent, containing only the agents which are relevant (strategy 1.1). This reduces the number of communication steps to $2(n-1)$. The main drawback for this improved communication protocol is that a node can only receive from (or send to) one other node at a time. It implies that in the meantime, the others are idle.

Strategy 1.2 therefore involves dedicating one node on the cluster, (called node 0 hereafter), solely to the role of intermediary, to ensure that it is always ready to send and receive, rather than engaged in an iteration. This should reduce the shortcoming of strategy 1.1, though it will not eliminate it entirely.

Even if node 0 is available and immediately receives the list from node i, that node will be idle while waiting for list i of all agents migrating to node i, as this list can only be prepared once all

incoming transmissions have been completed on node 0. Inclusion of an iteration between sending the first list, (agents leaving a node), and receiving the second, (agents arriving at that same node), prevents "computing nodes" from being idle, and gives time for node 0 to finish receiving every list and sorting the agents (strategy 1.3).

As scale increases, so too does the time that a given node has to wait before being able to send/receive. An alternative is to create more "intermediary nodes", and subnetworks. For instance, on a 16-node cluster, we could have four groups, each formed with three "computing nodes" to deal with modelling and one node only used for communication. The first three would run an iteration, send their list, compute another iteration, and receive the new list. The other one receives the lists, shares information with other similar nodes, and sends the new lists. With this configuration, any node finds a maximum of three nodes in the queue at the time it joins, and expectation for program speed-up is improved, (strategy 1.4).

To determine agent destination, the strategies, detailed so far, rely on a local function controlling inter-node cell mobility. The lymph network is not a "dense" network. Using terminology from graph theory, (see e.g. Diestel (2005)), it can be described as a directed finite graph which includes cycles, but is not a complete graph: if two lymph nodes are randomly chosen, it is likely that there will be no direct connections between them (incomplete), even though there is always a path from one to the other (connected). These properties can be used to implement the lymph network. A communication network can be created explicitly, rather than by a function as described

Table 4. Communication times for advanced strategies, compared to baseline strategy 1.1

Communication strategy	Relative communication time
1	2.68
1.1	1
1.2	0.97
1.3	0.92
1.4	N/A
3	0.89
3.1	0.76

above; communication can then be physically limited to this network (strategy 3). Without invalidating biological features, we can also impose the requirement that nodes have either two, (one incoming and one outgoing) or three connections (two incoming and one outgoing, or vice versa). This would imply that for any given node, at any stage of the simulation, there is a maximum of two nodes in front of the queue.

A further improvement is to design this network to satisfy two-colouring, (strategy 3.1), thus decreasing the communication load: during odd iterations, black nodes send data and white ones receive it, and vice versa during even iterations.

All advanced strategies were tested on the same local cluster. In Table 4, the results for 20,000 iterations and 16 nodes are shown. More advanced versions of strategy 1 provide improved performance, with the notable exception of strategy 1.4. Tested on 16 nodes, it is faster than other implementations of strategy 1. However, this configuration means that only 12 nodes are used for actual biological simulation, while the other 4 nodes focus solely on data transfer. A fair comparison with other strategies must take this into account and look at results on smaller node counts for these. In that case, strategy 1.4 does not show any particular improvement, and was not, therefore, included in Table 4. Strategy 1.3, on the other hand, demonstrated useful performance and is a candidate for large-scale implementation.

Another promising candidate is the new strategy, particularly in its 3.1 "coloured" version. It shows significant improvements compared to any other communication protocol. The fact that all 16 nodes are used for biological simulation is a further advantage: i.e. this strategy combines efficient communication and better resource usage.

A model size of hundreds of nodes should lead to enhanced realism, but large computing clusters to handle this are a limited resource, and an optimal communication strategy is vital. This also implies that it is preferable to determine the optimal strategy before using these clusters. In other words, small clusters can be used for development, but larger clusters should be kept for actual biological simulations.

An interesting property of MPI is its very good scalability, so that an implementation validated on a small configuration has a high probability of working on larger ones. This is still not enough in this case, however. What is required is an accurate estimate of communication time associated with each strategy, to satisfactorily assess the best one for this particular model. As noted for naive implementations 1 and 2, a communication strategy may have a critical model size above which performance is significantly reduced. This size was small for strategy 2, (which is only efficient on very small networks), but more advanced strategies may have a high critical threshold.

We considered that the best answer to this dilemma, (defining the best strategy without excessive physical testing), was to develop a performance simulator, with six key parameters:

- Communication algorithm. This corresponds to the communication strategy currently tested, i.e. which node sends to which, type of communication, and time of the transmission.
- Node size: the size of the matrix modelling the lymph node.
- Agent count. Each node is initialised with a particular number of agents. This number is kept up-to-date throughout the performance simulation by taking into account the agents that are sent or received. For simplicity, population variations due to other biological variations, (e.g. production of virions by infected immune agents), are neglected in this context.
- Length of local iteration: a function of the node size and agent count.
- Network latency: a property of the hardware architecture under virtual evaluation.
- Bandwidth, or rate of data transfer. Additional property of the simulated computing cluster.
- Length of data transfer. Transfer is obtained using the relation explained above, where data size is a probabilistic function of agent count.

This performance simulator was implemented, and validated, using the results obtained on the local computing cluster. Strategies were then repeatedly tested. Of main interest here were performance measures, such as the average total execution time and communication time. Maximal values were also monitored, in case a particular configuration proved able to block a communication strategy or lead to abnormal performance. Again, as found for small configurations, strategy 3.1 performs best.

These results are consistent with expectations based on the configuration of communication steps. Strategy 1.3 leads to $n(n-1)$ communication steps, and strategy 3.1 to the order of $3n/2$, but the distribution of these data transmissions is very different. Strategy 1.3 puts an increasing load on node 0 as the number of nodes increases, while strategy 3.1 conserves the limit of three connections per node. Since the latter does not correspond to an increase in the size of transferred data, evolution of communication time is more satisfactory.

Importantly, this strategy was designed with biological realism in mind, in the sense that there is no unnecessary communication between nodes that are not connected in the lymph network structure. Increased efficiency of communications is not, therefore, balanced by any inaccuracy introduced into the model. As a consequence, this is the strategy chosen for large-scale model implementation.

This implementation was then used on a 56-node cluster. Each "cluster node" has a dual-processor, and each processor has four cores. The cluster therefore offers 448 "computing cores" at 2.66 GHz, so that modelling hundreds of lymph nodes is a reasonable target, (and reflects a scale similar to that of the whole immune system).

MPI allows specification of the number of processes to be run on each "cluster node"; (these are then scattered over cores of this node). For the lymph network model, one process is equivalent to one lymph node. Of particular interest, therefore, is the evolution of the computation time:

- as a function of the number of agents at the start and of lymph nodes.
- as a function of the number of lymph nodes and of processes per "cluster node".

Table 5 displays the relative computation time as a function of initial agent count, (a.p.n), and the number of lymph nodes, (l.n.), for simulations using eight processes per cluster node, (p.p.n.). Optimisation of local iteration allows simulation of very large populations of agents at a relatively

Table 5. Model efficiency: relative computation time for several configurations of lymph nodes (l.n.) and agents per node at initialisation (a.p.n.)

a.p.n.	8 l.n.	16 l.n.	32 l.n.	64 l.n.	128 l.n.	256 l.n.
30000	1	1.04	1.08	1.12	1.16	1.21
150000	1.08	1.1	1.12	1.16	1.24	1.29
300000	1.21	1.22	1.23	1.24	1.27	1.31
600000	1.58	1.58	1.59	1.6	1.64	1.78
1500000	1.89	1.9	1.92	1.93	2.11	2.11

low computing cost. Significant increases in agent count only generate moderate overheads in terms of computation time. This is very important, as immune response and viral spread both lead to variations of specific agent counts.

Table 6 displays the relative computation time as a function of the number of lymph nodes and the number of processes per cluster node, for simulations starting with 150,000 agents per node. As expected from performance simulations, optimisation of the communication strategy allows simulation of a large lymph network, of size similar to that of the real system.

This is crucial to model realism, as confirmed by tests on cell mobility and its effects on immune activation, and on viral propagation throughout an organism. Implementation efforts, therefore, provide the opportunity for simulations reaching the scale of the immune system, both in terms of populations involved and "geographical" location of entities throughout the body. This is a significant increase in scale compared to existing models,

and permits a more detailed representation of the immune system. In particular, it allows inclusion of localised effects such as those related to the gastrointestinal tract, which are known to be crucial in the early stages of HIV infection,(see e.g. Guadalupe et al., 2003; Mattapallil et al., 2005), but which have not been modelled previously.

CONCLUSION

In this Chapter, we described how, using an optimal parallel implementation, it is possible to develop a large-scale model of the immune response to HIV infection. This model uses the agent-based paradigm, involving, in its current form, four agent types, which represent immune cells and virions.

This type of model requires large computing resources, and parallelisation was strongly indicated. Analysis of the biological system requirements and model form led to the identification

Table 6. Model efficiency: relative computation time for several configurations of lymph nodes (l.n.) and processes per "cluster node" (p.p.n.)

p.p.n.	8 l.n.	16 l.n.	32 l.n.	64 l.n.	128 l.n.	256 l.n.	512 l.n.
1	1	1.06	1.14	N/A	N/A	N/A	N/A
2	1.08	1.09	1.1	1.14	N/A	N/A	N/A
4	1.25	1.26	1.27	1.28	1.3	N/A	N/A
8	1.68	1.71	1.75	1.81	1.94	2	N/A
16	N/A	3.47	3.52	3.54	3.57	3.58	4.04

of spatial parallelisation as the most promising approach.

Data transfer was then considered, and several communication strategies tested. This was addressed through a combination of physical tests on smaller network sizes, and virtual experiments for larger configurations, using a purpose-built performance simulator. We optimised each communication strategy, and the "bio-inspired solution" was shown to be the most efficient, and implemented for large simulations.

These simulations confirmed the efficiency of the approach, and our model scaled up well. Using a 448-node cluster, it is now possible to run simulations with more than one billion cells, and approximately 500 lymph nodes. A better insight into the dynamics of HIV infection has been gained, both in terms of local interactions and overall progression, in particular through inclusion of the gastrointestinal tract in the model structure.

While providing for significant advance with respect to HIV modelling, this study is also important in the wider context of Computational Biology. Similar developmental frameworks can be postulated for other complex systems with multiple network layers. Examples include cell differentiation and cancer initiation, which involve complex genetic and epigenetic interactions, as well as epidemics, which features non-trivial dynamics and multiple potential intervention policies.

ACKNOWLEDGMENT

D. Perrin would like to acknowledge support for this study from the Irish Research Council for Science, Engineering and Technology (Embark Initiative).

REFERENCES

Anderson, R. M. (1988). The epidemiology of HIV infection: Variable incubation plus infectious periods and heterogeneity in sexual activity. *Journal of the Royal Statistical Society. Series A, (Statistics in Society)*, *151*(1), 66–98. doi:10.2307/2982185

Baldazzi, V., Castiglione, F., & Bernaschi, M. (2006). An enhanced agent based model of the immune system response. *Cellular Immunology*, *244*, 77–79. doi:10.1016/j.cellimm.2006.12.006

Benyoussef, A., HafidAllah, N. E., ElKenz, A., Ez-Zahraouy, H., Loulidi, M. (2003). Dynamics of HIV infection on 2D cellular automata. *Physica A*, *322*, 506–520. doi:10.1016/S0378-4371(02)01915-5

Berlekamp, E. R., Conway, J. H., & Guy, R. K. (2004). *Winning Ways for your Mathematical Plays* (2nd edition). Wellesley, MA: A. K. Peters Ltd.

Bernaschi, M., & Castiglione, F. (2001). Design and implementation of an immune system simulator. *Computers in Biology and Medicine*, *31*, 303–331. doi:10.1016/S0010-4825(01)00011-7

Bohringer, C., & Rutherford, T. F. (2008). Combining bottom-up and top-down. *Energy Economics*, *30*(2), 574–596. doi:10.1016/j.eneco.2007.03.004

Brehm, M., Pinto, A., Daniels, K., Schneck, J., Welsh, R., & Selin, L. (2002). T cell immunodominance and maintenance of memory regulated by unexpectedly cross-reactive pathogens. *Nature Immunology*, *3*, 627–634.

Burns, J. (2005). *Emergent networks in immune system shape space*. PhD thesis, Dublin City University, School of Computing.

Burns, J., & Ruskin, H. J. (2004). *Network topology in immune system shape space* (. *LNCS*, *3038*, 1094–1101.

Cahill, R. N. P., Frost, H., & Trnka, Z. (1976). The effects of antigen on the migration of recirculating lymphocytes through single lymph node. *The Journal of Experimental Medicine, 143*, 870–888. doi:10.1084/jem.143.4.870

Cammarata, S., McArthur, D., & Steeb, R. (1983). Strategies of cooperation in distributed problem solving. In *Proceedings of the Eighth International Joint Conference on Artificial Intelligence (IJCAI-83)*, Karlsruhe, Germany.

Castiglione, F., Poccia, F., D'Offizi, G., & Bernaschi, M. (2004). Mutation, fitness, viral diversity, and predictive markers of disease progression in a computational model of HIV type 1 infection. *AIDS Research and Human Retroviruses, 20*(12), 1314–1323. doi:10.1089/aid.2004.20.1314

Celada, F., & Seiden, P. E. (1992). A computer model of cellular interactions in the immune system. *Immunology Today, 13*(2), 56–62. doi:10.1016/0167-5699(92)90135-T

Cheynier, R., Henrichwark, S., Hadida, F., Pelletier, E., Oksenhendler, E., Autran, B., & Wain-Hobson, S. (1994). HIV and T cell expansion in splenic white pulps is accompanied by infiltration of HIV-specific cytotoxic T lymphocytes. *Cell, 78*(3), 373–387. doi:10.1016/0092-8674(94)90417-0

Corning, P. A. (2002). The re-emergence of "emergence": a venerable concept in search of a theory. *Complexity, 7*(6), 18–30. doi:10.1002/cplx.10043

Crutchfield, J. P. (1994). The calculi of emergence: computation, dynamics and induction. *Physica D. Nonlinear Phenomena, 75*(1-3), 11–54. doi:10.1016/0167-2789(94)90273-9

De Wolf, T., & Holvoet, T. (2005). *Emergence versus Self-Organisation: different concepts but promising when combined.* (. LNCS, 3464, 1–15.

Diestel, R. (2005). *Graph Theory* (Graduate Texts in Mathematics, Volume 173). New York: Springer.

Durfee, E. H. (1998). *Coordination of distributed problem solvers.* Boston: Kluwer Academic Publishers.

Gardner, M. (1970). Mathematical games: The fantastic combinations of John Conway's new solitaire game Life. *Scientific American, 223*, 120–123.

Goertzel, B. (1992). Self-organizing evolution. *Journal of Social and Evolutionary Systems, 15*(1), 7–53. doi:10.1016/1061-7361(92)90035-C

Gropp, W., Lusk, E., & Skjellum, A. (1999a). *Using MPI-2: Advanced Features of the Message Passing Interface.* Cambridge, MA: MIT Press.

Gropp, W., Lusk, E., & Skjellum, A. (1999b). *Using MPI: Portable Parallel Programming With the Message-Passing Interface,* (2nd Ed.). Cambridge, MA: MIT Press.

Grossman, Z., Feinberg, M. B., & Paul, W. E. (1998). Multiple modes of cellular activation and virus transmission in HIV infection: a role for chronically and latently infected cells in sustaining viral replication. *Proceedings of the National Academy of Sciences of the United States of America, 95*(11), 6314–6319. doi:10.1073/pnas.95.11.6314

Guadalupe, M., Reay, E., Sankaran, S., Prindiville, T., Flamm, J., McNeil, A., & Dandekar, S. (2003). Severe CD4+ T-cell depletion in gut lymphoid tissue during primary human immunodeciency virus type 1 infection and substantial delay in restoration following highly active antiretroviral therapy. *Journal of Virology, 77*(21), 11708–11717. doi:10.1128/JVI.77.21.11708-11717.2003

Hayes-Roth, B., Hewett, M., Washington, R., Hewett, R., & Seiver, A. (1989). Distributing intelligence within an individual. In L. Gasser & M. Huhns (Ed.), *Distributed Artificial Intelligence*, (Vol. 2, pp. 385-412). San Francisco: Pitman Publishing and Morgan Kaufmann.

Hecquet, D., Ruskin, H. J., & Crane, M. (2007). Optimisation and parallelisation strategies for Monte Carlo simulation of HIV infection. *Computers in Biology and Medicine, 37*(5), 691–699. doi:10.1016/j.compbiomed.2006.06.010

Helsinger, A., Kleinmann, K., & Brinn, M. (2004). A framework to control emergent survivability of multi agent systems. In *Proceedings of Third International Joint Conference on Autonomous Agents and Multiagent Systems, 1*, 28-35.

Horwitz, P., & Wilcox, B. A. (2005). Parasites, ecosystems and sustainability: an ecological and complex systems perspective. *International Journal for Parasitology, 35*(7), 725–732. doi:10.1016/j.ijpara.2005.03.002

Janssen, M. A., & Ostrom, E. (2006). Governing social-ecological systems. In L. Tesfatsion and K. Judd, (Ed.), *Handbook of Computational Economics*, (pp. 1465-1509).

Jennings, N., Sycara, K., Wooldridge, M. (1998). A roadmap of agent research and development. *Autonomous agents and multi-agents systems, 1*(1), 7-38.

Kunkel, E. J., & Butcher, E. C. (2002). Chemokines and the tissue-specific migration of lymphocytes. *Immunity, 16*, 1–4. doi:10.1016/S1074-7613(01)00261-8

Mackay, C. R., Marston, W. L., & Dudler, L. (1990). Naive and memory T cells show distinct pathways of lymphocyte recirculation. *The Journal of Experimental Medicine, 171*, 801–817. doi:10.1084/jem.171.3.801

Mannion, R., Ruskin, H. J., & Pandey, R. B. (2000). Effect of mutation on helper T-cells and viral population: a computer simulation model for HIV. *Theory in Biosciences, 119*, 145–155.

Mannion, R., Ruskin, H. J., & Pandey, R. B. (2002). A Monte-Carlo approach to population dynamics of cells in a HIV immune response model. *Theory in Biosciences, 121*, 237–245.

Mattapallil, J. J., Douek, D. C., Hill, B., Nishimura, Y., Martin, M., & Roederer, M. (2005). Massive infection and loss of memory CD4+ T cells in multiple tissues during acute SIV infection. *Nature, 434*, 1093–1097. doi:10.1038/nature03501

McCarthy, I. P., & Tan, Y. K. (2000). Manufacturing competitiveness and fitness landscape theory. *Journal of Materials Processing Technology, 107*(1-3), 347–352. doi:10.1016/S0924-0136(00)00687-7

Minar, N., Burkhart, R., Langton, C., & Askenazi, M. (1996). *The Swarm simulation system: A toolkit for building multi-agent simulations*. (Working Paper 96-06-042, Santa Fe Institute, Santa Fe, NM).

Mizoguchi, F., Nishiyama, H., Ohwada, H., & Hiraishi, H. (1999). Smart office robot collaboration based on multi-agent programming. *Artificial Intelligence, 114*, 57–94. doi:10.1016/S0004-3702(99)00068-5

Naresh, R., Tripathi, A., & Omar, S. (2006). Modelling the spread of AIDS epidemic with vertical transmission. *Applied Mathematics and Computation, 178*(2), 262–272.

Ndifon, W. (2005). A complex adaptive systems approach to the kinetic folding of RNA. *Bio Systems, 82*(3), 257–265. doi:10.1016/j.biosystems.2005.08.004

Pandey, R. B., Mannion, R., & Ruskin, H. J. (2000). Effect of cellular mobility on immune response. *Physica A*, *283*, 447–450. doi:10.1016/S0378-4371(00)00206-5

Perelson, A. S., & Nelson, P. W. (1999). Mathematical analysis of HIV-1 dynamics in vivo. *SIAM Review*, *41*(1), 3–44. doi:10.1137/S0036144598335107

Perelson, A. S., & Oster, G. F. (1979). Theoretical studies of clonal selection: Minimal antibody repertoire size and reliability of self-non-self discrimination. *Journal of Theoretical Biology*, *81*(4), 645–670. doi:10.1016/0022-5193(79)90275-3

Perrin, D., Ruskin, H. J., & Crane, M. (2006c). HIV modelling - a parallel implementation of a lymph network. In *Selected proceedings of Third International Conference on Cluster and Grid Computing Systems* (CGCS 2006), Venice, Italy.

Ruskin, H. J., & Burns, J. (2005). *Network emergence in immune system shape space.* (. *LNCS*, *3481*, 1254–1263.

Ruskin, H. J., & Burns, J. (2006). Weighted networks in immune system shape space. *Physica A*, *365*(2), 549–555. doi:10.1016/j.physa.2005.11.006

Ruskin, H. J., Pandey, R. B., & Liu, Y. (2002). Viral load and stochastic mutation in a Monte Carlo simulation of HIV. *Physica A*, *311*(1-2), 213–220. doi:10.1016/S0378-4371(02)00832-4

Seiden, P., & Celada, F. (1992). A model for simulating cognate recognition and response in the immune system. *Journal of Theoretical Biology*, *158*, 329–357. doi:10.1016/S0022-5193(05)80737-4

Sharkasi, A., Crane, M., Ruskin, H. J., & Matos, J. A. O. (2006). The reaction of stock markets to crashes and events: A comparison study between emerging and mature markets using wavelet transforms. *Physica A*, *368*(2), 511–521. doi:10.1016/j.physa.2005.12.048

Thurner, S., & Biely, C. (2007). The eigenvalue spectrum of lagged correlation matrices. *Acta Physiologica Polonica*, *38*(13), 4111–4122.

Uhrmacher, A. M., Tyschler, P., & Tyschler, D. (2000). Modeling and simulation of mobile agents. *Future Generation Computer Systems*, *17*, 107–118. doi:10.1016/S0167-739X(99)00107-7

Witherden, D. A., Kimpton, W. G., Washington, E. A., & Cahill, R. N. P. (1990). Non-random migration of CD4+, CD8+ and γδ+T19+ lymphocytes through peripheral lymph nodes. *Immunology*, *70*, 235–240.

Wooldridge, M., & Jennings, N. (1995). Intelligent agents: Theory and practice. *The Knowledge Engineering Review*, *2*(10), 115–152. doi:10.1017/S0269888900008122

Zhang, S., Yang, J., Wu, Y., & Liu, J. (2005). *An enhanced massively multi-agent system for discovering HIV population dynamics.* (. *LNCS*, *3645*, 988–997.

KEY TERMS AND DEFINITIONS

Adaptive Immune Response: In contrast to the innate response, the specific, or adaptive, immune response is based on the accurate recognition of foreign non-self antigens. Antigen-specific response has two arms, namely cell-mediated and antibody-mediated responses. The latter, also known as the humoral response, features B lymphocytes as effector cells, and mainly targets bacterial attacks. Humoral response is characterised by production, by these cells, neutralizing antibodies, following activation by CD4+ T helper cells through release of interleukin IL-4. Cell-mediated response is targeted more specifically at viral attacks and takes place in lymph nodes.

Agent: An intelligent agent is a modelling object with specific properties which include autonomy, social behaviour, reactivity, and pro-activity.

Agent-Based Model: A model in which the key abstraction elements are agents. When using several agents, such a model is often called a multi-agent system.

AIDS: Acquired ImmunoDeficiency Syndrome. Collection of symptoms and infections resulting from the specific damage to the immune system caused by the human immunodeficiency virus. This is the last phase of HIV infection progression.

HIV: Human Immunodeficiency Virus. Retrovirus targeting immune cells and using them as hosts. This results in massive depletion of immune cell populations. Infection progression is typically divided in three phases, ending with AIDS.

Innate Immune Response: A non-specific, or innate, response is based upon recognition of the pattern of the microbial surface components of the pathogens, rather than by a specific antigenic sequence. Innate response does not confer long-lasting immunity to the host, i.e. there is no memory of previous responses.

MPI: Message-Passing Interface, a communication protocol used for parallel implementation of programs. MPI provides support for point-to-point and collective communications, inquiry routines to query the execution environment, as well as constants and data-types.

Chapter 4
Computer Aided Tissue Engineering from Modeling to Manufacturing

Mohammad Haghpanahi
Iran University of Science and Technology, Iran

Mohammad Nikkhoo
Iran University of Science and Technology, Iran

Habib Allah Peirovi
Shaheed Beheshti University of Medical Science and Health Services, Iran

ABSTRACT

Computer aided tissue engineering integrates advances of multidisciplinary fields of biology, biomedical engineering, and modern design and manufacturing. It enables the application of advanced computer aided technologies and biomechanical engineering principles to derive systematic solutions for complex tissue engineering problems. After an introduction to tissue engineering, this chapter presents the recent development on computer aided tissue engineering, including computer aided tissue modeling, computer aided tissue scaffold informatics and biomimetic design, and computer aided biomanufacturing.

INTRODUCTION

Tissue engineering, a field of science which is approximately a decade old, has been labeled as one of the more promising domains within the broader field of biotechnology. In a simple definition, it is an interdisciplinary field that applies the principles and methods of bioengineering, material science, and life sciences toward the assembly of biologic substitutes that will restore, maintain, and improve tissue functions following damage either by disease or traumatic processes. The general principles of tissue engineering involve combining living cells with a natural/synthetic scaffold to build a three-dimensional living construct that is functionally, structurally and mechanically equal to the tissue that is to be replaced. Tissue engineering is founded on three principal components (Scaffolds, Cells, Growth Factor and Mechanical Stress), which may be used independently or incorporated in combinatorial form (Figure 1).

DOI: 10.4018/978-1-60566-768-3.ch004

Figure 1. Principal components of the tissue engineering

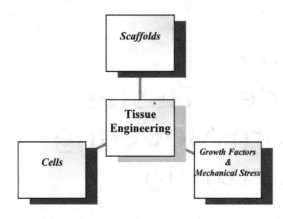

Scaffold materials are three-dimensional tissue structures that guide the organization, growth and differentiation of cells. There are several requirements in the design of scaffolds for tissue engineering. Many of these requirements are complex and not yet fully understood. In addition to being biocompatible both in bulk and degraded form, these scaffolds should possess appropriate mechanical properties to provide the correct stress environment for the tissues. Also, the scaffolds should be porous and permeable to permit the ingress of cells and nutrients, and should exhibit the appropriate surface structure and chemistry for cell attachment.

Cells are a key to tissue regeneration and repair due to their proliferation and differentiation, cell-to-cell signaling, biomolecule production, and formation of extracellular matrix. The functionality of an engineered tissue may be structural (e.g., bone, cartilage, and skin) or metabolic (e.g., liver, pancreas), or both. Cells may be a part of an engineered tissue, or alternatively, these cells may be recruited in vivo with the help of biomaterials or biomolecules.

Growth factors are soluble peptides capable of binding cellular receptors and producing either a permissive or preventive cellular response toward differentiation and proliferation of tissue. All cells and tissues of the organism are continually subject

to mechanical stresses. These forces have many various origins, from pressure forces linked to gravity to motion forces (i.e., blood circulation, loading on cartilage and bone during activity and etc.). Their range is a few Pascals in vascular wall shear stress and several mega Pascals in hip cartilage. It has now been accepted that these applied forces are likely to modify cellular behavior by affecting metabolism, paracrine or autocrine factor secretion and gene expression.

In early tissue engineering trials the aim of bioreactor culture was to provide nutrient perfusion to cells in the centre of a thick construct. There were two different bioreactors systems, in common usage; rotating wall or perfusion systems. In rotating wall bioreactors samples are suspended in culture medium in a cylindrical chamber and the cylinder rotated at a speed that allows the constructs to fall through the medium but not hit the sides. Perfusion bioreactors involve pumping of nutrient containing medium through the construct. Since these systems apply mechanical forces as well as allowing nutrient movement it is hard to separate the two effects. The stresses and strains applied to the cells are not measurable in these basic bioreactor systems, making it difficult to fine tune the system or understand the mechanism of increased tissue formation. While research in biorheology and biomechanics have in the last decades helped understanding the physical properties of cells and tissues, recent works have focussed on the physiological consequences of applied stresses and opened a new avenue for research that can be defined as mechanobiology, leading through its applications to a better understanding of a variety of diseases or pathological conditions and to novel therapeutical approaches using tissue engineering concept and the development of a new generation of biomaterials. In general mechanobiology is the study of how cells respond to mechanical forces and many cell types have been studied in vitro in order to tease out the signaling mechanisms by which a mechanical force results in a biochemical response. For instance, a layer of endothelial (blood vessel lining) cells can be subjected to fluid shear in vitro and will respond with changes in cell shape, cell alignment and production of signaling molecules. Bone cells can be stretched on a membrane and be shown to release factors involved in bone remodeling. Cartilage cells seeded in a gel and subjected to dynamic compression modulate production of the glycocaminoglycans that form the extracellular matrix. The mechanobiology approach is usually to apply precise loading conditions in an environment designed for the assays required.

By the way, the first step of tissue engineering process (such as bone, cartilage, intervertebral disc and etc.) begins with the design and fabrication of a porous three dimensional scaffold. In general, the scaffold should be fabricated from a highly biocompatible material which does not have the potential to elicit an immunological or clinically detectable primary or secondary foreign body reaction.

Scaffold characteristics are interconnectivity, pore size, pore curvature, microporosity, macroporosity and surface roughness which influence cellular responses and also control the degree of nutrient delivery, penetration depth of cells and metabolic waste removal. A tissue engineered scaffold must provide a germane environment for in vitro cell culturing in a bioreactor as well as providing a suitable environment once implanted in vivo.

The scaffolds should be designed in order to mimic the function of the natural extracellular matrix. The primary roles of scaffold are serving as an adhesion substrate for the cell, facilitating the localization and delivery of cells when they are implanted. Scaffolds should provide temporary mechanical support to the newly grown tissue by defining and maintaining a 3D structure and guide the development of new tissues with the appropriate function.

So an ideal scaffold for tissue engineering should bring about the desired biological response which is possible by considering following characteristics in fabrication process.

1. Having a porous three-dimensional with interconnected pore network for cell or tissue growth and flow transport of nutrients and metabolic waste
2. Having suitable surface chemistry for cell attachment, proliferation and differentiation
3. Being biodegradable or bioresorbable with a controllable degradation and resorption rate to match cell or tissue growth in vitro or in vivo
4. Having requisite mechanical properties and being easily processed to form a variety of shapes and sizes

The main scaffold materials which have been successfully investigated for tissue engineering are: (1) Natural polymers, such as collagens, glycosaminoglycan, starch, chitin and chitosan, (2) Synthetic polymers, based on polylactic acid (PLA), polyglycolic acid (PGA) and their copolymers (PLGA), and (3) Ceramics, such as hydroxyapatite (HA) and β-tricalcium phosphate (β-TCP).

Many research have defined scaffold pores based on size as either micro (pore diameter < 100μm) or macro (pore diameter > 100μm). The pore size employed may also be dependent on the tissue type desired. For example scaffolds with pore sizes less than 150μm have been successfully used for regeneration of skin and pore size about 400 μm is suitable for cartilage tissue.

A scaffold should provide an open porous networked structure allowing for easier vascularisation, which is important for the growth of cells. Approaches in scaffold design must be able to create hierarchical porous structures to attain desired mechanical function and mass transport (that is, permeability and diffusion) properties, and to produce these structures within arbitrary and complex three dimensional (3D) anatomical shapes. Hierarchical refers to the fact that features at scales from the nanometer to millimeter level will determine how well the scaffold meets conflicting mechanical function and mass transport

Figure 2. Principal steps of image based modeling

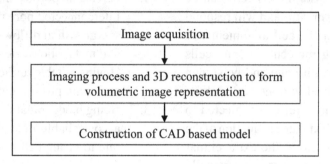

needs. Material chemistry together with processing determines the maximum functional properties that a scaffold can achieve, as well as how cells interact with the scaffold. However, mass transport requirements for cell nutrition, porous channels for cell migration, and surface features for cell attachment necessitate a porous scaffold structure. This porous structure dictates that achievable scaffold properties will fall between the theoretical maximum set by the material and the theoretical minimum of zero predicted by composite theories.

However, the objective of this chapter is to present how computer aided methods can facilitate this procedure, including an overview on computer aided tissue modeling, design and manufacturing.

COMPUTER AIDED TISSUE MODELING

Computer aided tissue modeling is including representation and modeling of 3D tissue and anatomic system, bio-modeling, 3D reconstruction and application to surgical planning and simulation. Construction of a computer aided modeling for a specific tissue often starts from the acquisition of anatomic data. This is referred to as image-based modeling in which the imaging modality must be capable of producing three dimensional views of anatomy, differentiating heterogeneous tissue types and displaying the vascular structure, and

generating computational tissue models for other down stream applications, such as analysis and simulation. In general, an image based modeling process involves three principal steps which are shown in Figure 2.

The current imaging modalities used in tissue modeling are CT, MRI, and optical microscopy.

Computed Tomography (CT) is an imaging modality that provides images generated by X-rays in a third dimension. In general, an X-rays beam is passed through a patient in the same location from a variety of angles. This action produces a single 'slice' of the patient. The CT computer displays the images of interest as a variation of a shade of gray. The structures outside the area of interest will be displayed as all black or all white. CT images have increased contrast resolution compared to two dimensional survey radiographic images. CT or μCT scans require exposure of a sample to small quantities of ionizing radiation, the absorption of which is detected and imaged. This results in a series of 2D images displaying a density map of the sample. Stacking these images creates a 3D representation of the scanned area. The main advantage of CT and μCT as an imaging modality for tissue engineering purposes is reasonably high resolution. The latest development of μCT technology has been successfully used to quantify the microstructure function relationship of tissues and the designed tissue structures, include characterizing micro-architecture of tissue scaffolds.

Magnetic Resonance Imaging (MRI) also provides cross sectional imaging without ionizing radiation. Aligning the water molecules of the body within a strong, stable external magnetic field generates the image. A radiofrequency pulse is applied which causes the water molecules to lose their alignment with the main magnetic field. As the molecules begin to realign themselves, they emit signals which are detected by the MRI machine. The signals are processed by a computer, which creates the image. In general, portions of the body that do not contain a lot of water (i.e., lung, bone) will not produce any significant signal. Contrast studies with the MRI utilize a paramagnetic substance injected intravenously, which alters the area in which it accumulates. It shows up as a whiter area in the image compared to the noncontrast study. So MRI provides images for soft tissues as well as for hard tissues, and as such is vastly superior in differentiating soft tissue types and recognizing border regions of tissues of similar density.

Optical microscopy has limited applications for tissue modeling due to the intensive data manipulation. In this case if we want to examine a sample with high resolution using optical microscopy, it must be physically sectioned to a thickness of between 5 and 50 microns and placed onto slides, providing a square sample for fine resolution. The division into these slides is a labor intensive process, and the resulting images of the target organ would be thousands of 2D images that must be both digitally stacked into 3D columns. This is so complicated and a memory intensive process. However, differentiating tissue down to the level of the individual cell may still be only possible by using optical microscopy.

Most CT and MRI units have the ability of exporting data in common medical file format (DICOM) digital imaging and communication in medicine. After saving CT or MRI image data, they should be transferred to computer aided design models. So the next step is processing these data, which is a very complex and important step that the quality of the final medical model depends on. For this step engineers need software packages such as Mimics, 3D Doctor, Simpleware, Amira,

Rhino and etc in which they can make segmentation of this anatomy image, achieve high resolution 3D rendering in different colors, make 3D virtual model and finally make possible to convert CT or MRI scanned image data from DICOM to IGES, STL or other requisite file format. These software packages allow making segmentation by threshold technique, considering the tissue density. In this way, at the end of image segmentation, there are only pixels with a value equal or higher than the threshold value. The virtual model of internal structures of human's body, which is needed for final production of 3D physical model, requests very good segmentation with a good resolution and small dimensions of pixels. This demands good knowledge in this field which should help engineers to exclude all structures which are not the subject of interest in the scanned image and choose the right region of interest (separate bone from tissue, include just part of a bone, exclude anomalous structures, noise or other problems which can be faced). Depending on complexity of the problem this step usually demands collaboration of engineers with radiologists and surgeons who will help to achieve good segmentation, resolution and a finally accurate 3D virtual model. As an example, figure 3 shows an overview of the stages of cervical spine computer aided modeling in our research laboratory.

On the basis of this computer modeling methods, we can develop the mechanobilogical models to gain the capability of optimizing the design parameters of porous scaffolds, role of electrochemical and mechanobiological parameters in tissue properties and prediction of the stress distribution in different stages of the tissue engineering.

COMPUTER AIDED SCAFFOLD MANUFACTURING

Several techniques have been developed for scaffold fabrication. These include fiber meshes

Figure 3. Overview of the procedure for developing 3D image based model

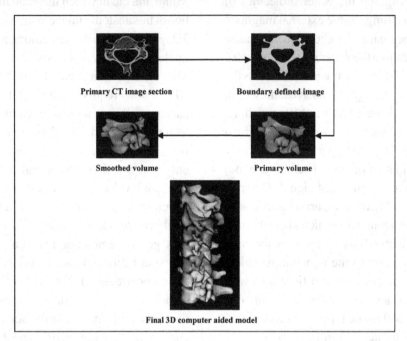

Figure 4. A SEM picture of our produced scaffold using freeze drying method (Bar indicates 100 μm)

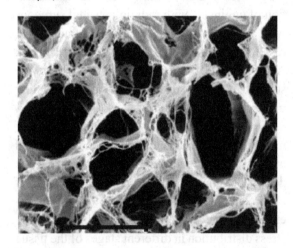

or fiber bonding, freeze drying, gas foaming, melt molding, phase separation, solution casting, solvent casting and particulate leaching (Figure 4 shows one of the scaffolds which was produced in our research center using freeze drying methods). Table 1 reviews the features of the mentioned

conventional methods. However, it is important to mention that we can not exactly judge that which methods are the best because it directly depends on the types of predicted tissue and also the experience of the researchers.

In general, convectional methods of fabricating scaffolds, through material processing and casting, have largely been unsuccessful in controlling the internal architecture to a high degree of accuracy or homogeneity, since the resulting interior architectures are determined by the processing technique. For example, particulate leaching is a process whereby the internal architecture is determined by embedding a high density of salt crystals into a dissolved polymer or ceramic matrix. The dissolved mixture is then poured into a mould and treated under heat and pressure to form the external shape. So it clear that conventional scaffold fabrication techniques are incapable of precisely controlling pore size, pore geometry, spatial distribution of pores and construction of internal channels within the scaffold. Also conventional fabrication techniques produce scaffolds

Table 1. Conventional scaffold fabrication methods

Method	Features	
	Advantage	Disadvantage
Fiber bonding	Can produce Highly porous structures	Limited range of polymers can be used Lack of mechanical strength
Freeze drying	Can produce Highly porous structures High pore interconnectivity	Limited to small pore sizes
High-pressure processing	No organic solvents	Causing nonporous external surface Interconnectivity is limited
Hydrocarbon templating	No thickness limitation Independent control of porosity and pore size	Problem with Residual solvents
Melt moulding	Independent control of porosity and pore size Macro shape control	High temperature required for nonamorphous polymer
Membrane Lamination	Independent control of porosity and pore size Macro shape control	Lack of mechanical strength Interconnectivity is limited
Phase separation	Can produce Highly porous structures Permits incorporation of bioactive agents	control over internal architecture is weakness range of pore sizes are limited
Polymer or ceramic fiber composite foam	Independent control of porosity and pore size Good mechanical strength	Problems with residual solvents
Solvent casting and particulate leaching	Can produce Highly porous structures Crystallinity can be tailored	membrane thickness and interconnectivity is limited control over internal architecture is weakness

that are foam structures. Cells are then seeded and expected to grow into the scaffold. However, this approach has resulted in the in vitro growth of tissues with cross-sections of less than 500μm from the external surface which is probably due to the diffusion constraints of the foam (Ishaug-Riley et al. and Freed et al.). The pioneering cells cannot migrate deep into the scaffold because of the lack of nutrients and oxygen and insufficient removal of waste products; cell colonization at the scaffold periphery is consuming, or acting as an effective barrier to the diffusion of, oxygen and nutrients into the interior of the scaffold. Furthermore, for bone tissue engineering, the high rates of nutrient and oxygen transfer at the surface of the scaffold promote the mineralization of the scaffold surface, further limiting the mass transfer to the interior of the scaffold (Martin et al.). Thus cells are only able to survive close to the surface. In this connection, it should be noted that no cell, except for chondrocytes, exists further than 25-100μm away from a blood supply. The low oxygen requirement of cartilage may be the reason why only this tissue

has been successfully grown in vitro to thick cross-sections i.e. greater than 1mm using conventional scaffold fabrication techniques. Skin is a relatively 2D tissue and thus thick cross-sections of tissue are not required, thereby explaining the success of producing this tissue with conventional scaffold fabrication techniques (Eaglstein et al.). However, most other 3D tissues require a high oxygen and nutrient concentration. The human body supplies its tissues with adequate concentrations of oxygen and nutrients via blood vessels. Tissue engineering scaffolds should embrace this approach and have some form of an artificial vascular system present within them to increase the mass transport of oxygen and nutrients deep within, and removal of waste products from, the scaffold. It is in this application that rapid prototyping fabrication method can optimize tissue-engineering scaffolds.

Consequently, researchers try to modify the conventional techniques to overcome the inherent process limitations. In recent years rapid prototyping method as an advance scaffold fabrication technique has been proposed. RP is a common name for a group of techniques that can generate a physical model directly from CAD data. It is an additive process in which each part is constructed in a layer-by-layer manner. This section presents the recent advanced technology on scaffold fabrication in tissue engineering.

Selective Laser Sintering (SLS)

In general, Selective Laser Sintering is an additive rapid manufacturing technique that uses a high power laser (for example, a carbon dioxide laser) to fuse small particles of plastic, metal, or ceramic powders into a mass representing a desired three dimensional object. This method uses a deflected laser beam selectively to scan over the powder surface following the cross sectional profiles carried by the slice data. The interaction of the laser beam with the powder elevates the powder temperature to reach the glass transition temperature. This process causes surfaces in contact to deform and fuse together. So this method has a great potential in fabrication process for producing complex porous ceramic matrices suitable for implantation in bone defect. Vail et al. and Tan et al. have been proposed their research results which confirm this subject. This fast processing method does not need support structure which is a good advantage.

Fused Deposition Modeling (FDM)

Fused deposition modeling, which is often referred to by its initials FDM, is a type of rapid prototyping or rapid manufacturing technology commonly used within engineering design. In this method, a filament of a suitable material is fed and melted inside a heated liquefier. Then the material is extruded through a nozzle. The system operates in an environment which is temperature controlled to maintain sufficient fusion energy between each layer. In recent years, researchers have demonstrated the feasibility of using FDM for scaffold fabrication. In a recent study, human

mesenchymal progenitor cells were seeded on PCL and PCL-hydroxyapatite scaffolds fabricated by FDM. Proliferation of cells toward and onto the scaffold surfaces was detected. Having good mechanical strength and being versatile in lay down pattern design is the most important advantage in this method.

3D Bioplotter (3DBP)

The 3D Bioplotter is as simple as useful and is based on dispensing a plotting material into a plotting medium to cause solidification of the material and to compensate gravity force through buoyancy. This method is based on the 3D dispensing of liquids and pastes into a liquid medium with matched density. The material leaves the nozzle and solidifies in the plotting medium after bonding to the previous layer. The liquid medium compensates for gravity and hence no support structure is needed. Hydrogel scaffolds with interconnected internal pore structure were prepared by Landers et al. However, the produced scaffolds had limited resolution and mechanical strength. According to the importance of having a suitable rigidity, Wong et al.. extended this research and it was shown that in this case, cells displayed a preference for stiffer regions, and tended to migrate faster on surfaces with lower compliance. As an advantage, in this method we can use enhanced range of materials and incorporate biomolecules.

3D Printing (3DP)

Three dimensional printing is a method of converting a virtual 3D model into a physical object. In this method, a stream of adhesive droplets is expelled through an inkjet print head, selectively bonding a thin layer of powder particles to form a solid shape. Kim et al. employed 3DP using a particulate leaching technique to create porous scaffolds using PLGA (polylactic glycolic acid) mix with salt particles and a suitable organic solvent. In this research the pores formed with 60% porosity. In vitro cell culture with hepatocytes showed ingrowth of these cells into the pore space. In an effort to render the system more biocompatible, Lam et al. have formulated a blend of starch-based polymer powders that can be bonded together using distilled water. Having a fast process and using enhanced range of material is the advantage of this method. It should be also mentioned that this method does not need support structure which is another advantage.

Stereolithography (SLA)

Stereolithography is the most widely used rapid prototyping technology. A UV laser traces out the first layer of photocurable resin, solidifying the model's cross-section while leaving the remaining areas in liquid form. The elevator then drops by a sufficient amount to cover the solid polymer with another layer of liquid resin A sweeper recoats the solidified layer with liquid resin and the laser traces the second layer atop the first. Chu et al. have produced HA-based porous implants using SLA-built epoxy moulds. A thermal curable HA–acrylate suspension was cast into the mould to obtain a scaffold with interconnected channels. The preliminary results showed that controlling the overall geometry of the regenerated bone tissue was possible through the internal architectural design of the scaffolds. The most important advantage of this method can be summarized as controlling of external and internal morphology in scaffold fabrication.

CONCLUSION

Tissue engineering is an interdisciplinary field that applies the principles and methods of bioengineering, material science, and life sciences toward the assembly of biologic substitutes that will restore, maintain, and improve tissue functions following damage either by disease or traumatic processes. The first step of tissue engineering process (such as

bone, cartilage, intervertebral disc and etc.) begins with the design and fabrication of a porous three dimensional scaffold and this chapter presents how computer aided methods can facilitate this procedure, including an overview on computer aided tissue modeling, design and manufacturing. The emergence of various different approaches in tissue engineering has highlighted the fact that the field of tissue engineering is still growing. So as it discussed, on the basis of computer aided modeling methods, we can develop the mechanobilogical models to gain the capability of optimizing the design parameters of porous scaffolds, role of electrochemical and mechanobiological parameters in tissue properties and prediction of the stress distribution in different stages of the tissue engineering. Also it is clear that computer aided manufacturing methods can improve the current limitations in tissue engineering. Looking towards the future, Rapid prototyping technologies hold great potential in the field of scaffold fabrication. This technology enables the tissue engineering researchers to have full control over the design, modeling and fabrication of the scaffold. On the other hand, these advanced fabrication methods can provide a systematic learning channel for investigating cell matrix structures.

REFERENCES

Atala, A. (2004). Tissue engineering and regenerative medicine: concepts for clinical application. *Rejuvenation Research*, 7, 15–31. doi:10.1089/154916804323105053

Audet, J. (2004). Stem cell bioengineering for regenerative medicine. *Expert Opinion on Biological Therapy*, 4, 631–644. doi:10.1517/14712598.4.5.631

Bhatia, S. N., & Chen, C. S. (1999). Tissue engineering at the micro scale. *Biomedical Microdevices*, 2, 131–144. doi:10.1023/A:1009949704750

Carter, D. R., & Beaupré, G. S. (2001). *Skeletal Function and Form: Mechanobiology of Skeletal Development, Aging, and Regeneration*. Cambridge, UK: Cambridge University Press.

Chapekar, M. S. (1996). Regulatory concerns in the development of biologic-biomaterial combinations. *Journal of Biomedical Materials Research, 33*, 199–203. doi:10.1002/(SICI)1097-4636(199623)33:3<199::AID-JBM10>3.0.CO;2-C

Chapekar, M. S. (2000). Tissue engineering: challenges and opportunities. *Journal of Biomedical Materials Research, 53*, 615–620. doi:10.1002/1097-4636(2000)53:6<617::AID-JBM1>3.0.CO;2-C

Cheah, C. M., Chua, C. K., Leong, K. F., Cheong, C. H., & Naing, M. W. (2004). Automatic algorithm for generating complex polyhedral scaffold structures for tissue engineering. *Tissue Engineering, 10*, 595–610. doi:10.1089/107632704323061951

Chu, T. M. G., Halloran, J. W., Hollister, S. J., & Feinberg, S. E. (2001). Hydroxyapatite implants with designed internal architecture. *Journal of Materials Science. Materials in Medicine, 12*, 471–478. doi:10.1023/A:1011203226053

Ciardelli, G., Chiono, V., Cristallini, C., Barbani, N., Ahluwalia, A., & Vozzi, G. (2004). Innovative tissue engineering structures through advanced manufacturing technologies. *Journal of Materials Science. Materials in Medicine, 15*, 305–310. doi:10.1023/B:JMSM.0000021092.03087.d4

Eaglstein, W. H., & Falanga, V. (1997). Tissue engineering and the development of Apligraf: a human skin equivalent. *Clinical Therapeutics, 19*, 894–905. doi:10.1016/S0149-2918(97)80043-4

Endres, M., Hutmacher, D. W., Salgado, A. J., Kaps, C., Ringe, J., & Reis, R. L. (2003). Osteogenic induction of human bone marrow derived mesenchymal progenitor cells in novel synthetic polymer hydrogel matrices. *Tissue Engineering, 9*, 689–702. doi:10.1089/107632703768247386

Fang, Z., Starly, B., & Sun, W. (2005). Computer aided characterization for effective mechanical properties of porous tissue scaffolds. *Computer Aided Design, 37*, 65–72. doi:10.1016/j.cad.2004.04.002

Freed, L. E., & Vunjak-Novakovic, G. (1998). Culture of organized cell communities. *Advanced Drug Delivery Reviews, 33*, 15–30. doi:10.1016/S0169-409X(98)00017-9

Fuchs, J. R., Nasseri, B. A., & Vacanti, J. P. (2001). Tissue engineering: a 21st century solution to surgical reconstruction. *The Annals of Thoracic Surgery, 72*, 577–591. doi:10.1016/S0003-4975(01)02820-X

Gooch, K. J., & Tennant, C. J. (1997). *Mechanical forces: their effects on cells and tissues*. Berlin: Springer Verlag.

Haghpanahi, M., & Miramini, S. (2008). Extraction of Morphological Parameters of Tissue Engineering Scaffolds using Two-Point Correlation Function. In *Proceedings of the 6th IASTED International Conference on Biomedical Engineering*, Austria

Haghpanahi, M., Nikkhoo, M., & Peirovi, H. (2008). Mechanobiological models for Intervertebral Disc Tissue Engineering. In *Proceedings of the Biomedical Electronics and Biomedical Informatics*, Greece.

Haghpanahi, M., Nikkhoo, M., Peirovi, H., & Ghanavi, J. (2007). Mathematical Modeling of the Intervertebral Disc as an Infrastructure for Studying the Mechanobiology of the Tissue Engineering Procedure. *Transactions on Applied and Theoretical Mechanics, 2*(12), 263–275.

Haghpanahi, M., Nikkhoo, M., Peirovi, H., & Ghanavi, J. (2007). A Poroviscoelastic Finite Element Formulation Including Transport and Swelling for Tissue Engineered Intervertebral Disc. In *Proceedings of the European Society of Biomechanics Workshop 2007, Ireland*, (pp. 46-47).

Hollister, S. J. (2005). Porous scaffold design for tissue engineering. *Nature Materials, 4*, 518–524. doi:10.1038/nmat1421

Hollister, S. J., Levy, R. A., Chu, T. M., Halloran, J. W., & Feinberg, S. E. (2002). An image based approach for designing and manufacturing craniofacial scaffolds. *International Journal of Oral and Maxillofacial Surgery, 29*, 67–71. doi:10.1034/j.1399-0020.2000.290115.x

Hutmacher, D. W. (2000). Scaffolds in tissue engineering bone and cartilage. *Biomaterials, 21*, 2529–2543. doi:10.1016/S0142-9612(00)00121-6

Hutmacher, D. W., Hurzeler, M., & Schliephake, H. (1996). A review of material properties of biodegradable and bioresorbable polymers and devices for GTR and GBR applications. *The International Journal of Oral & Maxillofacial Implants, 11*, 667–678.

Hutmacher, D. W., Sittinger, M., & Risbud, M. V. (2004). Scaffold-based tissue engineering: rationale for computer-aided design and solid free-form fabrication systems. *Trends in Biotechnology, 22*, 354–362. doi:10.1016/j.tibtech.2004.05.005

Ishaug-Riley, S. L., Crane, G. M., Gurlek, A., Miller, M. J., Yasko, A. W., Yaszemski, M. J., & Mikos, A. G. (1997). Ectopic bone formation by marrow stromal osteoblast transplantation using poly(DL-lactic-co-glycolic acid) foams implanted into the rat mesentery. *Journal of Biomedical Materials Research, 36*, 1–8. doi:10.1002/(SICI)1097-4636(199707)36:1<1::AID-JBM1>3.0.CO;2-P

Kim, S. S., Utsunomiya, H., Koski, J. A., Wu, B. M., Cima, M. J., & Sohn, J. (1998). Survival and function of hepatocytes on a novel three-dimensional synthetic biodegradable polymeric scaffold with an intrinsic network of channels. *Annals of Surgery, 228*, 8–13. doi:10.1097/00000658-199807000-00002

Lam, C. X. F., Mo, X. M., Teoh, S. H., & Hutmache, D. W. (2002). Scaffold development using 3D printing with a starch based polymer. *Materials Science and Engineering, 20*, 49–56. doi:10.1016/S0928-4931(02)00012-7

Landers, R., Pfister, A., Hübner, U., John, H., Schmelzeisen, R., & Mülhaupt, R. (2002). Fabrication of soft tissue engineering scaffolds by means of rapid prototyping techniques. *Journal of Materials Science, 37*, 3107–3116. doi:10.1023/A:1016189724389

Langer, R., & Vacanti, J. P. (1993). Tissue engineering. *Science, 260*, 920–926. doi:10.1126/science.8493529

Lin, C. Y., Kikuchi, N., & Hollister, S. J. (2004). A novel method for biomaterial scaffold internal architecture design to match bone elastic properties with desired porosity. *Journal of Biomechanics, 37*, 623–636. doi:10.1016/j.jbiomech.2003.09.029

Martin, I., Padera, R. F., Vunjak-Novakovic, G., & Freed, L. E. (1998). In vitro differentiation of chick embryo bone marrow stromal cells into cartilaginous and bone like tissues. *Journal of Orthopaedic Research, 16*, 181–189. doi:10.1002/jor.1100160205

Martin, I., Padera, R. F., Vunjak-Novakovic, G., & Freed, L. E. (1998). In vitro differentiation of chick embryo bone marrow stromal cells into cartilaginous and bone-like tissues. *Journal of Orthopaedic Research, 16*, 181–189. doi:10.1002/jor.1100160205

Mikos, A. G., Sarakinos, G., Lyman, M. D., Ingber, D. E., Vacanti, J. P., & Langer, R. (1993). Prevascularization of porous biodegradable polymers. *Biotechnology and Bioengineering, 42*, 716–723. doi:10.1002/bit.260420606

Murphy, W. L., Dennis, R. G., Kileny, J. L., & Mooney, D. J. (2002). Salt Fusion: An approach to improve pore interconnectivity within tissue engineering scaffolds. *Tissue Engineering, 8*, 43–52. doi:10.1089/107632702753503045

Nikkhoo, M., Haghpanahi, M., Peirovi, H., & Ghanavi, J. (2007). Mathematical model for tissue engineered intervertebral disc as a saturated porous media. *Proceedings of the 3rd WSEAS International Conference on Applied and Theoretical Mechanics, Spain*, (pp. 197-201).

Niknejad, H., Peirovi, H., Jorjani, M., Ahmadiani, A., Ghanavi, J., & Seifalian, A. M. (2008). Properties of the amniotic membrane for potential use in tissue engineering. *European Cells & Materials, 15*, 88–99.

O'Brien, F. J., Harley, B. A., Yannas, I. V., & Gibson, L. (2004). Influence of freezing rate on pore structure in freeze-dried collagen-GAG scaffolds. *Biomaterials, 25*, 1077–1086. doi:10.1016/S0142-9612(03)00630-6

Sachlos, E., & Czernuszka, J. T. (2003). Making tissue engineering scaffolds work Review on the application of solid freeform fabrication technology to the production of tissue engineering scaffolds. *European Cells & Materials, 5*, 29–40.

Stoltz, J. F., Wang, X., Muller, S., & Labrador, V. (1999). Introduction to mechanobiology of cells. *Applied Mechanics and Engineering, 4*, 177–183.

Sun, W., Darling, A., Starly, B., & Nam, J. (2004). Computer-aided tissue engineering: overview, scope and challenges. *Biotechnology and Applied Biochemistry, 39*, 29–47. doi:10.1042/BA20030108

Sun, W., Starly, B., Darling, A., & Gomez, C. (2004). Computer aided tissue engineering: application to biomimetic modelling and design of tissue scaffolds. *Biotechnology and Applied Biochemistry, 39*, 49–58. doi:10.1042/BA20030109

Taboas, J. M., Maddox, R. D., Krebsbach, P. H., & Hollister, S. J. (2003). Indirect solid free form fabrication of local and global porous, biomimetic and composite 3D polymerceramic scaffolds. *Biomaterials, 24*, 181–194. doi:10.1016/S0142-9612(02)00276-4

Tan, K. H., Chua, C. K., Leong, K. F., Cheah, C. M., Cheang, P., & Abu Bakar, M. S. (2003). Scaffold development using selective laser sintering of polyetheretherketone-hydroxyapatite biocomposite blends. *Biomaterials, 24*, 3115–3123. doi:10.1016/S0142-9612(03)00131-5

Tancred, D. C., Carr, A. J., & McCormack, B. A. O. (1998). Development of a new synthetic bone graft. *Journal of Materials Science. Materials in Medicine, 9*, 819–823. doi:10.1023/A:1008992011133

Tancred, D. C., McCormack, B. A. O., & Carr, A. J. (1998). A synthetic bone implant macroscopically identical to cancellous bone. *Biomaterials, 19*, 2303–2311. doi:10.1016/S0142-9612(98)00141-0

Tsang, V. L., & Bhatia, S. N. (2004). Three dimensional tissue fabrication. *Advanced Drug Delivery Reviews, 56*, 1635–1647. doi:10.1016/j.addr.2004.05.001

Vail, N. K., Swain, L. D., Fox, W. C., Aufdlemorte, T. B., Lee, G., & Barlow, J. W. (1999). Materials for biomedical applications. *Materials & Design, 20*, 123–132. doi:10.1016/S0261-3069(99)00018-7

Vats, A., Tolley, N. S., Polak, J. M., & Gough, J. E. (2003). Scaffolds and biomaterials for tissue engineering: a review of clinical applications. *Clinical Otolaryngology, 28,* 165–172. doi:10.1046/j.1365-2273.2003.00686.x

Whitaker, M. J., Quirk, R. A., Howdle, S. M., & Shakesheff, K. M. (2001). Growth factor release from tissue engineering scaffolds. *The Journal of Pharmacy and Pharmacology, 53,* 1427–1437. doi:10.1211/0022357011777963

Wintermantel, E., Mayer, J., Blum, J., Eckert, K. L., Luscher, P., & Mathey, M. (1996). Tissue engineering scaffolds using superstructures. *Biomaterials, 17,* 83–91. doi:10.1016/0142-9612(96)85753-X

Wong, J. Y., Velasco, A., Rajagopalan, P., & Pham, Q. (2003). Directed movement of vascular smooth muscle cells on gradient-compliant hydrogels. *Langmuir Journal, 19,* 1908–1913. doi:10.1021/la026403p

Xiong, Z., Yan, Y., Zhang, R., & Sun, L. (2001). Fabrication of porous poly(L-lactic acid) scaffolds for bone tissue engineering via precise extrusion. *Scripta Materialia, 45,* 773–779. doi:10.1016/S1359-6462(01)01094-6

Yamane, S., Iwasaki, N., Kasahara, Y., Harada, K., Majima, T., & Monde, K. (2007). Effect of pore size on in vitro cartilage formation using chitosan-based hyaluronic acid hybrid polymer fibers. *Journal of Biomedical Materials Research, 81,* 586–593.

Yang, S., Leong, K. F., Du, Z., & Chua, C. K. (2001). The design of scaffolds for use in tissue engineering: Part I. Traditional factors. *Tissue Engineering, 7,* 679–689. doi:10.1089/107632701753337645

KEY TERMS AND DEFINITIONS

CAD: Abbreviation of Computer Aided Design. General term referring to applications and the method to design things using computer. CAD is mainly used for detailed engineering of 3D models and/or 2D drawings of physical components, but it is also used throughout the engineering process from conceptual design, through strength and dynamic analysis of assemblies to definition of manufacturing methods of components.

DICOM: Abbreviation of Digital Imaging and Communications in Medicine. The DICOM image format is commonly used for transfer and storage of medical images.

Mechanobiology: Mechanobiology merges the older science of mechanics with the newer and emerging disciplines of molecular biology and genetics. Mechanobiology is the study of the influence of mechanics on biological processes.

RP: Abbreviation of Rapid Prototyping. It can be defined as a group of techniques used to quickly fabricate a scale model of a part or assembly using three-dimensional computer aided design (CAD) data. The first techniques for rapid prototyping became available in the late 1980s and were used to produce models and prototype parts. Today, they are used for a much wider range of applications and are even used to manufacture production quality parts in relatively small numbers.

Scaffold: A support either natural or artificial, which maintains tissue contour and guide the organization, growth and differentiation of cells.

SEM: Abbreviation of Scanning electron microscope. A microscope in which a finely focused beam of electrons is scanned across a specimen, and the electron intensity variations are used to construct an image of the specimen. This type of microscope can effectively achieve magnifications from 200 to 35,000 times.

Tissue Engineering: A multidisciplinary field involving biology, medicine, and engineering that is likely to revolutionize the ways we improve the health and quality of life for millions of people worldwide by restoring, maintaining, or enhancing tissue and organ function. The foundation of tissue engineering for either therapeutic or diagnostic applications is the ability to exploit living cells in a variety of ways.

Chapter 5
Electronic Health Records in a Tele-Ophthalmologic Application with Oracle 10g

Isabel de la Torre Díez
University of Valladolid, Spain

Roberto Hornero Sánchez
University of Valladolid, Spain

Miguel López Coronado
University of Valladolid, Spain

María Isabel López Gálvez
University of Valladolid, Spain

ABSTRACT

Electronic health record (EHR) refers to the complete set of information that resides in electronic form and is related to the past, present and future health status. EHR standardization is a key characteristic to exchange healthcare information. Health Level Seven (HL7) and Digital Imaging and Communications in Medicine (DICOM) are intensively influencing this process. This chapter describes the development and experience of a web-based application, TeleOftalWeb 3.2, to store and exchange EHRs in ophthalmology. We apply HL7 Clinical Document Architecture (CDA) and DICOM standards. The application has been built on Java Servlet and Java Server Pages (JSP) technologies. EHRs are stored in the database Oracle 10g. Its architecture is triple-layered. Physicians can view, modify and store all type of medical images. For security, all data transmissions were carried over encrypted Internet connections such as Secure Sockets Layer (SSL) and HyperText Transfer Protocol over SSL (HTTPS). The application verifies the standards related to privacy and confidentiality. TeleOftalWeb 3.2 has been tested by

from the University Institute of Applied Ophthalmobiology (IOBA), Spain. Nowadays, more than thousand health records have been introduced.

DOI: 10.4018/978-1-60566-768-3.ch005

INTRODUCTION

Telemedicine is a general concept including diagnoses, examinations, medical meetings, collaborative operations and nurseries (Xiang et. al, 2003). Telemedical information systems are necessary for the implementation of telemedicine applications (Horsch & Balbach, 1999). Amongst these systems are the Electronic Patient Records (EPRs) and the Electronic Health Records (EHRs) (Holle & Zahlmann, 1999). EPR can be defined as a set of relevant patient information stored in digital format that allows adequate medical assistance delivered to the patient even in distinct places and scenarios (Furuie et al., 2007). EHR is a secure, real-time, point-of-care and patient-centric information resource for physicians (HIMSS, 2003). EHRs include information such as observations, laboratory tests, diagnostic imaging reports, treatments, therapies, drugs administered, patient identifying information, legal permissions and allergies. Currently, this information is stored in all kinds of proprietary formats through a multitude of medical information systems available on the market (Eichelberg et. al, 2005). EHR must enable the communication of healthcare information to support shared patient care, improved quality of care and effective resource utilisation (IEEE, 1993). The primary purpose of EHR is the support of continuing, efficient and quality integrated health care. Amongst EHRs benefits are their universal access, coding efficiency and efficacy, easier and quicker navigation through the patient record (Smith & Newell, 2002). The potential advantages of an EHR over a traditional paper-based patient record involve: distributed and simultaneous access, high availability, fast information retrieval, better quality, higher confidence and possibility of reanalysis (Furuie et al., 2007). There are several barriers to their adoption such as training, costs, complexity and lack of a national standard for interoperability (Gans et. al, 2006).

Telemedicine applications and services often involve many institutions using different systems and technologies. This complicates the technical standardization. International and European institutions are concerned with EHR standardization such as the International Standards Organization Health Informatics Standards Technical Committee (ISO/TC) 215, European Committee for Standardization Technical Committee (CEN/TC) 251, openEHR, Health Level Seven (HL7), Extensible Markup Language (XML), Digital Imaging and Communication in Medicine (DICOM), American National Standards Institute (ANSI) and others (Bott, 2004). The development of HL7 and DICOM standards has also been of great benefit in the telemedicine services and applications.

HL7 standard is used for many different medical environments. It is a not-for-profit organization involved in development of international healthcare standards. It is used for many different medical environments. For example, there are mobile clinical information systems by using HL7 to integrate the patient data (Choi, 2006). HL7 Document is intended to be the basic unit of a document-oriented EPR. The patient medical record is represented as a collection of documents. HL7 Clinical Document Architecture (CDA) is a XML-based document markup standard that specifies the structure and semantics of EHR for the purpose of exchange. Clinical Document Architecture – Release One (CDA–R1), became an American National Standards Institute (ANSI)–approved HL7 Standard in November 2000, representing the first specification derived from the HL7 Reference Information Model (RIM). CDA – Release Two (CDA–R2), became an ANSI-approved HL7 Standard in May 2005 (Dolin et al., 2006). HL7/CDA is a XML-based document markup standard that specifies the structure and semantics of EHR for the purpose of exchange. CDA standard provides an exchange model for clinical documents. Many CDA documents comprise an individual EHR. A CDA document is a contextually complete information object that can include text, images, sounds and other multimedia content. CDA Level one sets a requirement for the structure of the header. It sets practically no requirements for the body. To transfer these structured level, one document XML can be employed. To display these in different formats, you can use DTDs (Data Type Definitions) or style sheets (XSL).

DICOM is a cooperative standard. It was developed from 1990 to 1996, mainly by the American College of Radiology (ACR) and National Electrical Manufacturers Association (NEMA) committee in the United States, with contributions from European standardization organizations, the Japanese Industry Radiology Apparatus (JIRA), the Institute of Electrical and Electronics Engineers (IEEE), HL7 and ANSI as well as from European manufacturers and societies. This standard allows the exchange of medical images and related information between systems from different manufacturers (NEMA, 2000). DICOM format has been recognized as the de facto standard for storage, transferring and sharing of images along different modalities like Magnetic Resonance Imaging, Nuclear Medicine, Computer Tomography (CT), Digital Angiography and Digital Radiology. In many medical environments there is a large need to have images available in formats (i.e. GIF, TIFF, BMP or JPEG), which are compatible with widely used office automation applications (Marcheschi, 2003).

EHR has a great potential to improve safety, quality and efficiency in medicine. However, adoption has been slow. Most previous studies addressing this issue have been done in primary care (Lo et. al, 2007). In our application, the EHRs are shared between ophthalmologists, endocrinologists and primary care physicians. We reviewed articles about EHR systems in different specialities such as pediatric (Ginsburg, 2007), emergency departments (Amouh et. al, 2005), ophthalmology (Chew et. al, 1998) and oncology (James, 2001). In these systems, EHR standardization in HL7 and DICOM was not presented. Moreover, we compare TeleOftalWeb 3.2 with other applications like Julius (Chen, 2007), OpenSDE (Los,

2004), CareWeb™ (Halamka, 1999), PHIMS (Kim, 2006), PedOne System (Ginsburg, 2007) and CipherMe (Hansen, 2006).

Ophthalmology is an ideal specialty for testing EHRs due to the use of images and objective measures during diagnosis of eye diseases. It is an ideal speciality for telemedicine (Constable et. al, 2000; Lamminen et. al, 2003; Murdoch, 1999; Yogesan et. al, 1998). In ophthalmology, text data entry is minimal as images and objective tests are more frequently used (Chew et. al, 1998). Moreover, within tele-ophthalmology the patient care concern has focused mostly on the use of telemedicine for retinal evaluations and more specifically for diabetic retinopathy. EHRs systems can assist ophthalmologists in improving the quality of care being provided as well as assist the ophthalmologists in building solid relationships with their patients. Physicians who have shared EHRs available yet fail to consult them before beginning treatment could face increasing liability in the future.

In this chapter, we present a web-based application to store and exchange Electronic Health Records (EHRs) in ophthalmology (TeleOftalWeb 3.2). We apply HL7/CDA and DICOM standards. The application has been built on Java Servlet and Java Server Pages (JSP) technologies. EHRs and fundus photographs are stored in the database Oracle 10g. Its architecture is triple-layered. The application server is Tomcat 5.5.9. The application is platform-independent thanks to using Extensible Markup Language (XML) and Java technologies. Physicians can view, store and modify all type of medical images. For security, all data transmissions were carried over encrypted Internet connections such as Secure Sockets Layer (SSL) and HyperText Transfer Protocol over SSL (HTTPS). The application verifies the standards related to privacy and confidentiality. It has been tested by ophthalmologists from the University Institute of Applied Ophthalmobiology (IOBA), Spain. Currently, more than thousand EHRs have been introduced.

Figure 1. Application architecture

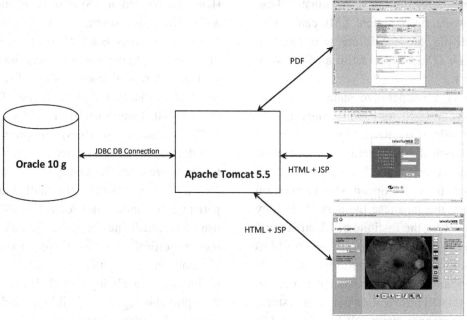

METHODS

Figure 1 shows the application architecture. It is a triple-layered with a database server (Oracle 10g) and one application server (Apache Tomcat 5.5.9). Oracle 10g provides high-performance, native XML storage and retrieval technology. An Oracle Java Database Connectivity (JDBC) driver was used to connect to the database instance. We chose a free open-source application server to process the requests. The web-based system was built on Java Servlet and JSP technologies, which enables rapid development of web-based applications. Tomcat was also chosen as the JSP engine. In Oracle database, we stored all the user data, access information to the web application and records with fundus photographs.

The development environment was NetBeans IDE 4.1 of Sun Microsystems. The application is platform-independent thanks to using XML and Java Technologies. Java was the basis application programming language. XML is an open standard that provides a unified model for data, content and metadata. It is being used to manage mission critical information.

We included all tools and Application Programming Interface (API) as Javascript, JSP, Java Servlets and Java Database Connectivity (JDBC). The evolution of Java Technology brings more features to the Java development tools. This facilitates the creation of telemedicine applications and reduces the time of developing programs (Fedyukin, et. al, 2002). We used XML technology to store and exchange EHRs. Some XML advantages are: easily readable, self-describing and interoperable. Moreover, there are several types of object-based parser components available for this language. Wherever Java programs can run, they can also access XML information. This enables Java and XML information to interoperate efficiently and effectively on different platforms. Combining Java and XML leads to the attractive dual portability of code and data (Fan, et. al, 2005).

Moreover, in Oracle database we stored all the user data and access information to the web application. It has two tables: "users" and "permis-

Figure 2. Data modeling

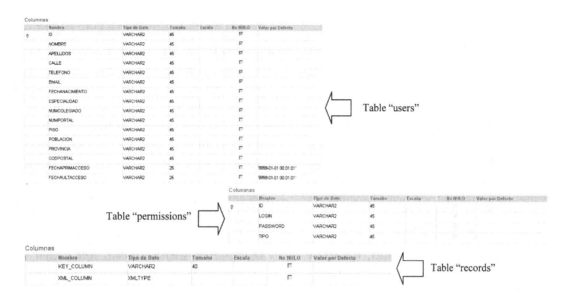

sions". The table "users" contains personal user data. The user identification, user name, password and user type appear in table "permissions". EHRs are stored in Oracle 10g database according to the ANSI/HL7 CDA R2.0-2005 template. The data modelling in the Oracle database is shown in Figure 2.

Oracle 10g introduced a new datatype, XM-LType, to facilitate native handling of XML data in the database. It supports comprehensive models (i.e., structured, unstructured and binary XML storage models) to server diverse XML use cases with different requirements. For table "records" (see Figure 2), we use this datatype. XMLType is stored in Large Objects (LOBs). LOB storage maintains content accuracy to the original XML (whitespaces and all). When we create an XMLType column without any XML schema specification, a hidden CLOB column is automatically created to store the XML data. The XMLType column itself becomes a virtual column over this hidden CLOB column. It is not possible to directly access the CLOB column. However, we can set the storage characteristics for the column using the XMLType storage clause. Oracle XML

database provides efficient support of SQL/XML XMLTable function and its COLUMNS clause for mapping XML data into relational views. By taking advantage of XQuery rewrite technology, storage models, indexing schemes, downstream processing on a relational view created with XMLTable function can approach pure-relational-performance (Oracle, 2007).

XPath language was used. It is a W3C Recommendation for navigating XML documents. XPath models an XML document as a tree of nodes. It provides a rich set of operations that walk this tree and apply predicates and node-test functions. Applying an XPath expression to an XML document can result in a set of nodes (Oracle, 2007).

CDA scope is the standardization of clinical documents for exchange. CDA R2 model is richly expressive, enabling the formal representation of clinical statements (such as observations, medication administrations and adverse events) such that they can be interpreted and acted upon by a computer (Dolin et al., 2006). As it was indicated in the introduction, CDA documents are encoded in XML. The CDA is only the first example of HL7's commitment to the advancement of XML-based e-healthcare technologies within the clinical, patient care domain. The CDA specification prescribes XML markup for CDA Documents: CDA instances must be validated against the CDA Schema and may be subject to additional validation. The document can be sent inside or outside a HL7 message. CDA also supports the semantically interoperable exchange of complex medical information between healthcare applications by virtue of its adherence to the HL7 V3 development methodology (Klein, 2005). CDA does not specify the creation or management of documents as such, only their exchange mark-up.

Technically, CDA Level One is specified by three components: the CDA Header, the CDA Level One Body and Reference Information Model (RIM) data type DTD. The CDA Header identifies and classifies the document and provides information on authentication, the encounter, the patient and the provider. The CDA Level One Body is specified in the CDA Level One DTD and is derived from document analysis. It contains the clinical-related information that we want to exchange. The CDA Level One Body is comprised of nested containers. There are four types of containers: sections, paragraphs, lists and tables. 'Containers' have contents and optional captions. Contents include plain text, links and multimedia.

The CDA Header identifies and classifies the document and provides information on the document authenticator, the patient, the encounter, provider and other service actors. Document-related information includes the *id, set id*, *version*, *type* and various *timestamps*. The *id* element uniquely identifies the specific clinical document,

while the *set id* identifies the TCF. The *type* and *version* elements identify the clinical document template. Encounter data include the *id, code, timestamps, service location* and *local header.* The *id* and *code* elements uniquely identify the relevant encounter and its type in the regional network, while attribute-value pairs in the *local header* facilitate interoperability with the local EHR system. The body of the clinical document consists of *section* elements. Sections correspond to reusable XML fragments. Each CDA *section* may contain CDA structures such as *paragraph, list* and *table* elements, nested CDA sections, or *coded_entry* elements. CDA structures contain CDA "entries" such as *content, link, coded_entry* and *local_markup. Sections* including only a *link* are used to refer to external multimedia objects such as an ECG (Chronaki et. al, 2001). RIM data type DTD is an XML implementation of the abstract data type specification. It used by both the CDA and the HL7 Version 3 message specifications. The CDA Level One Body is specified in the CDA Level One DTD and is derived from document analysis (HL7, 2000).

CDA distinguishes three different levels of granularity as shown in Table 1, where each level iteratively adds more markup to clinical documents, although the clinical content remains constant at all levels (Eichelberg, M, et. al, 2006). The CDA specification prescribes XML markup for CDA Documents: CDA instances must valid against the CDA Schema and may be subject to additional validation. The CDA Schema is shown in Figure 3.

The XML-based architecture described in the CDA v2.0 standard has been used to define

Table 1. Levels of document granularity in CDA Release One and Release Two

CDA Release One	CDA Release Two
CDA Level One	Unconstrained CD specification
CDA Level Two	CDA specification with section-level templates applied
CDA Level Three	CDA specification with entry-level templates applied

the health information format. Thanks to the use of XML-based technologies and HL7 specifications, our application fulfils the EHR standards. Its development methodology is a continuously evolving process that seeks to carry out specifications that facilitate interoperability between healthcare systems.

Results

Fistly, we present the two application modules: manager and user. Then, we describe the experience of introducing diabetic patient's health records from a screening program of diabetic retinopathy in a rural area of Spain (Hornero, et. al, 2003).

Manager Module

The manager can access the web platform with any browser. Login and password have to be introduced by users. In Figure 4, we can see several application users. The two user roles are: manager and user. The application manager allows to:

1. Create new users.
2. Show user information.
3. Erase users.

4. Modify physician's information.
5. Show user statistics.
6. Show the user patient records.
7. Search users by different criteria such as surname, identification number, type of user and member number.

User Module

The authorized users can access to this module. They have their login and password. The access is similar to the manager module. The users can do the following actions:

1. Create new records (see Figure 5). They have to introduce the necessary data: patient affiliation information, patient precedents, medical exploration and diagnostic.
2. Erase records and revisions.
3. Create new revisions in a record.
4. Search different EHRs and revisions,
5. Add new images in different records. Physicians can add new images in an EHR. These may be in different formats such as DICOM, JPEG, GIF, BMP, TIFF amongst others.
6. Edit and erase images. The images editor (see Figure 6) shows images and allows us

Figure 3. CDA schema

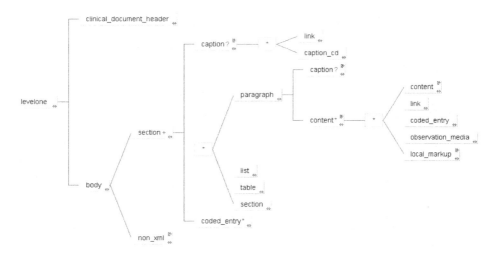

Figure 4. TeleOftalWeb 3.2 users

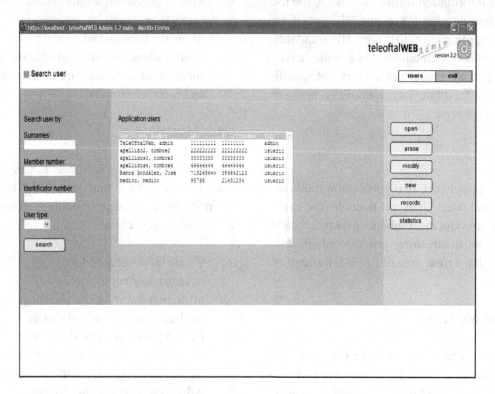

Figure 5. Exploration in a new record

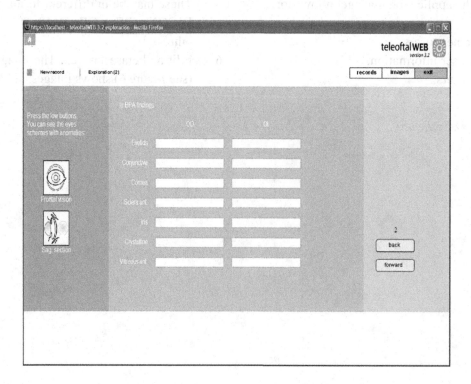

Figure 6. Images editor

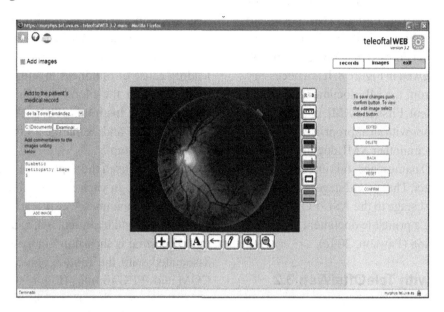

Figure 7. Searching images in an EHR

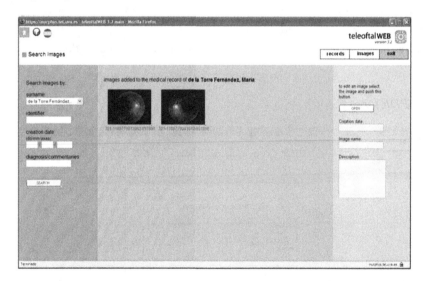

to change their shape and colour, zoom in or zoom out. It supports image editing of brightness and contrast. Other editor functions are: RGB (Red, Green, Blue) scale, add and delete text and arrows. It supports all type of images (DICOM, JPEG, GIF, etc.).

7. Search images according to different criteria: image identification number, surnames, image creation and comments. Figure 7 shows this action.

Furthermore, we use Extensible Stylesheet Language Formatting Objects (XSL-FO) to format XML data. XSL-FO is a complete XML vocabulary for laying out text on a page. An XSL-FO document is a well-formed XML document that uses this vocabulary. EHR output format is a Portable Document Format (PDF). An EHR in PDF format can be viewed in Figure 8. It is a necessary process to get from a XML document to a PDF printable document. First, the XML must be fed to an XSLT processor with an appropriate stylesheet in order to produce another XML document which uses the XSL-FO namespace. It is intended for an XSL-FO formatter. The second stage is to feed the output of the first stage to the XSL-FO formatter that can produce a printable document styled for visual presentation (Pawson, 2002).

Experience with TeleOftalWeb 3.2

TeleOftalWeb 3.2 has been tested by physicians from the University Institute of Applied Ophthalmobiology (IOBA), Spain. Nowadays, more than thousand health records were introduced during two months. All the patients were diabetic and

they participated in a telemedicine program for diabetic retinopathy screening in a rural area of Spain (Hornero, et. al, 2003). Diabetic retinopathy is the most common diabetic eye disease and a leading cause of blindness in adults. It is caused by changes in the blood vessels of the retina (National Eye Institute, 2007). Our application allows to store and exchange all the records and fundus photographs.

Physicians used the application with different web browsers. In each record there are the following parts: anamnesis, exploration (see Figure 5), diagnosis and treatment. An example of EHR in PDF format is shown in Figure 8. The EHR is associated with the fundus photographs in DICOM and JPEG formats (Figure 9). The process to introduce a patient record in the system was around 10 min by each record.

A survey with ten questions about the application usability was done (see Figure 10). We used SUS (System Usability Scale) to make the questionnaires. It is a Likert scale. SUS has proved

Figure 8. EHR in PDF format

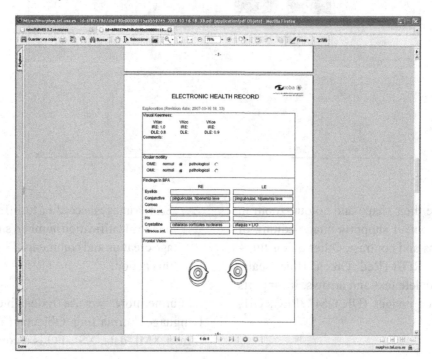

Figure 9. EHR in PDF format with DICOM fundus photographs

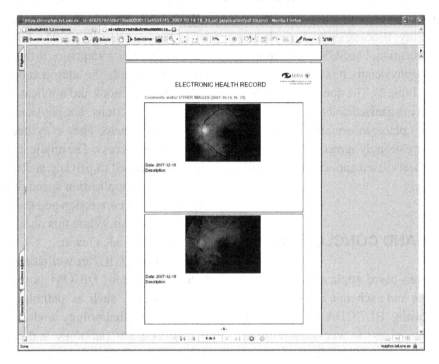

Figure 10. Satisfaction survey

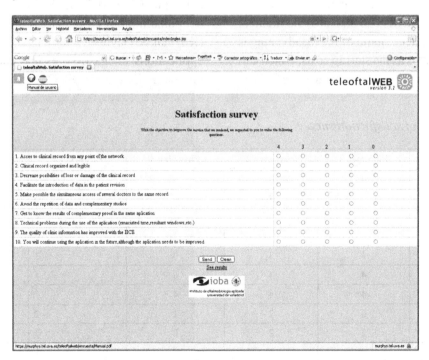

to be a valuable evaluation tool, being robust and, reliable. It correlates well with other subjectives measures of usability (Brooke, 1996). Five physicians used the application. The SUS score is major than fifty for all the physicians. Its average value is seventy four (see Table 2). The questions such as: clinical records are organised and legible, access to EHRs from any place, information quality in the application were strongly agreed. According to these results, our web-based application is useful for the physicians.

DISCUSSION AND CONCLUSIONS

In this study, a web-based application has been developed to store and exchange EHRs in Ophthalmology by using HL7/CDA and DICOM standards. The physicians can access and retrieve patient medical information and fundus photographs through any Web browser. EHRs have several distinct advantages and disadvantages over paper health records. One advantage is the fact that there are increased storage capabilities for longer periods of time. EHRs can also provide medical alerts and reminders. Some of the disadvantages include such items as the startup costs, which can be excessive. Another disadvantage to an EHR is

that there is a substantial learning curve and it is helpful when the users have some type of technical knowledge (Gurley, 2004). TeleOftalWeb 3.2 advantages are: its adaptation to the HL7/CDA and DICOM standards, the interoperability facilitation between institutions and applications and its security in transactions. The physicians can analyze EHRs everywhere. They only need a computer with Internet access. The application verifies the standards related to privacy and confidentiality. However, the application speed depends mainly on the Internet connection and the number of users in the system. When this number is high, the application speed is lower.

DICOM and HL7 are well-accepted healthcare industry standard. DICOM is used in diferent medical fields, such as pathology, endoscopy, dentistry, ophthalmology and dermatology. It is a success for radiology and cardiology and it is now beginning to be used for other clinical specialties. The US Department of Veterans Affairs has been instrumental in promoting this technological advancement. This work involved in extending DICOM to the clinical specialties such as ophthalmology (Kuzmak & Dayhoff, 2003). In (Kuzmak & Dayhoff, 2000), new colour imaging applications for gastrointestinal endoscopy and ophthalmology using DICOM are under develop-

Table 2. Results of satisfaction survey

Questions	Physician 1	Physician 2	Physician 3	Physician 4	Physician 5
1	4	4	4	4	4
2	3	4	4	3	4
3	3	4	4	4	3
4	3	3	3	4	3
5	4	3	3	3	4
6	4	4	4	4	4
7	4	4	4	4	4
8	2	2	1	2	2
9	2	2	2	2	2
10	4	4	3	4	4
SUS score	67,5	75	67,5	80	80

ment. In our application, we apply the DICOM standard in ophthalmology. The extensive use of HL7/CDA standard is desirable (Marcheschi, et. al, 2004). In this paper, these standards are applied in an ophthalmologic application.

According to our review, we analyzed several studies about EHR systems. In Paterson et al. (2002), these systems have been presented using XML-based Clinical Document Architecture to exchange discharge summaries. There are EHR applications in different specialities such as pediatric (Ginsburg, 2007), ophthalmology (Chew et. al, 1998), emergency departments (Amouh et. al, 2005) and oncology (James, 2001). In the telematic system for oncology, they use a data warehouse as EPRs server. The authors do not present a standardization process for the EHRs. In our application, we apply EHR standards. The information system designed for emergency department has been implemented by prototyping a web-based application. It is a multi-platform and multi-user system, using the Java language. It makes uses the XML-based *openEHR* standard. EHR systems are increasingly being adopted in pediatric practices (Ginsburg, 2007). The systems uses open standards to ensure interoperability and conform to stated the NHIH (National Health Information Network) goals. Some authors designed and developed a template based system (called Julius) that was integrated with existing EHR systems (Chen, 2007). The system has been implemented, tested and deployed to three health care units in Stockholm, Sweden. In the application OpenSDE, authors have expanded the traditional row modeling methodology with additional columns that allow structured representation of medical narrative (Los, 2004).

There are other web-based applications such as CareWeb™ that implements web-exposed HL7-based medical information servers at each participating institution in the health care delivery network (Halamka, 1999). A multicentre trial of a Web-based personal EHR service was conducted in three different European hospitals. The service was customised according to the needs of three groups of patients who had congenital heart disease (Karagiannis, et. al, 2007). Other EHRs applications are: PHIMS and CipherMe. PHIMS is a web-based repository of patient health information, which provides interfaces for storing structured and categorized patient information (Kim, 2006). CipherMe architecture enables individual entities to securely store private information about themselves and to manage access to selected items by other parties, according to needs or legal obligations. All CipherMe data are held in the form of XML objects (Hansen, 2006).

In Table 3, we compare different EHRs applications in order to aspects such as speciality, use of web technologies and whether the system

Table 3. Comparation of EHR applications

Name	Speciality	Uses web technology	Applies HL7 standard	Applies DICOM standard
CareWeb™	All	√	√	
CipherMe	All			
Julius	All	√		
OpenSDE	All	√		
OphthWeb	Ophthalmology	√		
PedOne System	Pediatry	√		
PHIMS	Elderly Population	√		
TeleOftalWeb	Ophthalmology	√	√	√

applies HL7, DICOM standards. TeleOftalWeb 3.2 is the only application that applies the both standards.

In summary, we have developed a web-based application to store and share EHRs in ophthalmology by using HL7/CDA and DICOM standards. Physicians can view electronic records and all type of fundus photographs. Our application treats to solve some of the barriers to the EHRs adoption in ophthalmology. It has several distinct advantages over paper health records. The records and the fundus photographs in all type of formats are continuously updated and are available concurrently for use everywhere. We verified that the application was useful for the physicians.

REFERENCES

Amouh, T., Gemo, M., Macq, B., Vanderdonckt, J., Wahed, A., & Reynaert, M. S. (2005). Versatile Clinical Information System Design for Emergency Departments. *IEEE Transactions on Information Technology in Biomedicine, 9*(2), 174–183. doi:10.1109/TITB.2005.847159

Bott, O. J. (2004). Electronic Health Record: Standardization and Implementation. In *2ⁿᵈ OpenECG Workshop*, Berlin, Germany, (pp. 57-60).

Brooke, J. (1996). *SUS: a "quick and dirty" usability scale.*

Chen, R., Enberg, G., & Klein, G. (2007). Julius--a template based supplementary electronic health record system. *BMC Medical Informatics and Decision Making*, 7–10.

Chew, S. J., Cheng, H. M., Lam, D. S. C., Cheng, A. C. K., Leung, A. T. S., & Chua, J. K. H. (1998). OphthWeb-cost-effective telemedicine for ophthalmology. *HKMJ, 4*(3), 300–304.

Choi, J., Yoo, S., Park, H., & Chun, J. (2006). MobileMed: A PDA-Based Mobile Clinical Information System. *IEEE Transactions on Information Technology in Biomedicine, 10*(3), 627–635. doi:10.1109/TITB.2006.874201

Chronaki, E., Lelis, P., Demou, C., Tsiknakis, M., & Orphanoudakis, S. C. (2001). An HL7/CDA framework for the design and deployment of Telemedicine services. *Proceedings of the 23ʳᵈ Annual EMBS International Conference*, October 25-28, Istanbul, Turkey, (pp. 3504-3507).

Constable, I., Yogesan, K., Eikelboom, R., Barry, C., & Cuypers, M. (2000). Fred Hollows lecture: digital screening for eye disease. *Clin. Exp. Ophthalmol., 28*, 129–132. doi:10.1046/j.1442-9071.2000.00309.x

Dolin, R. H., Alschuler, L., Boyer, S., Behlen, F. M., Biron, P. V., & Shabo, A. (2006). HL7 Clinical Document Architecture, Release 2. *Journal of the American Medical Informatics Association, 13*, 30–39. doi:10.1197/jamia.M1888

Eichelberg, M., Aden, T., & Riesmeier, J. (2005). A survey and Analysis of Electronic Healthcare Record Standards. *ACM Computing Surveys, 5*(14), 1–47.

Fan, R., Ceded, L., & Toser, O. (2005). Java plus XML: a powerful new combination for SCADA systems. *Computing & Control Engineering Journal, 16*(5), 27–30. doi:10.1049/cce:20050505

Fedyukin, I., Reviakin, Y. G., Orlov, O. I., Doarn, C. R., Harnett, D. M., & Merrell, R. C. (2002). Experience in the application of Java Technologies in telemedicine. *Ehealth International, 1*(3), 1–6.

Furuie, S. S., Rebelo, M. S., Moreno, R. A., Santos, M., Bertozzo, N., & Mota, G. (2007). Managing Medical Images and Clinical Information: InCor's Experience. *IEEE Transactions on Information Technology in Biomedicine, 11*(1), 17–24. doi:10.1109/TITB.2006.879588

Gans, D., Kralewski, J., Hammons, T., & Dowd, B. (2006). Medical groups' adoption of electronic health records and information systems. *Health Affairs (Project Hope)*, *24*(5), 1323–1333. doi:10.1377/hlthaff.24.5.1323

Ginsburg, M. (2007). Pediatric Electronic Health Record Interface Design: The PedOne System. In *Proceedings of the 40th Hawaii International Conference on System Sciences,* (pp. 1-10).

Gurley, L. (2004). *Advantages and disadvantages of the Electronic Medical Record.* Des Plaines, IL: American Academy of Medical Administrators.

Halamka, J. D., Osterland, C., & Safran, C. (1999). CareWeb™, a web-based medical record for an integrated health care delivery system. *International Journal of Medical Informatics*, *54*, 1–8. doi:10.1016/S1386-5056(98)00095-1

Hansen, I. B. (2006). CipherMe: personal Electronic Health Records in the hands of patients-owners. *Proceedings of the 1st Distributed Diagnosis and Home Healthcare (D2H2) Conference,* (pp. 148-151).

Health Level 7, HL7. (2000). *ANSI Standard HL7 V 2.4-2000.*

HIMSS Electronic Health Record Committee. (2003). *EHR Definition, Attributes and Essential Requirement.*

Holle, R., & Zahlmann, G. (1999). Evaluation of Telemedical Services. *IEEE Transactions on Information Technology in Biomedicine*, *3*(2), 84–91. doi:10.1109/4233.767083

Horsch, A., & Balbach, T. (1999). Telemedical Information Systems. *IEEE Transactions on Information Technology in Biomedicine*, *3*(3), 166–175. doi:10.1109/4233.788578

IEEE Computer Society (1993). *IEEE Recommended Practice for Software Requirements Specifications, Std. 830.*

James, A., Wilcox, Y., & Naguib, R. N. G. (2001). A Telematic System for Oncology Based on Electronic Health and Patient Records. *IEEE Transactions on Information Technology in Biomedicine*, *5*(1), 16–17. doi:10.1109/4233.908366

Karagiannis, G. E., Stamatopoulus, V. G., Rigby, M., Kotis, T., Negroni, E., Munoz, A., & Mathes, I. (2007). Web-based personal health records: the personal electronic health record (pEHR) multi-centred trial. *Journal of Telemedicine and Telecare*, *13*, 32–34. doi:10.1258/135763307781645086

Kim, E. H. (2006). Web-based Personal-Centered Electronic Health Record for Elderly Population. In *Proceedings of the 1st Distributed Diagnosis and Home Healthcare (D2H2) Conference,* (pp. 144-147).

Klein, J. (2005). Integrating Electronic Health Records Using HL7 Clinical Document Architecture.

Kuzmak, P. M., & Dayhoff, R. E. (2000). The use of digital imaging and communications in medicine (DICOM) in the integration of imaging into the electronic patient record at the Department of Veterans Affairs. *Digital Imaging*, *13*(2), 133–137.

Kuzmak, P. M., & Dayhoff, R. E. (2003). Experience with DICOM for the clinical specialities in the healthcare enterprise . *Proceedings of the Society for Photo-Instrumentation Engineers*, *5033*, 18–29. doi:10.1117/12.480668

Lamminen, H., Voipio, V., Ruohonen, K., & Uusitalo, H. (2003). Telemedicine in ophthalmology. *Acta Ophthalmologica Scandinavica*, *81*, 105–109. doi:10.1034/j.1600-0420.2003.00045.x

Lo, H. G., Newmark, L. P., Yoon, C., Volk, L. A., Carlson, V. L., & Kittler, A. F. (2007). Electronic Health Records in Specialty Care: A Time-Motion Study. *Journal of the American Medical Informatics Association*, *14*(5), 609–615. doi:10.1197/jamia.M2318

Los, R., van Ginneken, A. M., de Wilde, M., & van der Lei, J. (2004). OpenSDE: Row Modeling Applied to Generic Structured Data Entry. *Journal of the American Medical Informatics Association, 11*, 162–165. doi:10.1197/jamia.M1375

Marcheschi, P., Mazzarisi, A., Dalmiani, S., & Benassi, A. (2004). HL7 clinical document architecture to share cardiological images and structured data in next generation. *Computers in Cardiology, 617–620*. doi:10.1109/CIC.2004.1443014

Marcheschi, P., Positano, V., Ferdegnini, E. M., Mazzarisi, A., & Benassi, A. (2003). A open source based Application for integration and sharing of multi-modal cardiac image data in a heterogeneous environment. *Computers in Cardiology, 367–370*. doi:10.1109/CIC.2003.1291168

Murdoch, I. (1999). Telemedicine. *The British Journal of Ophthalmology, 83*, 1254–1256. doi:10.1136/bjo.83.11.1254

National Eye Institute [Online]. (2007). Retrieved from http://www.nei.nih.gov.

NEMA. (2000). *Digital Imaging and Communications in Medicine (DICOM): Version 3.0.* Hornero, R., López, M.I., Acebes, M. and Calonge, T. (2003). Teleophthalmology for diabetic retinopathy screening in a rural area of Spain, *Eighth Annual Meeting of the American Telemedicine Association (ATA'2003)*.

Oracle Database Online Documentation 10g Release 2 [Online]. (2007). Retrieved from http://youngcow.net/doc/oracle10g/index.htm.

Paterson, G. I., Shepherd, M., & Wang, X. Watters, C. & Zitner, D. (2002). Using the XML-based Clinical Document Architecture for Exchange of Structured Discharge Summaries. In *Proceedings of the 35th Hawaii International Conference on System Sciences,* (pp. 119-128).

Pawson, D. (2002). *XSL-FO Making XML Look Good in Print.*

Smith, D., & Newell, L. M. (2002). A Physician's Perspective: Deploying the EMR. *Journal of Healthcare Information Management, 16*(2), 71–79.

Xiang, Y., Gu, Q., & Li, Z. (2003). A Distributed Framework of Web-based Telemedicine System. In *Proceedings of the 16th IEEE Symposium on Computer-Based Medical Systems,* (pp. 108-113).

Yogesan, K., Constable, I., Eikelboom, R., & van Saarloos, P. (1998). Tele-ophthalmic screening using digital imaging devices. *Australian and New Zealand Journal of Ophthalmology, 26*(Suppl. 1), S9–S11. doi:10.1111/j.1442-9071.1998.tb01385.x

KEY TERMS AND DEFINITIONS

DTD: Document Type Definition. DTD is primarily used for the expression of a schema via a set of declarations that conform to a particular markup syntax and that describe a class, or type, of SGML or XML documents, in terms of constraints on the structure of those documents. A DTD may also declare constructs that are not always required to establish document structure, but that may affect the interpretation of some documents.

HTTPS: HyperText Transport Protocol Secure. It is the protocol for accessing a secure Web server. Using HTTPS in the URL instead of HTTP directs the message to a secure port number rather than the default Web port number of 80.

Java: It is an object-oriented applications programming language developed by Sun Microsystems in the early 1990s. Java applications are typically compiled to bytecode, although compilation to native machine code is also possible. The language itself derives much of its syntax from C and C++ but has a simpler object model and fewer low-level facilities.

JDBC: Java Database Connectivity. It is an API for the Java programming language that defines how a client may access a database. It provides methods for querying and updating data in a database.

JPEG: Joint Photographic Experts Group. JPEG is a *lossy compression* technique for color images. Although it can reduce files sizes to about 5% of their normal size, some detail is lost in the compression

JSP: Java Server Pages: It is a Java technology that allows software developers to dynamically generate HTML, XML or other types of documents in response to a web client request. The JSP syntax adds additional XML-like tags, called JSP actions, to be used to invoke the functionality. It lets you separate the dynamic part of your pages from the static HTML.

Oracle: An Oracle database system comprises at least one instance of the application, along with data storage. An instance comprises a set of operating-system processes and memory-structures that interact with the storage. It has become a major presence in database computing.

PDF: Portable Document Format. It is a file format developed by Adobe Systems. PDF captures formatting information from a variety of desktop publishing applications, making it possible to send formatted documents and have them appear on the recipient's monitor or printer as they were intended. To view a file in PDF format, you need Adobe Reader, a free application distributed by Adobe Systems.

RIM: Reference Information Model. It specifies the grammar of HL7 messages and, specifically, the basic building blocks of the language and their permitted relationships. The RIM is not a model of healthcare, although it is healthcare specific, nor is it a model of any message, although it is used in messages. At first site the RIM is quite simple.

SSL: Secure Sockets Layer, a protocol developed by Netscape for transmitting private documents via the Internet. SSL uses a cryptographic system that uses two keys to encrypt data, a public key known to everyone and a private or secret key known only to the recipient of the message.

XSL-FO: Extensible Style Language Formatting Objects is a markup language for XML document formatting which is most often used to generate PDFs. XSL-FO is a unified presentational language. It has no semantic markup in the way it is meant in HTML.

XSLT: Extensible Style Language Transformation, the language used in XSL style sheets to transform XML documents into other XML documents. An XSL processor reads the XML document and follows the instructions in the XSL style sheet, then it outputs a new XML document or XML-document fragment.

Chapter 6
High Performance Computing in Biomedicine

Dimosthenis Kyriazis
National Technical University of Athens, Greece

Andreas Menychtas
National Technical University of Athens, Greece

Konstantinos Tserpes
National Technical University of Athens, Greece

Theodoros Athanaileas
National Technical University of Athens, Greece

Theodora Varvarigou
National Technical University of Athens, Greece

ABSTRACT]

A constantly increasing number of applications from various scientific fields are finding their way towards adopting Grid technologies in order to take advantage of their capabilities: the advent of Grid environments made feasible the solution of computational intensive problems in a reliable and cost-effective way. This book chapter focuses on presenting and describing how high performance computing in general and specifically Grids can be applied in biomedicine. The latter poses a number of requirements, both computational and sharing / networking ones. In this context, we will describe in detail how Grid environments can fulfill the aforementioned requirements. Furthermore, this book chapter includes a set of cases and scenarios of biomedical applications in Grids, in order to highlight the added-value of the distributed computing in the specific domain.

INTRODUCTION

High performance distributed computing has emerged in the last years as a technology for large-scale, flexible and coordinated resource sharing. A successful example of high performance computing are Grids (Foster, 1999), which are increasingly considered as an infrastructure able to provide distributed and heterogeneous resources in order to deliver in a transparent way computational power to

DOI: 10.4018/978-1-60566-768-3.ch006

resource demanding applications (Foster, 2001), (Leinberger, 1999). The main objective of any distributed infrastructure is to serve as a means for providing resources for a set of purposes such as computational / processing, data storage / networking of file systems, communications and bandwidth, applications as services etc. Furthermore, a Grid-based environment enables the storage and distribution of data allowing access to various sources and analysis of them. Therefore, the information contained for example in medical records can be accessed and analyzed for various reasons (e.g. selection of the best treatment and prediction of its outcomes).

Exploitation of Grid technologies is imperative for medical applications due to a set of reasons such as the exponential increase of the required storage and computational resources, the heterogeneity of the required data (medical records, images, information obtained from sensors) with different preprocessing requirements and the large number of involved patients. Medical-related applications generally belong to those collaborative environments that are based on input from networked sensors and aggregation of acquired data under real-time conditions. With the simultaneous advent of technologies to support heterogeneous sources of information and computing resources (through Service Oriented Architectures - SOA (Sprott, 2004), Grids, etc), it is expected that in the years to come, there will be a great blooming in the development of infrastructures comprising multiplicity in resources both in number and nature.

To this end, a significant part of the value of Grid technology lies on the fact that Grids are in position to provide the fundamental management mechanisms for distributed data. This is one major reason that often many developed Grid-based systems were referenced as "data Grids", since the integration of data, infrastructures, digital libraries and persistent archives was a challenge forcing continued evolution of Grid technology. This challenge remains valid for medical applications, the

requirements of which range from the transition from data handling, sharing and aggregation to the provision of knowledge as utility. Therefore, besides the main property of Grids - referring to their computational power, we will also describe in detail their added-value in terms of data sharing and aggregation.

Nowadays, high performance systems are realized following the SOA principles: as Grids and Clouds (Boss, 2007) are considered to be the most successful examples; and are finding their way into the medical sector both for computational reasons and for storage and aggregation of medical data. A set of medical scenarios will also be described in this book chapter to demonstrate the added value of distributed computing in the aforementioned cases. These refer to the use of Grids for computational reasons for a clinical trial simulation and for data aggregation and analysis for the simulation of possible therapeutic schemes and personalized medicine proposals.

The remainder of this chapter is structured as follows: the first section (namely *Background*) includes information related to high performance computing as realized through Service Oriented Infrastructures (SOIs) (Wikipedia, 2008) and Grids, while the following section focuses on Grid environments and the way these can be used either as computational infrastructures (Computational Grids) or as distributed, networking and sharing ones (Data Grids). Thereafter, in the subchapter called *"Biomedical Applications: Cases and Scenarios in Grids"* we discuss various cases and scenarios of biomedical applications in Grids. Finally, the last subchapter concludes with a discussion on future research and potentials for the current study.

BACKGROUND

As already mentioned, research efforts of the past years have led to the realization of high performance computing in environments that follow the

SOA principles, such as Grids, SOIs and Clouds. In this subchapter the main concepts of the aforementioned infrastructures are discussed, while we also present related work in the research area of enabling medical applications to be executed in Grids.

Starting from distributed computing, it refers to any environments in which the service provision is done through a federation of resources. It can be seen as a case of parallel computing, with the main difference lying on the fact that in parallel computing usually a task (programming part) is split into many parts and executed in parallel in various nodes while in distributed computing these nodes are heterogeneous resources and distributed doesn't necessarily means splitting the code into many parts but executing it in a remote resource or a set of resources (e.g. in cases of workflows a process of it may be executed in a specific resource while another process may be executed in a different one). Distributed computing offers a set of benefits such as fault tolerance, data aggregation, cooperative processing and scalability through the distributed resources, and cost effectiveness since the service providers act in an open market by advertising their services and the end users are able to select providers based on their offerings. One can of course name much more advantages of distributed computing, however these can be found in literature (e.g. (Waldo, 1994)) and it is not within the scope of this subchapter to analyze them in detail but provide a general overview.

A service oriented architecture is an architecture allowing the collection and provision of services including description both of the offered services and of the way these services interact. These services communicate through specific interfaces and are loosely coupled with the operating systems on which they run and the programmable languages, allowing flexibility, replaceablity, scalability, risk mitigation and fault tolerance. Furthermore, one of the main SOA principles refers to its business nature and the alignment

with the business industry in terms of robustness, cost effectiveness and reduction of complexity, while its agility enables the automation of business processes and the orchestration of composite services. The most well-known / used connection technology for SOA is Web services - which can be seen as distributed software components. While the web services are application oriented, they are platform and language independent, usually built on XML, SOAP and WSDL (w3c, 2008). Figure 1 (http://mrwebservice.files.wordpress.com/2008/10/soa-detailed-diagram.png) depicts the main concept of SOA - offering resources, network, applications, storage as a service.

Based on the above, Service Oriented Infrastructures are a realization of SOA providing distributed databases, virtualized storage, application services and virtualized servers. The resources may not only refer to computational resources but also to network and storage allowing them to be provided as a service. Current Grid middleware implementations follow the SOA principles and in many cases allow the deployment of service oriented infrastructures and therefore called "*Service Oriented Grids*". Grid is the foundation of collaborative high performance computing and data sharing, which allows the creation of virtual computing systems that are sufficiently integrated to deliver non-trivial Quality of Service by using standard, open, general-purpose protocols and interfaces.

While Grids are beginning to find their way into real-world applications in many industrial and scientific sectors (such as automobile industry or biomedicine), a set of standards have been developed to allow their deployment, mainly OGSA (Open Grid Services Architecture), which aims to define a common, standard and open architecture for Grid-based systems, and WSRF (Web Services Resource Framework), which improves aspects of Wed services with the main improvement being the approach to statefullness. In that context, Service Oriented Grids are becoming a reality providing

Figure 1. Conceptual approach of a service oriented architecture

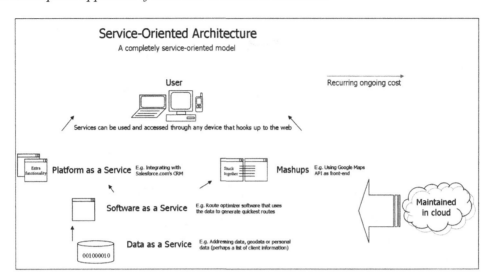

virtualized resources through the Grid middleware. An example of this approach is depicted in figure 2 (Srinivasan, 2005):

In general, enabling applications execution on Grid environment has been a research topic since the distributed nature of a Grid-based infrastructure makes feasible the solution of computational intensive problems in a reliable and cost-effective way. To this direction, literatures (E. Katsaloulis, 2006), (Ardaiz, 2004) and (Glatard, 2005) present the work performed for various application domains (biocomputational, learning and medical). Moreover, Parameter Sweep Applications (PSAs) are a class of applications that deal with the analysis of a specific simulation for a range of parameter values. PSAs are a very common class of applications that are met in computational Grids. High performance parametric modeling has been identified as a killer application for Grids (Abramson, 2000) and Grid-enabled PSAs have been recently developed in Bioinformatics (Aloisio, 2005). Following this direction, there are many projects dealing with the Grid technology in the Biology and Medicine sectors. The BRIDGES project (Biomedical Research Informatics Delivered by Grid Enabled Services) (BRIDGES Project) aimed at developing Grid-enabled bioinformatics tools to

support biomedical research. WISDOM initiative (Wide In Silico Docking On Malaria) (WISDOM Initiative) aims to demonstrate the relevance and the impact of the Grid approach to address drug discovery for neglected and emergent diseases with the use of the EGEE infrastructure (EGEE Project), while, the Akogrimo Project (AKOGRIMO Project) specified a mobile Grid infrastructure and evaluated it through a heart monitoring and emergency management scenario with the use of mobile devices. The Grid Relational Catalog (GRelC) project is working towards ubiquitous, integrated, seamless and comprehensive data Grid management solutions to fully address application specific requirements. Such an environment was used in the bioinformatics sector as described in (Fiore, 2008).

Furthermore, interesting works in the area of Grid computing with regard to biology are described in (Mirto, 2008) and (Tirado-Ramos, 2004). Authors in (Mirto, 2008) discuss a Grid-based infrastructure that performs ingestion into a relational DBMS for data integration of biological data sources, which reduces the data redundancy of biological flat files. Literature (Tirado-Ramos, 2004) describes an approach for using Grid resources are used for access to medical image

Figure 2. Service oriented grid

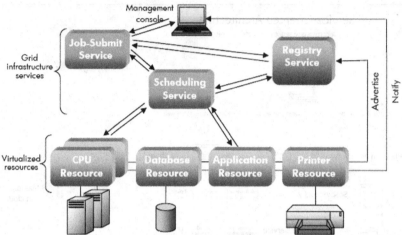

repositories, segmentation services, simulation of blood flow, and visualization in virtual environments of the simulated results.

Specific approaches for Grid services, such as an architecture for data management in Grids focused on the bioinformatics domain is presented in (Aloisio, 2006), while (Snel, 2006) includes a workflow management system description that supports software components for image import/export, caching, processing and notification. In that context, works on data aggregation are presented in (Reynaud, 2006) and (Tzung-Shi Chen, 2007).

The research efforts presented in the previous paragraphs confirm that work on the specific field is underway and many approaches are already implemented allowing applications to be executed in Grid environments.

HIGH PERFORMANCE COMPUTING: COMPUTATIONAL AND DATA GRIDS

The use of a Grid infrastructure for the execution of simulations is both a necessity for the involved researchers and an opportunity to make biomedical applications available to a wider research community. What follows is a deeper analysis on how Grids can serve biomedical applications either through computational / processing power provision or through networking / data sharing and aggregation.

Given that the exploitation of the vast resources provided by a Grid may lead to a significant decrease of the processing times, computer simulation may be employed in order to optimize treatment of diseases, by conducting a number of simulations for different therapeutic schemes based on the individual data of a patient. A restraining factor is that simulations need to be conducted in clinically accepted computational time. As the number of possible therapeutic schemes and consequently the number of simulations increases, the time required for evaluating and comparing the effects of the different schemes may become forbiddingly high especially since there is usually a large number of involved (real or virtual) patients. Exploiting Grid computing is a very attractive solution, as the resources provided in a Grid infrastructure may be efficiently used to reduce overall required execution time in a cost-effective and efficient manner, while one has also to take into account that there is considerable heterogeneity of required data (e.g. imaging, histopathologic, genetic) with different pre-processing requirements.

The resources are provided to the end-users / applications through the Grid middleware. The basic services of it refer to File Management (for transferring the input files to the Grid resources and obtaining the results), Execution Management (for submitting a "job" for execution to the infrastructure) and Monitoring (for monitoring the "jobs" status), while other services may also be part of the middleware, such as accounting and billing services or user management services for managing different classes of users: end-users, administrators, application workflow providers, etc.

An important module in many environments refers to the end-user's interface towards it. In Grids, what the mostly used user interface is a web-based Grid portal for enabling interactions in a simple and user-friendly way. This portal is based on a multi-tier architectural approach, defining different layers for the operations and functions of the application framework, and thus simplifying the installation and the maintenance processes. These layers are: the presentation layer (which includes all the functionality for the interaction with the end users and is presented to the users through web pages), the portal services layer (which includes all the functionality of the application and establishes the connection between the presentation and the middleware and database layers), the middleware layer (which includes all the functionality for the communication with the Grid services and resources) and the database layer (which includes a database storing the data regarding the application and user management).

The first step towards Grid-enabling an application is to bring it to a form appropriate for execution on the Grid. This involves adapting the source code in order to be externally parameterizable and also creating the required scripts and description files which are used by the grid workload management system for job execution on the grid. The source code has to be modified in order to be able to get input from standard parameter files, which are created at the user session

and are transferred along with the executable code to the Grid resource for execution. Moreover, the required wrapper-scripts are developed for setting up and executing the application on a Grid resource and the job description files are created. These files are used by the Grid workload management systems for the job submission and for the job match-making processes. A job description file describes various characteristics of the job to be submitted (such as executable name, input arguments, input and output files, etc) in a standard job description language. Obviously, each job in a parametric simulation corresponds to a different job description file and it is necessary to provide mechanisms for the automatic creation of these files according to user input. A user job is described by a Job Description Language (JDL) (JDL Attributes Specification). A JDL file specifies, for example, which executable to run and its parameters, files to be moved to and from the worker node and input files needed. For each one of the simulation applications, abstract JDL files are created with respect to its particular requirements and the end-user needs. These abstract files are used as a JDL template for the execution of simulation jobs and prior to the submission were instantiated with specific values for each parameter. The application porting process results to the design and development of bundles that include the Grid-enabled simulation code, the wrapper scripts and the JDL templates. These bundles, with the appropriate initialization in a manual or automatic way, are the only software prerequisites for the execution of applications in a Grid environment.

On the other hand, a Grid infrastructure may be used as a means for networking of resources and data sharing and aggregation. Medical-related applications generally belong to those collaborative environments that are based on input from various sources (e.g. sensors, cameras) and aggregation of acquired data under real-time conditions. Ubiquity renders data management issues of major importance since the nature of data will be changed in order to include high quality input from sources of all kinds, assuring privacy and aggregation associated and cross-checked with incomplete and inconsistent information. In this frame, any new achievements and directions in nanotechnologies, networks and biosensors are expected to achieve performance and interconnectedness. To this end, a significant part of the value of Grid technology lies on the fact that Grids are in position to provide the fundamental management mechanisms for distributed data. This is one major reason that often many developed Grid-based systems were referenced as "Data Grids", since the integration of data, infrastructures, digital libraries and persistent archives was a challenge forcing continued evolution of Grid technology. This challenge remains valid for medical applications, the requirements of which range from the transition from data handling, sharing and aggregation to the provision of knowledge as utility. Through data virtualization techniques (allowing representation of data as a service) and data management services, the Grid environments enable access to data regardless of the data source or physical location. Taking advantage of this feature, Grids can be used for aggregating data formats coming from a variety of sources, while the components deployed in the middleware are able to aggregate the data deriving from heterogeneous resources in a time sensitive manner. Moreover, there are cases of distributed data, such as medical records that are stored in local systems of institutions and updated regularly, while these institutions may be distributed worldwide. Nevertheless, a patient's treatment may require aggregation of data stored in different institutions and since the number of locations and size of the records tends to grow over time, an infrastructure able to provide the functionality of aggregating the shared data in real (or close to real) time is considered to be of major advantage - Grids allow the linking and aggregation of these data sources.

One should also take into consideration that these sources may not only refer to medical data /

records but also to information obtained by other sources referring to biosensors, cameras, digital inventories and knowledge libraries (e.g. pharmaceutical inventories or diagnosis databases). The challenging part in this design, is both the collection of the data as a bytestream and the configuration of the Grid components so as to "comprehend" the information and make sure the critical information is transmitted to the end nodes interested in the data - other Grid provided services. Roughly speaking, this implies that in order to resolve a query, the combinatory use of various pieces of information is needed. These pieces are defined by the application developer along with the appropriate way to present the results (e.g. store it, post it on the web, push it to another component).

In many cases, the concepts of event-driven SOAs are followed. Messages that are propagated by input resources, e.g. sensors, cameras, etc, are received by dedicated management systems that process them per category. The result of this process is a report that contains all the necessary information that needs to be relayed: the sender's ID, the receiver's ID, lifecycle information (e.g. a timestamp) and the message itself. These reports are generated and "wrapped" in the form of an event message, following the WS-Eventing specification of the W3C standardization body. Once data aggregator components receive an Event, the respective policy for that event is triggered and the information contained in the report is treated accordingly. This means that the information is stored to an intermediary repository and consequently combined with other pieces of information generally defined by the policy. Then the data aggregator components send the information extracted by the aforementioned process to the components that it is configured to (for example data analyzer components or specific databases). Again, the information is usually wrapped up in a WS-Event message so as to trigger the data analyzer components' policies in order to reach to a final meaningful conclusion regarding the patient. In an implementation level, this architecture is based on Service Oriented Computing concepts, inheriting the OGSA (http://www.ggf.org/documents/GFD.30.pdf) and Web Services Architecture principles through standards and specifications such as WS-Eventing (each component identified in the overall design is exposed as a Web Service).

In conclusion, a similar architecture for a service oriented data aggregator for a sensor network is described in (Kang 2004). There are two main differences between the proposed model and the one described in (Kang 2004). The first one is conceptual and lies to the fact that Kang is using a single notification mechanism as a Service Data Element broker. The second one is related to the design which is based on the Grid Web Services Description Language (GWSDL) that allows the specification of the properties of a Grid service's data. In our case, the properties of the service data are irrelevant to the system and are incorporated in the report. Therefore, it falls to the application developer's responsibility to properly configure the data aggregator components, allowing more flexibility in a business level.

BIOMEDICAL APPLICATIONS: CASES AND SCENARIOS IN GRIDS

What we discuss in this subchapter are two (2) scenarios of biomedical applications that have been applied to Grid-based solutions. These refer to radiotherapy simulations for in silico oncology and personalized healthcare and medicine demonstrating the added value of Grids in computational / processing and data aggregation and storage accordingly.

In silico ("on the computer") oncology is an emerging interdisciplinary field aiming at mathematically describing and computationally simulating the multi-scale biological mechanisms that constitute the phenomenon of cancer and its response to therapeutic techniques. Within

this framework, the "In Silico Oncology Group (ISOG)", National Technical University of Athens, has already developed a four-dimensional simulation model of glioblastoma multiforme (GBM) (Stamatakos 2002), (Dionysiou, 2004) response to radiotherapy. ISOG has adopted an essentially "top-down" modeling approach and developed a number of hybrid discrete Monte Carlo / cellular automata and continuous differential equation simulation models of tumour growth and response to therapeutic modalities. The aim is a better understanding of biological mechanisms concerning cancer and related therapeutic interventions and, in the long term, a contribution to the design of patient individualized therapies. For the purposes of the specific work, the ISOG simulation model of glioblastoma multiforme response to radiotherapy has been used in order to perform a virtual clinical trial. The model is based on the clinical, imaging, histopathologic, and molecular data of the patient and numerous fundamental biological mechanisms are incorporated and explicitly described. A prototype system of quantizing cell clusters included within each geometrical cell of a discretizing mesh covering the anatomic area of interest lies at the heart of the proposed simulation approach. The simulation environment for performing radiotherapy parameter sweep simulations mainly consists of a real Grid infrastructure provided by the EGEE project (EGEE Project). The simulation environment builds on the gLite middleware (gLite middleware) and provides a web-based Grid portal for enabling interactions with it in a simple and user-friendly way. In order to enable the execution of radiotherapy simulations on the Grid, the legacy code has been suitably migrated to the operating system used on Grid nodes and several scripts have been developed in order to automatically conduct parameter sweep radiotherapy simulations. The application – portal is enhanced with added functionality in order to simplify the job submission process and automate the interaction with the Grid services. In that way more users are able to access the computational resources while the administrators manage the application and monitor the operational status.

In the table that follows (Table 1), we present execution results in terms of the time needed for the simulations. The results indicate that by performing the parameter sweep simulations on the Grid, a considerable speedup may be achieved. This is very important in the context of in silico oncology, since the computational requirements of the simulations become overwhelmingly large as the required detail of simulation grows and because of the large number of potentially involved patients. Grid computing is a very appealing solution in the context of in silico oncology, since the vast resources provided by a Grid may be efficiently used for providing timely and accurate results.

The second application scenario refers to personalized healthcare and medicine. This scenario demonstrates the added value of Grid environments in terms of data aggregation, sharing and analysis. Aiming at providing healthcare and medicine proposals, all relevant data needs to be collected and analyzed for each patient. Moreover, prior to the medicine proposal, a simulation has to be performed so as to conclude which therapeutic method fits better to the specific patient. The historical data of the patient are available through various sources (usually distributed across medical institutions) as well as other relevant medical data (such as clinical, demographic, etc), which are obtained for a specific medical case using pattern-matching algorithms in order to identify patients that are similar.

The information is obtained through a data aggregator component that dynamically changes its functionality based on a set of policies that are acquired by a repository storing the data aggregation policies. These policies, which are associated with specific medical treatments or events from sensors, alter the data stream sources for the data aggregator and at the same time apply the algorithms that will be used on the streams so as to generate a set of patient specific information and relay it to the analyzer. The data streams are separated in two categories, the real-time data from the sensors and the historical data either the patient and or for the similar medical conditions and statistics. The analyzer component exploits this set of information and performs an analysis in order to produce the personalized healthcare and medicine proposal. The analysis process is also customized based on the real-time and historical data but in all cases includes a simulation of the available treatments and its parameters. For the simulation process, dedicated services are used that combine the mass data and computational capabilities of Grid environments. During the analysis process, the simulation results are evaluated and if they are not satisfying in terms of conformity with the historical data and statistical models, additional information are requested from the Data Aggregator and the simulation process for the particular treatment starts again.

This approach is presented in the following sequence diagram (Figure 3). The process is initiated by the end-user / doctor that requests an analysis

Table 1. Execution times of radiotherapy parameter sweep simulations

Scheme #	Mean Job Execution Time	Overall Schema Execution Time
1	~32 mins	~58 mins
2	~31 mins	~59 mins
3	~34 mins	~59 mins
4	~38 mins	~72 mins
5	~36 mins	~66 mins
6	~21 mins	~47 mins

for a particular patient. Thereafter the analyzer component interacts with the data aggregator in order to acquire information for the analysis and more specifically about the simulation. The aggregator loads the policies that are associated with the specific analysis and then establishes the data transfer connections with the sensors and the historic data services. The patient specific information is forwarded to the analyzer that performs simulations using the corresponding simulation services and based on the simulation results; a patient-specific medicine proposal is produced. Furthermore, the the process may be triggered by a real-time event from a sensor / camera that is used to monitor some parameters of a patient.

Due to the exponential increase in the complexity of the simulation and the heterogeneity of required data and their preprocessing needs, as well as the large number of potentially involved patients, the large scale execution capabilities and vast computational capacities offered by Grids may prove exceptionally beneficial. Execution times prove that a considerable speedup may be achieved by using the Grid and that the Grid can also provide solutions in case that comparative results for therapeutic schemas are needed in real time. Moreover, the proposed set of components along with their interconnections / interfaces can be implemented to any service-based Grid middleware (such as GRIA (Surridge, 2005), (http://www.gria.org)) since they can be deployed as web services.

CONCLUSION

High performance computing as realized nowadays through Service Oriented Infrastructures, Clouds and Grids can serve as a means to provide resources of any kind to biomedical applications. These resources refer to computational / processing, storage, networking and other ones such as cameras / sensors, which combined with the services offered by the Grid middleware allow the

Figure 3. Sequence diagram of the personalized medical healthcare & medicine scenario

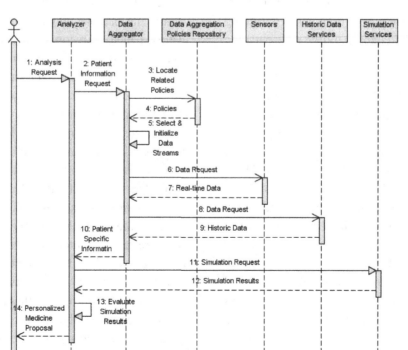

medical applications to become Grid-enabled and benefit from the added value the latter offers.

We have described the main concepts and principles of Service Oriented Architectures and Infrastructures as well as Grids, describing both computational and data Grids. Furthermore, two scenarios have been selected to demonstrate how the aforementioned Grid classification can be applied to biomedicine. The first one refers to in silico oncology, a multidisciplinary field that aims to model the multi-scale biological mechanisms that constitute the phenomenon of cancer and evaluate its response to therapeutic techniques by computer simulations. Due to the exponential increase in the complexity of the simulation as the density of discretization of the 4D Grid of the biological model increases and the heterogeneity of required data and their preprocessing needs, as well as the large number of potentially involved patients, the large scale execution capabilities and vast computational capacities offered by a Grid may prove exceptionally beneficial. Some of the experiments that have been performed demonstrate the potential of the exploitation of Grid technologies for the simulation of clinical trials, with the long-term aim of both better designing clinical studies and understanding their outcome based on basic biological science. Exploitation of grid resources has made possible the simulation and comparison of different therapeutic schemes that would be extremely time consuming in case of execution on a conventional computer. Execution times prove that a considerable speedup may be achieved by applying Grid technologies. These can also provide solutions in case that comparative results for therapeutic schemas are needed in real time (as described with the use of Historical Data). The second one refers to personalized healthcare and medicine, which poses the requirement of a complex infrastructure that will allow for aggregation of data from different sources and significant speed up of the data analysis and simulation process

Notwithstanding, one of the main challenges in both scenarios refers to a workflow management service, which based on a patient's record, will obtain information from specific providers since for example different hospitals - following their expertise - hold different kind of records for the same patient (e.g. heart records). Furthermore, future work on the data analyzer component may be focused on identifying the simulation with the longest execution time within a generation in order to avoid cases of low resources' utilization (occurring when the rest of the generation's simulations are completed). Another challenge would be to port additional treatment modalities to the Grid-enabled environment and use the Grid in order to perform similar comparative simulations. It is expected that during the next few years, such simulation models and mechanisms supported by new generation algorithms are likely to be used. The computer infrastructure needed for these simulations will be significant and to this direction, Grid infrastructures can serve as a means to support this endeavor

Concluding, Grids form networks of resources along with monitoring and diagnosis facilities around the patient, which in combination with historical medical records, diagnosis and analysis services, allow for the realization of therapeutic schemes. A prerequisite to produce the aforementioned schemes is a component able to obtain information from different sources (e.g. cameras, sensors, etc) and aggregate this kind of data along with historical data in order to be consumable by data analysis and simulation services. The outcome of these simulations will enable medicine to become more personalized and patient-oriented, targeting the optimal individual treatment. Furthermore, any computational intensive biomedical application can benefit from the distributed high performance nature of Grid environments, since as the number of possible therapeutic schemes and consequently the number of simulations increases, the time required for evaluating and comparing the effects of the different schemes may become forbiddingly high. Exploiting Grid computing is a very attractive solution, as the resources provided in a Grid infrastructure may be efficiently used to reduce the overall required time for simulations.

REFERENCES

Abramson, D., Giddy, J., & Kotler, L. (2000), "High Performance Parametric Modeling with Nimrod/G: Killer Application for the Global Grid?" IPDPS'2000, Mexico, USA.

G. Aloisio, M. Cafaro, S. Fiore, M. Mirto, (2005), "ProGenGrid: A Workflow Service Infrastructure for Composing and Executing Bioinformatics Grid Services"

Aloisio, G., Cafaro, M., Fiore, S., & Mirto, M. (2006), "A Split & Merge Data Management Architecture for a Grid Environment", 19th IEEE Symposium on Computer-Based Medical Systems.

Ardaiz, L., Diaz de Cerio, L., Gallardo, A., Messeguer, R., & Sanjeevan, K. (2004), "ULab-Grid Framework for Computationally Intensive Remote and Collaborative Learning Laboratories", IEEE International Symposium on Cluster Computing and the Grid.

Boss, G., Malladi, P., Quan, D., Legregni, L., & Hall, H. (2007), "Cloud Computing", http://www.ibm.com/developerworks/websphere/zones/hipods/

David Sprott and Lawrence Wilkes. (2004), "Understanding Service-Oriented Architecture", CBDI Forum

Dionysiou, D., Stamatakos, G., & Uzunoglu, N. (2004). A four-dimensional simulation model of tumour response to radiotherapy in vivo: parametric validation considering radiosensitivity, genetic profile and fractionation . *Journal of Theoretical Biology*, *230*(Issue 1), 1–20. doi:10.1016/j.jtbi.2004.03.024

Fiore, S., Mirto, M., Cafaro, M., Vadacca, S., Negro, A., & Aloisio, G. (2008), "A GRelC based Data Grid Management Environment," 21st IEEE International Symposium on Computer-Based Medical Systems.

Foster, I., & Kesselman, C. (1999), "The Grid: Blueprint for a Future Computing Infrastructure", Morgan Kaufmann Publishers, USA.

Foster, I., Kesselman, C., & Tuecke, S. (2001). The Anatomy of the Grid: Enabling Scalable Virtual Organizations . *The International Journal of Supercomputer Applications*, *15*(3).

Glatard, T., Montagnat, J., & Pennec, X. (2005), "Grid-enabled workflows for data intensive medical applications", Computer Based Medical Systems, Special Track on Grids for Biomedicine and Bioinformatics.

GRIA. Grid Resources for Industrial Applications, (2008), http://www.gria.org

http://mrwebservice.files.wordpress.com/2008/10/soa-detailed-diagram.png

Job Description Language (JDL) Attributes Specification. (2008), https://edms.cern.ch/document/590869/1/

Katsaloulis, E., Floros, A., Provata, Y., & Cotronis, T. (2006), "Gridification of the SHMap Biocomputational Algorithm", International Special Topic Conference on Information Technology in Biomedicine.

Latha Srinivasan and Jem Treadwell (2005), "An Overview of Service-oriented Architecture, Web Services and Grid Computing" HP Software Global Business Unit, 2005

Leinberger, W., & Kumar, V. (1999). Information Power Grid: The new frontier in parallel computing? *IEEE Concurrency*, 7(4), 75–84. doi:10.1109/MCC.1999.806982

Mirto, M., Fiore, S., Cafaro, M., Passante, M., & Aloisio, G. (2008), "A Grid-Based Bioinformatics Wrapper for Biological Databases,", 21st IEEE International Symposium on Computer-Based Medical Systems.

Open Grid Services Architecture (OGSA). (2008), http://www.ggf.org/documents/GFD.30.pdf

Reynaud, S., Mathieu, G., Girard, P., & Hernandez, F. (2006), "LAVOISIER: A Data Aggregation and Unification Service", Proceedings of Computing in High Energy and Nuclear Physics (CHEP06), Mumbai, India, February 2006.

Snel, J., Olabarriaga, S., Alkemade, J., Andel, H., Nederveen, A., Majoie, C., et al. (2006), "A Distributed Workflow Management System for Automated Medical Image Analysis and Logistics", 19th IEEE International Symposium on Computer-Based Medical Systems.

Stamatakos, G., Dionysiou, D., Zacharaki, E., et al. (2002), "In silico radiation oncology: combining novel simulation algorithms with current visualization techniques", Proceedings of the IEEE, Special Issue on "Bioinformatics: Advances and Chalenges", Volume 90, Issue 11, pp. 1764-1777.

Surridge, M., Taylor, S., De Roure, D., & Zaluska, E. (2005), "Experiences with GRIA-Industrial Applications on a Web Services Grid", in Proceedings of the First International Conference on e-Science and Grid Computing, pp. 98-105. IEEE Press.

The AKOGRIMO Project. http://www.mobileGrids.org/

The BRIDGES Project. http://www.brc.dcs.gla.ac.uk/projects/bridges/

The EGEE Project. http://www.eu-egee.org/

The gLite middleware, (2008), http://glite.web.cern.ch/glite

The WISDOM Initiative, http://wisdom.eu-egee.fr/

A. Tirado-Ramos, P.M.A. Sloot, A.G. Hoekstra, M. Bubak, (2004), "An Integrative Approach to High-Performance Biomedical Problem Solving Environments on the Grid", Parallel Computing, Special issue on High-Performance Parallel Biocomputing.

Tzung-Shi Chen. Yi-Shiang Chang, Hua-Wen Tsai, Chih-Ping Chu, (2007), "Data Aggregation for Range Query in Wireless Sensor Networks", IEEE Wireless Communications & Networking Conference (WCNC 2007), Hong Kong.

Waldo, J., Wyant, G., Wollrath, A., & Kendall, S. (1994), "A Note on Distributed Computing", http://research.sun.com/techrep/1994/smli_tr-94-29.pdf

Web Services Description Language (WSDL) 1.1, (2008), http://www.w3.org/TR/wsdl

Wikipedia, Service Oriented Infrastucture Definition, (2008), http://en.wikipedia.org/wiki/Service_Oriented_Infrastructure

YunHee Kang. (2004), "An Extended OGSA Based Service Data Aggregator by Using Notification Mechanism", Grid and Cooperative Computing, Springer.

Chapter 7
Pervasive Healthcare Services and Technologies for Memory Loss Diseases Support

Mata Ilioudi
University of Peloponnese, Greece

Dimitrios Karaiskos
Athens University of Economics and Business, Greece

Athina Lazakidou
University of Peloponnese, Greece

ABSTRACT

With an increasingly mobile society and the worldwide deployment of mobile and wireless networks, the wireless infrastructure can support many current and emerging healthcare applications. This could fulfill the vision of "pervasive healthcare" or healthcare to anyone, anytime, and anywhere by removing locational, time and other restraints while increasing both the coverage and the quality. In this chapter the authors present applications and requirements of pervasive healthcare, wireless networking solutions and several important research problems. The pervasive healthcare applications include pervasive health monitoring, intelligent emergency management system, pervasive healthcare data access, and ubiquitous mobile telemedicine. On top of the valuable benefits new technologies enable the memory loss patients for independent living and also reduce the cost of family care-giving for memory loss and elder patients.

INTRODUCTION

Pervasive computing (also referred to as ubiquitous computing or ambient intelligence) aims to create environments where computers are invisibly and seamlessly integrated and connected into our everyday environment. Pervasive computing and intelligent multimedia technologies are becoming increasingly important, although many potential applications have not yet been fully realized. These key technologies are creating a multimedia revolution that will have significant impact across a wide spectrum of consumer, business, healthcare, and governmental domains.

DOI: 10.4018/978-1-60566-768-3.ch007

Pervasive healthcare has the potential to reduce long-term costs and improve quality of service, but it also faces many technical and administrative obstacles. The healthcare industry has introduced wireless technologies on a limited scale for many simple tasks. However, researchers and advocates must overcome many challenges before a truly pervasive healthcare environment can become a reality.

Pervasive healthcare would improve the productivity of healthcare practitioners and greatly facilitate the delivery of a wider range of medical services. The rapidly increasing use of handheld devices and the deployment of wireless-based solutions should accelerate the development of such services, especially in areas where a wire line infrastructure is minimal or impractical. Faster and more accurate communication would result in substantial savings that could be used to expand basic healthcare to everyone, thereby reducing costs in the long run.

DEFINITION OF MEMORY LOSS

Many studies have been made by neuroscientists in order to analyse the structure of human memory, to assign how many kinds of memory there are, and how the brain selects, stores and retrieves information in each memory's part. In general, memory divided into three parts:

- Immediate or working memory which refers to the structures and processes used for temporarily storing and manipulating information. The Immediate memory includes information about the current task such as the name of a person you met moments ago or a phone number just to place the call.
- Short-term memory for holding in mind information of the recent past. This memory's part refers to memories which last for a few minutes such as events that happened

in several seconds or minutes ago, and what you ate for breakfast.
- Long-term memory can record the remote past. It is the part which contains everything we know about the world, semantic information as well as autobiographical experience, and memories of childhood (Familydoctor.org Editorial Staff, 2006).

Each memory part can be affected by different reason and can lead us in memory loss.

Memory loss disease can appear after brain damage due to physical trauma or disease, as well as, due to emotional trauma. Memory loss can affect memories partially or totally, slowly or suddenly, temporary or permanent. Anything that affects cognition - the process of thinking, learning, and remembering - can affect memory (FDA, 2009). In many cases, alternative names are used for memory loss, such as forgetfulness, amnesia or impaired memory.

Some causes which affect the memory's parts and enforce memory loss are listed below:

- *Depression*: Depressed patients performed significantly worse on measures of both processing speed and working memory (Nebes, Butters, et al, 2000). The major depression significantly affects working memory (Pelosi, Slade, et al, 2000).
- *Alzheimer disease*: Alzheimer's destroys brain cells, which control short-term memory (Alzheimer's Association, 2009), causing problems with memory, thinking and behaviour severe enough to affect work, lifelong hobbies or social life, and difficulty in remembering recently learned facts. Usually, Alzheimer's disease is diagnosed in people over 65 years of age (*Alzheimer's disease*, 2009). In addition, many other known chronic diseases such as Parkinson, Diabetes can lead in memory loss as well.
- *Dementia*: Short-term memory deficits in Dementia can be summarised as follows:

there is a reduction in memory capacity and an impairment in the ability to retain verbal material in memory for short durations (Morris, 1986). Dementia is a severe debilitating condition and, people with dementia experience a global decline in cognitive function leading to problems with memory, perception, reasoning and temporal and spatial awareness (Whitehouse, 1993).

- *Amnesia*: Retrograde amnesia gradients are often interpreted as revealing the time needed for the formation of long-term memories (Gold, 2006). The anterograte amnesia, the inability to lay down new memory records is associated with memories formed prior to brain damage are impaired, but the effect depends on the age of the memory trace at the time the damage occurs, with most recently formed memories suffering the most (Brown, 2002).
- *Emotional trauma*: Also, emotional trauma leads the brain to deny recalling from long-term memory past events, which are not pleasant for the patient.
- *Aging*: A person loses nerve cells at the rate of 1%per year and stops growing new nerve cells after age 25 (Memory loss, 2006). Aging may affect memory by changing the way the brain stores information and by making it harder to recall stored information. The immediate and long-term memories are not usually affected by aging, but the short-term memory may be affected. Also, aging is commonly accompanied by an increased likelihood of developing memory loss and around 4000 of people aged 65 or older have age-associated memory impairment and some 15% develop Alzheimer's disease each year (Small, 2002)
- *Reduction of oxygen*: In addition, reduction of oxygen to the brain can cause the

death of the nerves cells and finally memory loss.

- *Alcohol & Chemical products*: Large amounts of alcohol and chemical products such as drugs destroys brain cells and can produce partial or complete blackouts, which are periods of memory loss.

PERVASIVE HEALTHCARE: EXISTING DEVICES AND SYSTEMS

There already exist many applications in the domain of healthcare. For example, Geer (Lorincz, Fulford-Jones, et al, 2004) describes the technology involved in robotically-enhanced surgery, and various wearable devices. Motes are described: these are very small devices, usually composed of a processor, memory, low-power radio, antenna and power supply. A project, CodeBlue, is currently looking at use of motes for medical applications such as real-time triage and large-population studies.

Other applications described by Geer include the Pluto mote, which is similar to a wristwatch. This device measures heart rate and respiration, and can help study limb movement in cases of stroke. Such information helps clinicians predict tremor episodes and adjust medication dosages as needed.

There is also the 'Vital Dust' mote, which measures heart rate and blood oxygen saturation. Other research, looking at kidney failure in children, uses Bluetooth enabled scales and blood pressure cuffs. These allow nurses to manage fluid levels and monitor many more children at once. Finally, diabetic monitoring has been examined, using IBM's Personal Care Connect technology. This technology uses Johnson and Johnson's Lifescan glucose meter, and sends graphs of the device's readings to doctors.

Other applications include a system to ask unobtrusive questions of its users at random times, to predict declines in health (Intille, Larson, & Kukla, 2002); packaged medicine which alerts relevant people when it is not taken at an expected time (Floerkemeier & Siegemund, 2003); a medicine cabinet which can monitor conditions, provide medication reminders, interact with healthcare staff, and access health information; and systems to pass relevant information to paramedics at an emergency scene (Varshney, 2007).

On a similar note, Lorincz et al (Lorincz, Fulford-Jones, et al, 2004) propose a protocol and framework for integrating devices into disaster-response scenarios. Such technologies have been put into active use in the assisted living complex run by Elite Care (Stanford, 2002) in which residents reportedly have 'as much autonomy and responsibility for themselves and their environment as possible'. Sensors help staff identify residents who may need immediate care – for example, when a resident wanders off disoriented, or is wakeful in the night – helping to concentrate efforts where they are needed. This is achieved via the use of locator badges: dual-channel infrared RF tags, which act as apartment keys and can be detected when within 90 feet of an IR sensor. Embedded weight sensors in the beds help detect tossing and turning, and frequent bathroom trips. In-apartment touch screen computers allow communications with family and friends. Personalized databases allow monitoring of long-term trends, avoidance of extensive manual record keeping, and communication with families and physicians about any changes in their elderly relations.

PERVASIVE COMPUTING FOR MEMORY LOSS PATIENTS

"The most profound technologies are those that disappear. They weave themselves into the fabric of everyday life until they are indistinguishable from it". With this prompt Mark Weiser, in 1991, introduced a new computing paradigm to the academic community, which he baptized it *Ubiquitous Computing* and proclaimed it as the successor of the current paradigm, that of *Personal Computing* (Weiser, 1991). Ubiquitous computing appears into the literature with different titles such as pervasive computing, calm technology, ambient intelligence, among others all defining the same. Pervasive or ubiquitous computing involves the availability of many effectively invisible computers throughout the physical environment (Weiser, 1989).

Pervasive IS may support both personal and business activities. Kourouthanassis and Giaglis (Kourouthanassis & Giaglis, 2007) provide a taxonomy of pervasive IS and their features by identifying four pertinent application types: personal, domestic, corporate, and public:

- *Personal Pervasive IS* relies on wearable hardware elements to provide a fully functional computing experience wherever the user might be. Typical examples include biomedical monitoring systems (Jafari, Dabiri, Brisk, & Sarrafzadeh, 2005), human detection systems (Smith, Fishkin, Jiang, Mamishev, Philipose, Rea, Roy, & Sundara-Rajan, 2005), and remote plant operation systems (Najjar, Thompson, & Ockerman, 1997).
- *Domestic Pervasive IS* primarily automate tasks that otherwise require human supervision in the household (e.g. heating and lightning control, monitoring the home inventory, and so on). Typical examples include MIT's Home of the Future initiative (Intille, 2002) and the Aware Home (Kidd, Orr, Abowd, Atkeson, Essa, MacIntyre, Mynatt, Starner, & Newsetter, 1999).
- *Corporate Pervasive IS* may support enterprise-wide activities, such as supply chain management [e.g. warehouse and logistics management (Karkkainen, 2003),(Prater, Frazier, Reyes, 2005), workforce management [e.g. sales force automation (Walters, 2001) and office support (Greenberg &

Rounding, 2001), (Churchill, Nelson, & Denoue, 2003), and customer relationship management (Fano & Gershman, 2002), (Kourouthanassis, 2004).

- *Public Pervasive IS* may craft interactive environments in public places. Examples include wireless museum guides (Hsin & Liu, 2006) and mobile information devices in hospitals (Ziao, Lasome, Moss, Mackenzie, & Faraj, 2001), (Liszka, Mackin, Lichter, York, Pillai, & Rosenbaum, 2004) to name but a few popular applications.

Pervasive healthcare has been defined as 'the application of pervasive computing technologies for healthcare, health and wellness management' and 'as making healthcare available everywhere, anytime, pervasively (Korhonen & Bardram, 2004).

Patients with memory loss need care giving into everyday life, at home and everywhere. Pervasive tools can help patients manage memory loss disease, until a cure is found. Many applications which use pervasive technologies have been developed for supporting people with memory loss in different areas.

The Gloucester Smart House (Glo*ucestor smart house*, n.d.), is specifically targeted to support people with dementia with relatively simple solutions, such as a bath and basin monitor to prevent overflows, and a locator for finding lost items such as keys and glasses.

The application which is called Autominder uses pervasive technologies to help people compensate for memory loss by prompting them to take medications (Pollack, et al, 2003).

Another relevant application is called Smart Blister Pack which monitors the medication consumption unobtrusively, transmits the sensed data continuously to facilitate analysis by medical staff during the treatment and reminds patients when they missed a dose on their mobile phone (Floerkemeier & Siegemund, 2003).

In a more specific application area, Tran and Mynatt (Floerkemeier & Siegemund, 2003) have developed the Cook's Collage as an aid for people with memory loss, in which cameras record events and display snapshots to show patients where they are in the cooking process. It replays a series of digital still images in a comic strip reel format depicting people's cooking actions *in situ*, intended to help them remember if they have forgotten a step (e.g., adding a particular ingredient) after being distracted (Tran & Mynatt, 2002).

Pervasive technology could play an important role by providing a Smart Environment, which will help patients with memory loss to become more independent. Matilda's Smart House is equipped with various sensors and devices, and supports applications for Alzheimer Disease patients as well as for patients with Dementia (Tran, Calcaterra, & Mynatt, 2005). Especially, the Mobile Patient Care-Giving Assistant (mPCA) application is a cognitive assistant designed to improve the independence of live at home for Alzheimer Disease patients. The mPCA assistant is a smart phone that interacts with a set of sensors in a smart space, in which most of the computation, decisions, and events take place. The mPCA will assist individual Alzheimer Disease patients in daily activities by means of reminders, orientation, and context-sensitive teaching, and monitoring. Matilda's Smart House, supports a General Reminder System (GRS), which is a reminder application targeted to patients with Dementia. The GRS reminds the patients to perform critical tasks such as medication intake and doctor appointments.

SenseCam is a wearable digital camera that is designed to take photographs passively, without user intervention, while it is being worn. This device has been developed by Microsoft Research Cambridge, for patients with amnesia. The Sense-Cam is a device fitted with sensors that automatically records key moments in daily life and helps patients to stimulate recovery and retention of autobiographical memory (Helal, Mann, Giraldo, Kaddoura, Lee, & Zabbadani, 2003).

Several "Smart Home" platforms have been developed to implement and assess pervasive technologies in every day life. Especially, for the aging population which is characterized by memory loss and communication difficulties. The Homecare Hub (HCH) concept addresses this issue by providing an in-home care rich-data repository accessible by care providers and family members. Using a wireless handheld device, users can review text, images and audio files stored by previous carers and add their own observations for subsequent carers to access (Knies, 2007).

BENEFITS FOR MEMORY LOSS PATIENTS

- Provide an environment that is constantly monitored to ensure the patient with memory loss is safe (activity monitoring)
- Automate specific activities that a patient is unable to perform (medication)
- Provide a safe and secure environment (alerting the patients of potentially dangerous activities)
- Enable and empower the patient for independent living (by giving prompts that be auditory and/or visual)
- Reduce the cost of family care-giving for memory loss and elder patients.

CONCLUSION

The integration of multi-modal sensors, smart phones and smart spaces, provide a unique opportunity to create tools that would enhance the performance of the daily activities done by an elder or a memory loss patient. Using location sensors, the smart home will be able to locate the elder and infer important context information. The new technologies will hopefully reduce the cost of family care-giving for memory loss and elder patients.

REFERENCES

Alzheimer's Association. (2009). *What is Alzheimer's*. Retrieved from http://www.alz.org/alzheimers_disease_what_is_alzheimers.asp

Alzheimer's disease. (2009). Retrieved from http://en.wikipedia.org/wiki/Alzheimer%27s_disease

Brown, A. S. (2002). Consolidation theory and retrograde amnesia in humans. *Psychonomic Bulletin & Review, 9*, 403–425.

Churchill, E. F., Nelson, L., & Denoue, L. (2003). Multimedia fliers: Information sharing with digital community bulletin boards. In *Proceedings of the Communities and Technologies* (pp. 19-21). Amsterdam: Kluwer.

Familydoctor.org Editorial Staff. (2006). *Memory loss with aging: What's normal, what's not*. Retrieved from http://familydoctor.org/online/famdocen/home/seniors/common-older/124.html

Fano, A., & Gershman, A. (2002). The future of business services in the age of ubiquitous computing. *Communications of the ACM, 45*(12), 83–87. doi:10.1145/585597.585620

FDA. (2009). *For consumers. U.S. food and drug administration*. Retrieved from http://www.fda.gov/consumer/features/memoryloss0507.html

Floerkemeier, C., & Siegemund, F. (2003). Improving the effectiveness of medical treatment with pervasive computing technologies. In *Proceedings of the UbiHealth 2003: The 2nd International Workshop on Ubiquitous Computing for Pervasive Healthcare Applications. Gloucestor smart house*. (n.d.). Bath Institute of Biomedical Engineering. Retrieved from http://www.bath.ac.uk/bime/projects/smart/index.htm

Gold, P. E. (2006). The many faces of amnesia. *Learning & Memory (Cold Spring Harbor, N.Y.)*, *13*(5), 506–514. doi:10.1101/lm.277406

Greenberg, S., & Rounding, M. (2001). The notification collage: Posting information to public and personal displays. *CHI Letters*, *3*(1), 515–521.

Helal, A., Mann, W., Giraldo, C., Kaddoura, Y., Lee, C., & Zabbadani, H. (2003). Smart phone based cognitive assistant. In *Proceedings of the 2nd International Workshop on Ubiquitous Computing for Pervasive Healthcare Applications*, Seattle, WA.

Hsin, C., & Liu, M. (2006). Self-monitoring of wireless sensor networks. *Computer Communications*, *29*(4), 462–476. doi:10.1016/j.comcom.2004.12.031

Intille, S., Larson, S. K., & Kukla, C. (2002). Just-in-time context-sensitive questioning for preventative health care. In *Proceedings of the AAAI' 2002 Workshop on Automation as Caregiver: The Role of Intelligent Technology in Elder Care*.

Intille, S. S. (2002). Designing a home of the future. *IEEE Pervasive Computing / IEEE Computer Society [and] IEEE Communications Society*, *1*(2), 76–82. doi:10.1109/MPRV.2002.1012340

Jafari, R., Dabiri, F., Brisk, P., & Sarrafzadeh, M. (2005). Adaptive and fault tolerant medical vest for life-critical medical monitoring. In *Proceedings of the ACM symposium on Applied Computing*, Santa Fe, New Mexico. New York: ACM Press.

Karkkainen, M. (2003). Increasing efficiency in the supply chain for short life goods using RFID tagging. *International Journal of Retail & Distribution Management*, *31*(10), 529–536. doi:10.1108/09590550310497058

Kidd, C. D., Orr, R., Abowd, G. D., Atkeson, C., Essa, I., MacIntyre, B., et al. (1999). The aware home: A living laboratory for ubiquitous computing research. In *Proceedings of the Second International Workshop on Cooperative Buildings*. Berlin, Germany; Springer-Verlag.

Knies, R. (2007). *Memorable support for Sense-Cam memory-retention research*. Retrieved from http://research.microsoft.com/en-us/news/features/sensecam.aspx

Korhonen, I., & Bardram, J. E. (2004). Guest editorial: Introduction to the special section on pervasive healthcare. *IEEE Transactions on Information Technology in Biomedicine*, *8*(3), 229–234. doi:10.1109/TITB.2004.835337

Kourouthanassis, P. (2004). Can technology make shopping fun? *ECR Journal*, *3*(2), 37–44.

Kourouthanassis, P., & Giaglis, G. M. (Eds.). (2007). *Pervasive information systems*. New York: M. E. Sharpe Inc.

Liszka, K. J., Mackin, M. A., Lichter, M. J., York, D. W., Pillai, D., & Rosenbaum, D. S. (2004). Keeping a beat on the heart. *IEEE Pervasive Computing / IEEE Computer Society [and] IEEE Communications Society*, *3*(4), 42–49. doi:10.1109/MPRV.2004.10

Lorincz, K. M., & Fulford-Jones, D. J. (2004). Sensor networks for emergency response: Challenges and opportunities. *IEEE Pervasive Computing / IEEE Computer Society [and] IEEE Communications Society*, *3*, 16–23. doi:10.1109/MPRV.2004.18

Memory loss. (2006). *Encyclopedia of alternative medicine*. Gale Cengage. Retrieved from http://www.enotes.com/alternative-medicine-encyclopedia/memory-loss

Morris, R. G. (1986). Short-term forgetting in senile dementia of the Alzheimer's type. *Cognitive Neuropsychology*, *3*(1), 77–97. doi:10.1080/02643298608252670

Najjar, L., Thompson, J. C., & Ockerman, J. J. (1997). A wearable computer for quality assurance in a food-processing plant. In *Proceedings of the 1st International Symposium on Wearable Computers*. Los Alamitos, CA: IEEE Press.

Nebes, R. D., & Butters, M. A. (2000). Decreased working memory and processing speed mediate cognitive impairment in geriatric depression. *Psychological Medicine*, *30*, 679–691. doi:10.1017/S0033291799001968

Osman, K. A., Ashford, R. L., & Oldacres, A. (2007). Homecare Hub - A Pervasive Computing Approach to Integrating Data for Remote Delivery of Personal and Social Care. In *Proceedings of the 2ⁿᵈ International Conference on Pervasive Computing and Applications* (pp. 348-353).

Pelosi, L., & Slade, T. (2000). Working memory dysfunction in major depression: An event-related potential study. *Clinical Neurophysiology*, *11*(9), 1531–1543. doi:10.1016/S1388-2457(00)00354-0

Pollack, M. E. (2003). Autominder: An intelligent cognitive orthotic system for people with memory impairment. *Robotics and Autonomous Systems*, *44*(3-4), 273–282. doi:10.1016/S0921-8890(03)00077-0

Prater, E., Frazier, G. V., & Reyes, P. M. (2005). Future impacts of RFID on e-supply chains in grocery retailing. *Supply Chain Management: An International Journal*, *10*(2), 134–142. doi:10.1108/13598540510589205

Small, G. W. (2002). What we need to know about age related memory loss. *British Medical Journal*, *324*, 1502–1505. doi:10.1136/bmj.324.7352.1502

Smith, J. R., Fishkin, K., Jiang, B., Mamishev, A., Philipose, M., & Rea, A. D. (2005). RFID-based techniques for human-activity detection. *Communications of the ACM*, *48*(9), 39–44. doi:10.1145/1081992.1082018

Soerensen, R. A., & Nygaard, J. M. (2008). Distributed zero configuration base station. In *Proceedings of the 2nd International Conference on Pervasive Computing Technologies for Healthcare*, Tampere, Finland.

Stanford, V. (2002). Using pervasive computing to deliver elder care. *IEEE Pervasive Computing / IEEE Computer Society [and] IEEE Communications Society*, *1*, 10–13. doi:10.1109/MPRV.2002.993139

Tran, Q., Calcaterra, G., & Mynatt, E. (2005). Cook's collage: Deja vu display for a home Kkitchen. In *. Proceedings of HOIT*, *2005*, 15–32.

Tran, Q., & Mynatt, E. (2002). *Cook's collage: Two exploratory designs*. Paper presented at the New Technologies for Families Workshop, CHI 2002, Minneapolis, MN.

Varshney, U. (2007). Pervasive healthcare and wireless health monitoring. *Mobile Networking Applications*, *2*(2-3), 113–127. doi:10.1007/s11036-007-0017-1

Wagner, S. (2008a). Towards an open and easily extendible home care system infrastructure. In *Proceedings of the 2nd International Conference on Pervasive Computing Technologies for Healthcare*, Tampere, Finland.

Wagner, S. (2008b). Zero-configuration of pervasive healthcare sensor networks. In *Proceedings of the 3rd International Conference on Pervasive Computing and Applications (ICPCA' 2008)*, Alexandria, Egypt.

Walters, G. J. (2001). Privacy and security: An ethical analysis. *Computers & Society*, *31*(2), 8–23. doi:10.1145/503345.503347

Weiser, M. (1989). The computer for the 21st century. *ACM SIGMobile Mobile Computing and Communications Review*, *3*, 3–11. doi:10.1145/329124.329126

Weiser, M. (1991). The computer of the 21st century. *Scientific American*, *265*(3), 66–75.

Whitehouse, P. J. (Ed.). (1993). *Dementia*. Philadelphia: F. A. Davis Company.

Xiao, Y., Lasome, C., Moss, J., Mackenzie, C., & Faraj, S. (2001). Cognitive properties of a whiteboard: A case study in a trauma center. In *Proceedings of the 7th European Conference on Computer Supported Cooperative Work*, Bonn, Germany (pp. 16-20). Amsterdam: Kluwer.

KEY TERMS AND DEFINITIONS

Health Information: The term 'health information' means any information, whether oral or recorded in any form or medium, that is created or received by a health care provider, health plan, public health authority, employer, life insurer, school or university, or health care clearinghouse; and relates to the past, present, or future physical or mental health or condition of an individual, the provision of health care to an individual, or the past, present, or future payment for the provision of health care to an individual.

Memory Loss Disease: Memory loss can have many causes such as Alzheimer's disease, Parkinson's disease or Huntington's disease. It is sometimes a side effect of chemotherapy in which cytotoxic drugs are used to treat cancer. Certain forms of mental illness also have memory loss as a key symptom, including fugue states and the much more famous Dissociative Identity Disorder. Stress-related activities are another factor which can result in memory loss. It can also be caused by traumatic brain injury, of which a concussion is a form.

Ubiquitous Computing: It is a post-desktop model of human-computer interaction in which information processing has been thoroughly integrated into everyday objects and activities. In the course of ordinary activities, someone "using" ubiquitous computing engages many computational devices and systems simultaneously, and may not necessarily even be aware that they are doing so. This paradigm is also described as pervasive computing, ambient intelligence, or, more recently, everyware.

Chapter 8
Statistical Models in Bioinformatics

Stelios Zimeras
University of Aegean, Greece

Anastasia N. Kastania
Athens University of Economics & Business, Greece

ABSTRACT

In recent years, biological research has been witness of a sea change mainly spearheaded by the advent of novel high throughput technologies that can provide unprecedented amounts of valuable data. This has given rise to novel field sharing the popular suffix 'omics'. Genomics/transcriptomics, proteomics, metabolomics, interactomics/regulomics and numerous other terms have been coined to categorize this ever increasing number of new fields. Biomarkers comprise the most critical tools for the early detection, diagnosis, prognosis and prediction of diseases providing key clues for drug development processes. A significant challenge is to define appropriate levels of specificity and sensitivity of new biomarkers in detecting complex diseases. The establishment of new biomarkers is not only an issue of optimizing wet lab experiments but also of designing appropriate and robust data analysis methods. Various approaches, like multivariate analysis methods as well as standard statistical tests have been applied to search for the important features in 'omics' data. Likewise, several methods, e.g. FDA, SVM, CART, nonparametric kernels, kNN, boosted decision stump and genetic algorithms, have been reported. However, it still remains an unsolved challenge to analyze and interpret the enormous volumes of 'omics' data.

INTRODUCTION

Bioinformatics derives knowledge from computer analysis of biological data. These can consist of the information stored in the genetic code, but also experimental results from various sources, patient statistics, and scientific literature (Figure 1) (Makalowski, 2009). Research in bioinformatics includes method development for storage, retrieval, and analysis of the data.

A statistician – bioinformatician uses a collection of statistical methods for dealing with large biological data sets. In a computer science department - bioinformatics is the marriage of Computer

DOI: 10.4018/978-1-60566-768-3.ch008

Figure 1. Bioinformatics as interaction with wide areas of subjects

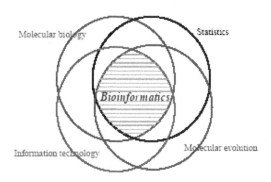

Science and Molecular Biology. An artificial intelligence researcher – bioinformatician uses the application of machine learning to biological data. A physicist – bioinformatician uses a collection of methods to solve a protein structure.

Bioinformatics is a rapidly developing branch of biology and is highly interdisciplinary, using techniques and concepts from informatics, statistics, mathematics, chemistry, biochemistry, physics, and linguistics. It has many practical applications in different areas of biology and medicine.

The history of computing in biology goes back to the 1920s when scientists were already thinking of establishing biological laws from data analysis by induction (Lotka, 1925). Practical applications of bioinformatics are readily available through the World Wide Web, and are widely used in biological and medical research.

Although bioinformatics is a new term developed in the early 1990s, bioinformatics research started before 1970. Over the past four decades, bioinformatics emerged gradually from a hardly noticeable area to a mainstream discipline in science (Ouzounis and Valencia, 2003).

Analyses in bioinformatics focus on three types of datasets: genome sequences, macromolecular structures, and functional genomics experiments (e.g. expression data, yeast two–hybrid screens). But bioinformatics analysis is also applied to various other data, e.g. taxonomy trees, relation-

ship data from metabolic pathways, the text of scientific papers, and patient statistics. A large range of techniques are used, including primary sequence alignment, protein 3D structure alignment, phylogenetic tree construction, prediction and classification of protein structure, prediction of RNA structure, prediction of protein function, and expression data clustering. Algorithmic development is an important part of bioinformatics, and techniques and algorithms were specifically developed for the analysis of biological data. A number of popular software packages and servers developed in the 1990s are widely used, as indicated by their large numbers of citations (see Figure 2) (Dong Xu, et.al, 2009; Baxevanis and Ouellete, 2001; Higgins and Taylor, 2000).

National health organizations in most developed nations allocate a good share of their public research funding to Bioinformatics, often establishing self-contained Bioinformatics institutions (Figure 3).

Regardless of the definitions, the scope of bioinformatics could address all bio-related issues, the current scope of bioinformatics is mainly at the biomolecular level, particularly on macro-molecules (DNA, RNA, and proteins), biological complexes/modules involving a group of genes/proteins, and biomolecular networks/pathways that control various interactions among genes/proteins. A demanding task for bioinformatics is to extract useful biological information and pat-

Figure 2. Popular bioinformatics packages (The number of journal citations was based on the "ISI Web of Knowledge" (http://nadc.isiknowledge.com) on August 4, 2006.)

Name	Functionality	URL	Reference	Citations
BLAST	Pairwise sequence alignment	http://www.ncbi.nlm.nih.gov/BLAST/	Altschul *et al.*, 1990	20,495
CLUSTAL-W	Multiple sequence alignment	http://www.ebi.ac.uk/clustalw/	Thompson *et al.*, 1994	18,837
SignalP	Signal peptide prediction	http://www.cbs.dtu.dk/services/SignalP/	Nielsen *et al.*, 1997	3002
DALI	Protein structure comparison	http://www.ebi.ac.uk/dali/	Holm and Sander, 1993	g
MODELLER	Protein tertiary structure prediction	http://www.salilab.org/modeller/	Sali and Blundell, 1993	1817
PHD	Protein secondary structure prediction	http://www.predictprotein.org/	Rost and Sander, 1993	1795
SEQUEST	Protein identification using mass-spec data	http://fields.scripps.edu/sequest	Eng *et al.*, 1994	1324
MFOLD	RNA secondary structure prediction	http://www.bioinfo.rpi.edu/applications/mfold/	Mathews *et al.*, 1999	1228
PHRED	DNA sequencing	http://www.phrap.org/	Ewing *et al.*, 1998	1162
GENESCAN	Gene identification in DNA	http://genes.mit.edu/GENSCAN.html	Burge and Karlin, 1997	1139

Figure 3. Worldwide Bioinformatics Centres (Martin Frith)

North America

- Portal to bioinformatics activities in Canada
- University of Washington Department of Genome Sciences
- LBNL Genome Sciences Department
- Bioinformatics at Stanford University
- UCSC Center for Biomolecular Science & Engineering
- UCSD Bioinformatics Program
- Washington U. St. Louis Center for Computational Biology
- Whitehead Institute / MIT Center for Genome Research
- Boston University Bioinformatics
- Harvard-Lipper Center for Computational Genetics
- Bioinformatics at Rensselaer and the Wadsworth Center
- Columbia Genome Center
- Genome Resources at Cold Spring Harbor Laboratory
- Penn. State Center for Comparative Genomics and Bioinformatics
- U. Pennsylvania Center for Bioinformatics
- The Institute for Genomic Research
- National Center for Biotechnology Information
- Nitrogen Fixation Research Center, National Autonomous University of Mexico

Australia / Oceania

- U. Queensland Institute for Molecular Bioscience

Africa

- South African National Bioinformatics Institute

Asia

- Weizmann Institute Bioinformatics & Biological Computing
- Institute of Cytology and Genetics, Siberia
- University of Pune Bioinformatics Center
- National University of Singapore Bioinformatics Centre
- Genome Institute of Singapore
- Keio University Institute for Advanced Biosciences
- University of Tokyo Human Genome Center
- Computational Biology Research Center, Tokyo
- RIKEN Genomic Sciences Center, Yokohama
- Center for Information Biology and DNA Data Bank of Japan
- Kyoto University Bioinformatics Center

Europe

- Virtual Institute of Bioinformatics (Éire)
- European Bioinformatics Institute
- Sanger Institute
- Research Group on Biomedical Informatics, IMIM, Barcelona
- Karolinska Institute Center for Genomics and Bioinformatics
- Technical University of Denmark Center for Biological Sequence Analysis
- Helix Group, Lyon
- Structural and Genetic Information Laboratory, Marseille
- Bielefeld University Center for Biotechnology
- EMBL Heidelberg
- GSF, München
- Max Planck Institute for Molecular Genetics
- Swiss Institute of Bioinformatics

terns from noisy data produced by high-throughput technologies. For example, one can compare sequences of multiple genomes to identify interesting evolutionary patterns. Analyzing microarray data can lead to the discovery of the genes that are associated with a particular disease (Dong Xu, et.al, 2009).

As massive biological data have become fundamentally important resources during discovery of new biological knowledge, a key task for bioinformatics is to identify meaningful information (or statistically significant patterns) from data and correlate such information with biological knowledge. However, such a task is highly challenging in many cases: (1) the data size is large with high dimensionality, with a complexity much higher than those typically handled by traditional computational sciences; (2) the information-rich data are heterogeneous in nature, noisy, and incomplete, as well as containing misleading outliers; and (3) biological systems, due to adaptability, evolution, redundancy, robustness, and emergence, are extremely complex. Many biological data are generated by biological processes which are not well understood. Interpretation of such data requires discovery of convoluted relationships hidden in the data.

The following are some most notable applications of computational and statistical methods in bioinformatics (Dong Xu, et.al, 2009):

1. Dynamic programming
2. Neural networks
3. Hidden Markov Models
4. Hypothesis test
5. Bayesian statistics
6. Clustering
7. Sampling search (Gibbs, Monte Carlo, etc)
8. Maximum likelihood methods
9. Information theory
10. Support Vector Machines

One of the main challenges of the 21st century biology will be to integrate data from several of the above mentioned sources such as genomic, proteomic, and metabolomic information to give a more complete picture of living organisms. Neural Networks: Artificial Neural Networks, often just called Neural Networks (NN), are modeled on biological neural networks and represent a major branch of Artificial Intelligence. A great number of bibliographic sources cover network architectures and training algorithms. Emphasis should be given in all approach to performance evaluation of neural learning, testing and evaluation; 'omics' neural networks can be developed for new complex diseases biomarkers detection.

METHODOLOGY IN BIOINFORMATICS

Genomics is the study of the entire genome of an organism. Here, we are mostly concerned with its subfield known as functional genomics which mainly involves patterns of gene expression during various conditions. Another subfield, known as *transcriptomics*, deals with all messenger RNA (mRNA) molecules, or transcripts, produced in a

Figure 4. DNA base pairing

cell or population of cells (Figure 4). Unlike the genome, which is roughly fixed for a given cell line (excluding mutations), the transcriptome can vary with external environmental conditions and can be influenced by phenomena such as mRNA degradation via transcriptional attenuation. The main source of data for functional genomics and transcriptomics are cDNA microarrays, where mRNA is converted to cDNA and this is in turn hybridized with thousands of the genes of the under question organism, revealing the expression profile of certain types of cells under certain conditions possible over time. Today, new genomics experimental assays, such as gene expression microarrays, are generating data for thousands of genes simultaneously (Hardiman G., 2006; Mount, 2001)

Most papers today report experimental results on a single gene or protein, though more and more large-scale experiments are being reported. Increasingly, the scientific literature is becoming available online in electronic format, raising the possibility of facile computational analysis.

PubMed abstracts are increasingly available for most biologically relevant articles (Figure 5), while electronic publishers such as High-Wire press and PubMed Central permit access to full text articles (Hutchinson 1998; Roberts 2001).

Automatic DNA sequencers allow high-throughput reading of fragments of DNA sequences with lengths of several hundred base pairs. The lengths of the genomes of organisms are much longer. Therefore, the basic strategy used in genome sequencing, called shotgun sequencing, involves assembling long DNA sequences from short overlapping pieces. In shotgun sequencing, large numbers of overlapping DNA fragments several hundred base pairs long are read randomly from the basic DNA strand and their sequences are recorded. Then on the basis of the overlaps between the reads, the whole DNA sequence is reconstructed. This strategy is illustrated in Figure 6 (Polanski and Kimmel, 2007)

Proteomics is the large-scale study of proteins, particularly their structures and functions, while the proteome of the organism corresponds to the

Figure 5. Popular molecular biology journals

	Journal	Total Articles 2003
1	American Journal of Human Genetics	330
2	Cancer Cell	134
3	Cancer Research	1311
4	Cell	356
5	Developmental Cell	208
6	European Molecular Biology Organization Journal	653
7	Genes and Development	288
8	Genome Research	291
9	Human Molecular Genetics	390
10	Immunity	173
11	Journal of Cell Biology	457
12	Journal of Experimental medicine	493
13	Molecular Cell	325
14	Molecular Cell Biology	803
15	Nature	2408
16	Nature Cell Biology	238
17	Nature Genetics	308
18	Nature Immunology	236
19	Nature Medicine	412
20	Nature Neuroscience	264
21	Nature Structure Biology	210
22	Neuron	421
23	Plant Cell	256
24	Proceedings of the National Academy of Science USA	2955
25	Science	2377

Figure 6. Illustration of the shotgun method sequencing of a DNA strand

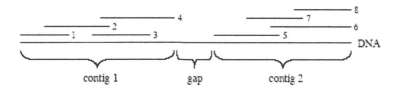

entirety of proteins in existence in an organism. in existence in an organism throughout its life cycle. Thus, proteomics is the study of the composition, structure, function, and interaction of the proteins directing the activities of each living cell. Proteome analysis is rendered particularly challenging because of a number of issues. First, gene transcription levels often do not correlate exactly with gene expression, since mRNA may be degraded rapidly, or translated inefficiently, so the amount of protein produced is reduced. Second, many proteins experience posttranslational modifications, such as phosphorylation, that have a profound effect on their activities. Furthermore, many transcripts give rise to more than one protein, through alternative splicing or alternative post-translational modifications. Finally, many proteins form complexes with other proteins or RNA molecules, and they only function in the presence of these other molecules. The current technological advances in the proteomics field such as the 2D gel electrophoresis and once proteins are separated and quantified, they can via mass spectrometry, specifically matrix-assisted laser desorption-ionization time-of-flight (MALDI-TOF) mass spectrometry system (Mount, 2001)

Proteins are molecules with complicated three-dimensional structures, but they always have an underlying linear chain of amino acids as their primary structure. On the basis of criteria, such as functional aspects the activity of a protein or the part of an organism or compartment of a cell where a protein appears as a building element, proteins are classified into protein families. An example of a protein, the enzyme trypsin, obtained by using information from the Protein Data Bank (PDB) and internet accessible molecular graphics program Ras Mol, is shown in Figure 7.

Metabolomics is the study of the unique chemical fingerprints that specific cellular processes leave behind especially, the study of their small molecule metabolite profiles. The term metabo-

Figure 7. Graphical presentation of the enzyme trypsin obtained with the use of spatial coordinates of atoms from Protein Data Bank (accession symbol 2ptn), and the molecular-graphics program Ras Mol

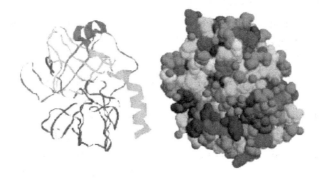

lome refers to the complete set of small molecule metabolites (such as metabolic intermediates, hormones and other signaling molecules) to be found within a biological sample, such as a single organism. Metabolite analysis includes separation and detection of the analytes (Fell, 1996). For separation, gas chromatography, especially when interfaced with mass spectrometry (GCMS), is one of the most widely used methods. However, some large and polar metabolites cannot be analyzed by GC and need therefore be analyzed by high performance liquid chromatography (HPLC), which has lower chromatographic resolution, but can separate a much wider range of analytes. The last option for analysis is capillary electrophoresis.

Interactomics/Regulomics: Interactome is the whole set of molecular interactions in cells, while the term regulome refers to the whole set of regulation components in a cell. Hence, the regulome is a subset of the interactome and includes components such as genes, mRNAs, proteins, and metabolites. Interactomics includes the interplay of regulatory effects between these components, and their dependence from variables such as subcellular localization, tissue, developmental stage, and pathological state, thus forming complex interaction networks. In regulomics, transcription factors play a major role in affecting gene expression. Other proteins that bind to transcription factors to form transcriptional complexes might modify their activity. The same effect can be achieved through signaling pathways, where a signal affecting gene expression is transmitted through a path involving more than two participants. Interactomics / regulomics take advantage of highthroughput technologies which allow measuring the levels of thousands of biological components such as mRNAs, proteins, or metabolites. Chromatin immunoprecipitation of transcription factors can be used to map transcription factor binding sites in the genome. Other methods for mapping the interactome include affinity purification, mass spectrometry and the yeast twohybrid method. Regarding detection methods, mass spectrometry

(MS) is used to identify and to quantify metabolites after separation by GC, HPLC, or CE, while nuclear magnetic resonance (NMR) spectroscopy can detect metabolites without prior separation at the price of much lower sensitivity. Technologies to measure endogenous metabolites accurately and quantitatively already exist, as reported by (Newman and Mason, 2006; Castrillo et. al., 2003)

STATISTICAL MODELING IN BIOINFORMATICS

The aim of this work is to provide different approaches to complex diseases biomarker prediction and classification using advanced statistical techniques.

A large number of algorithms make use of these theories in biological sequences, which have been developed to deal with the large amounts of biological and clinical data that play such an important role in the fields of bioinformatics, biomedicine, and so on. The main goals in the analysis are: (a) to introduce mathematical methods as they are used in the fields of molecular biology and bioinformatics, (b) to discuss the potential mathematical requirements in the study of molecular biology and bioinformatics, which will drive the development of new theories and methods and (c) to propose a framework within which bioinformatics may be combined with mathematics.

Generally the following description of a biological sequence could be introduced: A= {a_1, a_2, a_3,..., a_m) and B= {b_1, b_2, b_3,..., b_n), where the capital letters A,B represent the sequences, and a_i, b_i represent the basic units of the sequence, at positions i, whose elements are obtained from the set $\Omega = V_q = \{0, 1, \cdots, q-1\}$, m, n are the lengths of sequences A, B.

An alignment of two sequences is obtained by first appropriately inserting spaces (which we represent with gaps or dashes −), either into or at the ends of the two sequences, and then placing the two resulting sequences one above the other so that every character or space in one sequence is aligned with a character or space in the other sequence. Practically, two sequences are aligned by writing them across a page in two rows. Identical or similar characters are placed in the same column, and no identical characters can either be placed in the same column as a mismatch or opposite a gap in the other sequence (e.g. given two words A = accggta, B = aggctg then one possible alignment are accgg – ta and a – –ggctg). Organisation of the sequences could be achieved applying Suffix Tries and Suffix Trees. A suffix trie (or suffix tree) for a string S is a search trie (or compact search trie) as in Figure 8 constructed for all suffixes of the string S. A suffix of S is a trailing part of S. If we employ more a expanded notation for string of length n, namely S(1: n), where the range of the indices of the characters of the string, 1: n, is included, then a suffix of S is every substring of the form S(i: n), $1 \leq i \leq n$. Consider a string S = CACTAACTGA, defined over the alphabet of letters A, C, G, T, which can symbolize the nucleotides in DNA (Polanski and Kimmel, 2007)

In molecular biology, when consider nucleotide sequences, we have N = 4, Ω = V4 = {a, c, g, t}. Each site xi in a DNA sequence of length n x = x_1 . . . x_n is one of the four bases {a, g, c, t}, chosen randomly to some probabilities π_a, π_g, π_c, π_t. These probabilities follow $\pi_a + \pi_g + \pi_c + \pi_t = 1$, since each of the bases is certain to appear. The initial vector $p_0 = (\pi_a, \pi_g, \pi_c, \pi_t)$ describes the ancestral base distribution. Specifically, fix two DNA sequences, where the first one S_0 is called an ancestral sequence and the other the descendent after one time step,

S_0 = acttgtcggatgatcagcggtccatgcacctgacaacggt,

S_1 = acatgttgcttgacgacaggtccatgcgcctgagaacggc.

The mutation process over one time step maybe modelled, assuming that only base substitutions can occur - no deletions, insertions, or inversions are considered. The 16 conditional probabilities of observing a base substitution could be defined $p_{i,j} := P[S_1 - i \mid S_0 = j]$ where i, j = a, g, c, t. The $p_{i,j}$ form a Markov matrix P with ordering a, g, c, t,

$$P = [p_{i,j}] = \begin{bmatrix} p_{a,a} & p_{a,g} & p_{a,c} & p_{a,t} \\ p_{g,a} & p_{g,g} & p_{g,c} & p_{g,t} \\ p_{c,a} & p_{c,g} & p_{c,c} & p_{c,t} \\ p_{t,a} & p_{t,g} & p_{t,c} & p_{t,t} \end{bmatrix},$$

Figure 8. Suffix trie and Suffix tree for the string S = CACTAACTGA

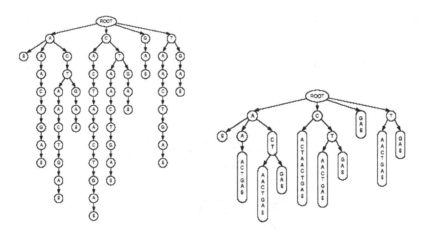

in each column of the matrix are entries referring to the same ancestral base, and in each row are entries referring to the same descendent base. Then all these probabilities can be estimated from the data. Data in the frequency array A of size 4 by 4 can be found by counting frequencies of bases in ancestral sequence S_0 to the columns, then split frequencies to bases of the descendent S_1 along each column. Next, computing $Pp_0 = p_1$, called the vector of probabilities for various bases occurring in the sequence S_1. By the addition rule, the probability that each site of descendent S_1 receives a base j is the probability of the union of the four events:

$$P\left[S_1 = j\right] = \sum_i p_{i,j} \pi_i$$

Calculating transition probabilities p_{ij} easily could be illustrated as a connected graph (Figure 9) (Isaev, 2006).

Hidden Markov Models (HMMs) can be used for searching in a way similar to that for Markov models, with the purpose of detecting potential membership of an unannotated sequence in a sequence family used as training data. An HMM is an ordinary discrete-time finite Markov chain (that we always assume to be non-trivially connected) with states $G_1 \ldots G_{N,}$ transition probabilities p_{0j}, p_{ij}, p_{j0}, i, j = 1, . . ., N, that, in addition, at each state emits symbols from an alphabet Q (the DNA alphabet in our examples). For each state G_k and

each symbol a ∈ Q an emission probability q_{Gk} (a) = q_k (a) is specified. For every k = 1. . . N, the probabilities q_k (a) sum up to 1 over all a ∈ Q (Figure 10) (Birney, 2001).

The Markov chain will be referred to as the underlying Markov chain of the HMM. The two-block DNA model can be thought of as an HMM where at each state one letter is emitted with probability 1 and the rest of the letters are emitted with probability 0. As in the case of Markov chains, any *a priori* connectivity defines a family of HMMs parameterized by the transition probabilities corresponding to the arrows present in the connectivity graph (subject to constraints dictated by the condition of non-trivial connectedness) and the emission probabilities. Analogously to Markov chains, HMMs are often derived as *models* for particular *training data*, from which the transition and emission probabilities can be *estimated*. This process of parameter estimation is more complicated than that for Markov chains. Let $x = x_1 \ldots x_L$ be a sequence of letters from Q and $\pi = \pi_1 \ldots \pi_L$ be a path of the same length. We will now define the probability $P(x, \pi)$ of the pair (x, π) as follows $P(x, \pi) = p_0\pi_1 q\pi_1 (x_1) p\pi 1\pi 2 q\pi 2 (x_2) * \ldots * p\pi_{L-1}\pi_L q\pi_L (x_L) p\pi L$ 0. An HMM will consist of the following five components: (1) A set of N states $S_1, S_2, \ldots, S_N$., (2) An alphabet of M distinct observation symbols A = {$a_1, a_2, \ldots, a_M$}.(3) The transition probability matrix P = (p_{ij}), where p_{ij} = Prob($q_{t+1} = S_j \mid q_t = S_i$), (4) The emission probabilities: For each state S_i and a in

Figure 9. Connectivity graph between DNA sequence based on Markov chain modelling

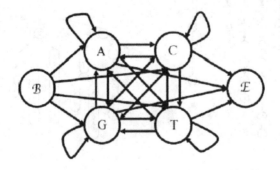

Figure 10. A graphical model representation for the standard Hidden Markov Model. The set of hidden (latent) states St form a sequence which evolves under a first-order Markov process. Each state generates an observation, Y_t, according to its observation model.

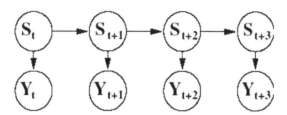

A, $b_i(a)$ = Prob(S_i emits symbol a).(5) An initial distribution vector $\pi = (\pi_i)$, where π_i = Prob(q_1 = S_i) (Warren Ewens and Gregory Grant, 2005)

In neural networks supervised learning, training algorithms are devised to minimize the prediction error made by the network by adjusting the network weights and thresholds. The training process aims at fitting the model represented by the network to the training data available. The most popular example of a neural network training algorithm is back propagation. Back propagation is the easiest algorithm to understand and has advantages in some circumstances, although modern secondorder algorithms such as conjugate gradient descent and LevenbergMarquardt are substantially faster for many problems. Furthermore, there are heuristic modifications of back propagation which work well for some problem domains, such as DeltaBarDelta and quick propagation. In supervised learning, the training data set contains cases featuring input variables together with the associated outputs. The initial weights are randomly chosen. The training cases are then submitted to the network, target and actual outputs are compared and the error is calculated. This error, together with the error surface gradient, is used to adjust the weights. Weight adjustment can happen immediately after each pattern is presented to the network or wait until a full iteration of the training set is presented (epoch training). If a given number of epochs has elapsed or the error has reached an acceptable level or the error has stopped improving, the training stops and the network is ready to be tested. Testing is a similar process, only simpler; the forward pass is only performed since no learning takes place. Unsupervised learning: In unsupervised neural learning the training data set contains only input variables. The self organizing network has only two layers: the input layer, and an output layer known as the topological map layer. The units in the topological map layer are laid out in space typically in two dimensions. The self organizing network then attempts to learn the structure of the data, using an iterative algorithm. Starting with an initially random set of radial centers, the algorithm gradually adjusts them to reflect the clustering of the training data. The iterative training procedure also arranges the network so that units representing centers close together in the input space; are also situated close together on the topological map. This technique can be useful in allowing the user to visualize data which might otherwise be impossible to understand. The self organizing network is most commonly used for exploratory data analysis and novelty detection. In the first case, the network can learn to recognize clusters of data and relate similar classes to each other. The user can develop an understanding of the data, which is then used to refine the network. The network becomes capable of classification tasks, as the classes of data that are recognized can be labeled. Such networks can also be used for classification when output classes are immediately available the advantage in this case is their ability to highlight similarities between classes. Another use of the self organizing network is in novelty detection. The network can learn to recognize clusters in the training data, and respond to it. If new data, unlike previous cases, is encountered, the network fails to recognize it and this indicates novelty (Wu and McLarty, 2000).

A typical neural network (shown in Figure 11) is composed of input units $X_1, X_2,...$ corresponding to independent variables, a hidden layer known as the first layer, and an output layer (second layer) whose output units $Y_1,...$ correspond to dependent variables (expected number of accidents per time period) (Wu C. H. and McLarty J., 2000; Rabindra Ku et. al, 2009)

In between are hidden units H_1, H_2, ... corresponding to intermediate variables. These interact by means of weight matrices $W^{(1)}$ and $W^{(2)}$ with adjustable weights. The values of the hidden units are obtained from the formulas: $H_j = f\left(\sum_k W^{(1)}_{jk} X_k\right)$ and $Y_i = f\left(\sum_j W^{(2)}_{ij} H_j\right)$, with proposed function $f = 1/{1 + e^{-u}}$. The accuracy of the estimated output is improved by an iterative learning process in which the outputs for various input vectors are compared with targets (observed frequency of accidents) and an average error term E is computed: $E = \frac{1}{N} \sum_{i=1}^{N} \left(Y^{(n)} - T^{(n)}\right)^2$, where N is the number of highway sites or observations, $Y^{(n)}$ is the estimated number of accidents at site n for $n = 1, 2,..., N$ and $T^{(n)}$ is the observed number of accidents at site n for $n = 1, 2,..., N$. After one pass through all observations (the training set), a gradient descent method may be used to calculate improved values of the weights $W^{(1)}$ and $W^{(2)}$, values that make E smaller. After revaluation of the weights with the gradient descent method, successive passes can

be made and the weights further adjusted until the error is reduced to a satisfactory level. The computation thus has two modes, the mapping mode, in which outputs are computed, and the learning mode, in which weights are adjusted to minimize E. Although the method may not necessarily converge to a global minimum, it generally gets quite close to one if an adequate number of hidden units are employed. The most delicate part of neural network modelling is generalization, the development of a model that is reliable in predicting future accidents. Over fitting (i.e., getting weights for which E is so small on the training set that even random variation is accounted for) can be minimized by having two validation samples in addition to the training sample. According to Smith and Thakar (Rabindra Ku et. al, 2009), the data set should be divided into three subsets: 40% for training, 30% to prevent over fitting, and 30% for testing.

This task will be achieved by: (a) Decomposition of the individual model response to its components and a systematic evaluation of the correlations between the related 'omics' input data and the output biomarkers. (b) Identification of the common (if any) components between the obtained models. This will provide useful insight in terms of the complementarily of the evolutionary computing approaches used. (c) Analysis of the model complexity versus obtained performance with training data and generalization ability. It is envisaged that the required mapping could be sufficient performed using more than one model in each category /approach. This step will permit the systematic treatment and understanding of each model suitability and 'fit' to the problem under investigation. (d) Investigation of the robustness of the acquired computational models in the presence of noise.

Complex diseases biomarkers investigation with fuzzy rule based systems: It is finally proposed that new methods will be developed, which will combine the powerful computational model of a Machine Learning approach, which is able

Figure 11. A simplified artificial neural network (Rabindra Ku et. al, 2009)

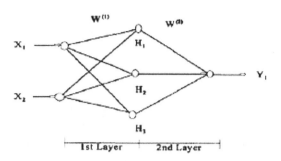

to solve the problem of combinatorial explosion, with the transparent classification process of Fuzzy Rule based Inference System. This can be accomplished using a support vector learning method for the construction of a fuzzy rule based pattern classification system. Instead of the standard SVM model as candidate Machine Learning Model, it is proposed that more recently proposed models e.g. Knowledge Based SVMs, SVM with Recursive Feature Elimination, Bayesian SVMs or Multicategory Proximal SVM are used (Bishop, 1994; 1995)

Pattern classification typically is a supervised process where, based on set of training samples with known classifications, a classifier is derived that performs automatic assignment to classes based on unseen data. Let us assume that our pattern classification problem is an n-dimensional problem with C classes (in microarray analysis C is often 2) and m given training patterns $x_p = (x_{p1}, x_{p2}, \ldots, x_{pn})$, $p = 1, 2, \ldots, m$. Without loss of generality, we assume each attribute of the given training patterns to be normalised into the unit interval $[0,1]$; that is, the pattern space is an n-dimensional unit hypercube $[0,1]^n$ with fuzzy rule:

Rule R_j: If x_1 is A_{j1} and $\ldots$ and x_n is A_{jn}

then Class C_j with CF_j, $j = 1, 2, \ldots, N$,

where R_j is the label of the j-th fuzzy if-then rule, $A_{j1}, \ldots, A_{jn}$ are antecedent fuzzy sets on the unit interval $[0,1]$, C_j is the consequent class (i.e. one of the C given classes), and CF_j is the grade of certainty of the fuzzy if-then rule R_j (Figure 12) (Schaefer et. al., 2008)

Detailed insilicovalidation of the identified complex diseases biomarkers: The results of the above described methodology can be evaluated with the usage of bioinformatics tools both publicly available and in-house built in the light of biomedical information and will be prioritized according to their importance (Figure 13)

CONCLUSION

Bioinformatics is a rapidly developing branch of biology and is highly interdisciplinary, using techniques and concepts from informatics, statistics, mathematics, chemistry, biochemistry, physics, and linguistics. It has many practical applications in different areas of biology and medicine. Analyses in bioinformatics focus on three types of datasets: genome sequences, macromolecular structures, and functional genomics experiments (e.g. expression data, yeast two–hybrid screens).

The aim of this work was to provide different approaches to complex diseases biomarker prediction and classification using advanced statistical techniques. 'Omics' Bayesian networks combined

Figure 12. Fuzzy rule

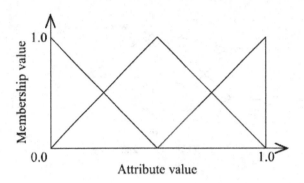

Figure 13. Applications of the neural networks in bioinformatics

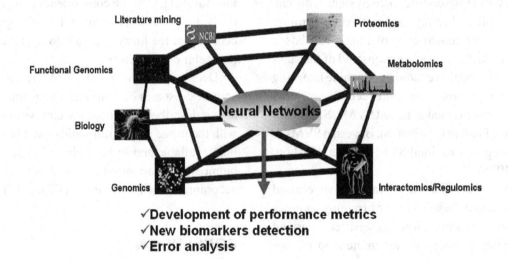

✓Development of performance metrics
✓New biomarkers detection
✓Error analysis

with Markov chains models were presented for new complex diseases biomarkers detection. Neural Networks can also be effectively adopted. A general research methodology related with neural networks is proposed with the following parts: (a) 'Omics' neural models development, analysis and interpretation using various neural network methodologies (b) Development of performance metrics for complex diseases biomarker investigation with fuzzy rules based systems. (c) Detailed in-silico validation of the identified complex diseases biomarkers. The described methodology is expected to be implemented in different complex diseases types with ultimate goal the identifica-

tion of the highest quality of complex diseases biomarkers which is of tremendous importance worldwide. Ultimately, unnecessary costs will be saved, patient suffering will be minimized, personalized health care will be sought and increase of quality of life will be achieved.

REFERENCES

Baxevanis, A. D., & Francis Ouellete, B. F. (2001). *Bioinformatics: Practical Guide to the Analysis of Genes and Proteins*. Hoboken, NJ: Wiley Interscience Press.

Birney, E. (2001). Hidden Markov Models in Biological Sequence Analysis. *IBM Journal of Research and Development, 45*(3/4).

Bishop, C. M. (1994). Novelty Detection and Neural Network Validation. *IEE Proceedings. Vision Image and Signal Processing, 141*, 217–222.

Bishop, C. M. (1995). *Neural Networks for Pattern Recognition*. Oxford, UK: Clarendon Press.

Castrillo, J. I., Hayes, A., Mohammed, S., Gaskell, S. J., & Oliver, S. G. (2003). An optimized protocol for metabolome analysis in yeast using direct infusion electrospray mass spectrometry. *Phytochemistry, 62*, 929–937.

Clote, P. (2000). *Computational Molecular Biology*. Hoboken, NJ: Wiley.

Des Higgins & Taylor. W. (2000). *Bioinformatics: Sequence, structure, and databanks*. Oxford, UK: Oxford University Press.

Ewens, W., & Grant, G. (2005). Statisical Methods in Bioinformatics: An Introduction. Berlin: Springer Science+Business Media, Inc.

Fell, D. (1996). *Understanding the Control of Metabolism, (Frontiers in Metabolism series)*. Aldershot, UK: Ashgate Publishing.

Frith, M. (n.d.). *Worldwide Bioinformatics Centres*. Retrieved from http://zlab.bu.edu/~mfrith/BioinfoCenters.html

Hardiman, G., (2006). Microarrays Technologies. an overview. *Pharmacogenomics, 7*(8, December), 1153-8.

Hutchinson, D. (1998). Medline for health professionals: how to search PubMed on the Internet. Sacramento, CA: New Wind.

Isaev, A. (2006). *Introduction to Mathematical Methods in Bioinformatics*. Berlin: Springer-Verlag.

Jena, R. K., Aqel, M. M., Srivastava, P., & Mahanti, P. K. (2009). Soft Computing Methodologies in Bioinformatics. *European Journal of Scientific Research, 26*(2), 189–203.

Lotka, A. J. (1925). *Elements of physical biology*. Baltimore: Williams and Wilkins.

Makalowski, Wojciech (n.d.). *web site notes*. Retrieved from http://www.compgen.uni-muenster.de/

Mount, D. W. (2001). Bioinformatics: Sequence and genome Analysis. Cold Spring Harbor, NY: CSHL Press.

Newman, K., & Mason, R. S. (2006). Organic mass spectrometry and control of fragmentation using a fast flow glow discharge ion source. *Rapid Communications in Mass Spectrometry, 20*(14), 2067–2073.

Ouzounis, C. A., & Valencia, A. (2003). Early bioinformatics: the birth of a discipline— a personal view. *Bioinformatics (Oxford, England), 19*(17), 2176–2190.

PDB. *protein data bank* (n.d.). Retrieved from http://www.rcsb.org/pdb/

Polanski, A., & Kimmel, M. (2007). *Bioinformatics*. Berlin: Springer-Verlag.

Ras Mol (n.d.). molecular visualization freeware. Retrieved from http://www.umass.edu/microbio/rasmol/

Roberts, R. J. (2001). PubMed Central: The GenBank of the published literature. *Proceedings of the National Academy of Sciences of the United States of America, 98*(2), 381–382.

Schaefer, G., Nakashima, T., & Yokota, Y. (2008). Fuzzy Classification for Gene Expression Data Analysis. In A. Kelemen, A. Abraham, & Y. Chen (Eds.), *Computational Intelligence in Bioinformatics*. Berlin: Springer-Verlag.

Sen, P. K. (2002). Computational sequence analysis: Genomics and statistical controversies. In Y. P. Chaubey (ed.), *Recent Advances in Statistical Methods* (pp. 274-289). London: World Scien. Publ.

Wu, C. H., & McLarty, J. (2000). *Neural Networks and Genome Informatics*. New York: Elsevier Science.

Xu, D., Keller, J. M., Popescu, M., & Bondugula, R. (2009). Applications of Fuzzy Logic in Bioinformatics. *Series on Advances in Bioinformatics and Computational Biology,* (Vol. 9).

Chapter 9

Computational Analysis and Characterization of Marfan Syndrome Associated Human Proteins

K. Sivakumar

Sri Chandrasekharendra Saraswathi Viswa Maha Vidyalaya University, India

ABSTRACT

Novel computational procedures and methods have been used to analyze, characterize and to provide more detailed definition of some Marfan syndrome associated human Fibrillin 1 proteins retrieved from NCBI Entrez protein database. Primary structure analysis reveals that the Marfan syndrome associated proteins are rich in cysteine and glycine residues. Extinction Coefficients of Marfan syndrome associated proteins at 280nm is ranging from 1490 to 259165 M^1 cm^{-1}. Expasy's ProtParam classifies most of the Marfan syndrome associated human Fibrillin 1 proteins as unstable on the basis of Instability index (II>40) and few proteins (AAB25244.1, 1EMO_A, Q504W9) as stable (II<40) proteins in the room temperature. The aliphatic index infers that the Fibrillin 1 proteins may become unstable at high temperature. GRAVY index of all the proteins indicates that all these proteins may interact equally and easily with water. The number of basic and acidic amino acids in each Marfan syndrome associated human Fibrillin 1 proteins correlates well with the corresponding pI computed. Secondary structure analysis shows that human Fibrillin 1 proteins are found to be with mixed secondary structural content. The average molecular weight of Marfan syndrome associated proteins calculated is 134086 Da. Scanprosite server identified EGF-like domain, TGF-beta binding domain and extracellular sushi domain profiles in Marfan syndrome associated proteins.

INTRODUCTION

Marfan syndrome (OMIM, 2000) [OMIM Number 154700 - http://www.ncbi.nlm.nih.gov/sites/ entrez?db=omim] is a systemic, rare connective tissue disorder that affects primarily the bones, joints, heart, blood vessels and the eyes. The main characteristic features of the Marfan syndrome are tall and thin stature with long arms, slender legs, elongated fingers and toes disproportionately long

DOI: 10.4018/978-1-60566-768-3.ch009

in relation to rest of the body and exceeds the body height. Marfan syndrome often leads to the eye problems where one or both lenses of the eye may dislocate, heart and blood vessel problems that may result in heart murmurs, irregular heartbeat, or in severe cases aortic aneurysm. Marfan syndrome affects numerous other structures and organs including the lungs, eyes, dural sac surrounding the spinal cord, and hard palate. Marfan syndrome affects males and females equally, and the mutation shows no geographical bias. According to National Marfan Foundation (http://www.marfan.org) approximately 1 in 5000 people have Marfan syndrome, including men, women and children of all races and ethnic groups. The parents with Marfan syndrome have high chance (50:50) of passing Marfan syndrome on to a child due to its autosomal dominant nature. Signs and symptoms of Marfan syndrome vary from one person to another, even within the same family. Some people have mild signs and symptoms, while others may have severe problems and discomfort, which occur, in many parts of the body. Genetic mutation studies reveal that the Marfan syndrome is due to the mutations in gene located on chromosome 15, locus 15q, 21.1 (Magenis, R.E., 1991) (http://www.ncbi.nlm.nih.gov/projects/mapview/). This gene is involved in the production and processing of Fibrillin 1 (OMIM, 2008) [OMIM Number 134797 - http://www.ncbi.nlm.nih.gov/sites/entrez?db=omim] protein. Mutations in this gene disrupt the production, processing, or assembly of Fibrillin 1 protein. Many researchers have reported clinical and genetics studies on Marfan syndrome. However, sequence analysis and physicochemical characterization of Marfan syndrome associated human Fibrillin 1 proteins have not been done so far. In this chapter we report the computational analysis and characterization studies on Marfan syndrome associated 14 human Fibrillin 1 proteins retrieved from NCBI Entrez protein database to provide more detailed description of Marfan syndrome associated proteins.

MATERIALS AND METHODS

Protein Sequence Databases

Protein sequence databases consist of protein sequences and information about proteins. These databases consist of protein sequences that have been translated from the nucleotide sequences and also sequenced by methods like N-terminal sequencing. Many protein sequence databases are available today and all of these databases allow free download of full content. Any researcher from all over the world can download these protein sequences to study the properties of encoded proteins and utilize them for healthcare, disease identification, drug discovery and development. Protein sequence databases contain the amino acid sequence of proteins; the constituent amino acids are represented by single letter amino acid code. Short descriptions of various protein databases used in this study are given in the following sections.

UniProtKB/Swiss-Prot

SwissProt (http://www.expasy.org/sprot) is a manually curated protein sequence database that provides a high level of annotation (e.g. the description of the function of a protein, its domain structure, etc), while maintaining a minimal level of redundancy and a high level of integration with other databases. Release 56.5 of (Nov-2008) contained 402,482 entries.

UniProtKB/TrEMBL

UniProtKB/TrEMBL (http://www.uniprot.org/database/knowledgebase.shtml) consists of computer-annotated entries, which are derived from the translation of coding sequences present in the EMBL/GenBank/DBJ Nucleotide Sequence Databases and also protein sequences extracted from the literature or submitted to UniProtKB. Release 39.5 (Nov-2008) contained 6796,837 entries. The Swiss Institute of Bioinformatics and the European Bioinformatics Institute maintain Swiss-Prot and TrEMBL collaboratively.

EMBL

The EMBL (www.ebi.ac.uk/embl/), Nucleotide Sequence Database is Europe's primary nucleotide sequence resource. The main sources of the DNA and RNA sequences in the database are direct submissions from individual researchers, genome sequencing projects and patent applications. Release 96 (Sep-2008) contained 97,703,661 entries.

GenBank

GenBank (http://www.ncbi.nlm.nih.gov/Genbank/) is a comprehensive database that contains publicly available nucleotide sequences for more than 240,000 named organisms, obtained primarily through submissions from individual laboratories and batch submissions from large-scale sequencing projects. GenBank is accessible through NCBI's retrieval system, Entrez, which integrates data from the major DNA and protein sequence databases along with taxonomy, genome mapping, protein structure and domain information. Release 168 (Nov-2008) contained 96,400,790 entries.

NCBI Entrez Protein Database

The National Center for Biotechnology Information, NCBI Entrez protein database (http://www.ncbi.nlm.nih.gov/entrez/) entries have been compiled from a variety of sources, including Swiss-Prot, PIR, PRF, PDB and Translations from annotated coding in GenBank and RefSeq. Protein sequence records in Entrez have links to protein structures, conserved protein domains, nucleotide sequences, pre-computed protein BLAST alignments, genomes and genes.

PDB

The Protein Data Bank, PDB (http://www.pdb.org/pdb/) is the single worldwide archive of structural data of biological macromolecules. The Protein Data Bank was established at Brookhaven National Laboratories in 1971 as an archive for biological macromolecular crystal structures. The PDB distributes coordinate data (3D structure data), structure factor files and NMR constraint files. It provides documentation and derived data. The coordinate data are distributed in PDB format. Release 08-12-08 (Dec-2008) contained 54,559 entries.

PRF

The sequence database of Protein Research Foundation, PRF (http://www.prf.or.jp/en) consists of amino acid sequences of peptides and proteins, including sequences predicted from genes. Sequences not included in EMBL, GenBank (Gene Products Data Bank) and Swiss-Prot is also found in PRF. The latest release 123 (Sep-2008) of PRF contains 1,010,213 entries.

DDBJ

DNA Data Bank of Japan, DDBJ (http://www.ddbj.nig.ac.jp/) was established in 1986. DDBJ has been functioning as one of the International DNA Databases, including European Bioinformatics Institute; responsible for the EMBL database in Europe and NCBI responsible for GenBank database in the USA as the two other members. DDBJ is the sole DNA data bank in Japan, which is officially certified to collect DNA sequences from researchers and to issue the internationally recognized accession number to data submitters. DDBJ provides worldwide many tools for data retrieval and analysis.

PMD

Protein Mutant Database, PMD ((http://pmd.ddbj.nig.ac.jp/) consists of information on functional and / or structural influences brought about by amino acid mutation at a specific position of protein. PMD covers natural as well as artificial mutants, including random and site-directed ones, for all proteins except members of the globin and immunoglobulin families. The PMD is based on literature, not on proteins. i.e, each entry in the database corresponds to one article which may describe one, several or a number of protein mutants. Each database entry is identified by a serial number and is defined as either natural or artificial, depending on the type of the mutation. PMD can be searched using keywords and by submitting protein sequences through similarity search.

Protein Sequence Analysis Tools and Servers

Sequence Retrieval Systems

Sequence Retrieval Systems, SRS is a user-friendly database search interface with flexible search procedure to the various sequence databases. The user can search the databases by submitting their queries (i.e., sequence similarity search) or by key words (i.e., protein name, function etc.). Sequence retrieval systems provide scientist to retrieve protein sequences and other useful details related to the proteins.

BioEdit

BioEdit (http://www.mbio.ncsu.edu/BioEdit/BioEdit.html) is a biological sequence alignment editor written for Windows 95/98/NT/2000/XP. An intuitive multiple document interfaces with convenient features makes alignment and manipulation of sequences relatively easy on desktop computer. Several sequence manipulation and analysis options and links to external analysis programs facilitate a working environment, which allows viewing and manipulating sequences with simple point-and-click operations.

CLC Free Workbench

CLC Free Workbench (http://www.clcbio.com) creates a software environment enabling users to make a large number of bioinformatics analyses, combined with smooth data management, and excellent graphical viewing and output options. The software package is a community edition, available for Windows, Mac OS X, and Linux. CLC Free Workbench is very versatile when it comes to importing and exporting both data and graphics files.

Statistical Analysis of Protein Sequences

The program Statistical Analysis of Protein Sequences, SAPS (http://www.isrec.isb-sib.ch/software/SAPS-form.html) computes extensive physical and chemical statistical information for any submitted protein sequence. The output consists of the results of compositional analysis, charge distribution analysis, including the locations of positively and negatively charged clusters, high-scoring charged and uncharged segments, and charge runs and patterns. The final sections present information on high-scoring hydrophobic and transmembrane segments, repetitive structures and periodicity analysis. Statistically significant sequence features highlighted by SAPS in the input sequence may suggest promising regions for experimental investigation. The program also finds application in the description of conserved features of families of proteins, as well as in the inverse problem of deriving protein groupings based upon sequence features. Short sequences are subject to larger statistical fluctuations than longer sequences. The statistical evaluations of SAPS are reliable only for sequences of at least about 200 residues and more.

Compute pI/Mw

Compute pI/Mw (http://www.expasy.ch/tools/pi_tool.html) is a tool, which allows the computation of the theoretical pI (isoelectric point) and Mw (molecular weight) of a protein.

ProtParam

ProtParam (http://us.expasy.org/tools/protparam.html) tool computes various physico-chemical properties that can be deduced from a protein sequence. No additional information is required about the protein under consideration. The protein can either be specified as a Swiss-Prot/TrEMBL accession number or the protein can be submitted in the form of a raw sequence.

Parameters Computed using ProtParam Tool

The parameters computed using ProtParam include the molecular weight, theoretical pI, amino acid composition, atomic composition, extinction coefficient, estimated half-life, instability index, aliphatic index and Grand average of hydropathy (GRAVY). All the parameters are explained in the following sections in detail.

Extinction Coefficient

The extinction coefficient indicates how much light a protein absorbs at a certain wavelength. It is useful to have an estimation of this coefficient for following a protein with a spectrometer when purifying it. From the molar extinction coefficient of tyrosine, tryptophan and cystine at a given wavelength, the extinction coefficient of the native protein in water can be computed using the following equation:

E(Prot)=Numb(Tyr)*Ext(Tyr)+Numb(Trp)*Ext(Trp) + Numb(Cystine)*Ext(Cystine)

where (for proteins in water measured at 280 nm): Ext(Tyr) = 1490, Ext(Trp) = 5500, Ext(Cystine) = 125;

The Absorbance (optical density) can be calculated using the following formula:

Absorb(Prot) = E(Prot) / Molecular Weight

Protparam computes the extinction coefficient and absorbance values based on the above equations at 280 nm.

In Vivo Half-Life

The half-life is a prediction of the time it takes for half of the amount of protein in a cell to disappear after its synthesis in the cell. The program gives an estimation of the protein half-life.

Instability Index

The instability index provides an estimate of the stability of the protein in a test tube. ProtParam computes the instability index using the formula,

i=L-1

$$II = (10/L) * \sum DIWV(x[i]x[i+1])$$

i=1

where - L is the length of sequence
   DIWV(x[i]x[i+1]) is the dipeptide instability weight value starting at a position i. A protein whose instability index is smaller than 40 is predicted as stable, a value above 40 predicts that the protein may be unstable.

Aliphatic Index

The aliphatic index of a protein is defined as the relative volume occupied by aliphatic side chains (alanine, valine, isoleucine, and leucine). It may be regarded as a positive factor for the increase of thermostability of globular proteins. The aliphatic index of a protein is calculated according to the following formula:

Aliphatic index = X(Ala) + a * X(Val) + b * (X(Ile) + X(Leu))

where, X(Ala), X(Val), X(Ile), and X(Leu) are mole percent (100 X mole fraction) of alanine, valine, isoleucine, and leucine. The coefficients a and b are the relative volume of valine side chain (a = 2.9) and of Leu/Ile side chains (b = 3.9) to the side chain of alanine.

GRAVY (Grand Average of Hydropathy)

The GRAVY value for a peptide or protein is calculated as the sum of hydropathy values of all the amino acids, divided by the number of residues in the sequence.

SOPM

SOPM (http://npsa-pbil.ibcp.fr/cgi-bin/npsa_au-tomat.pl?page=npsa_sopm.html) is a self-optimized method for protein secondary structure prediction tool. The first step of the SOPM is to build sub-databases of protein sequences and their known secondary structures drawn from 'DATABASE.DSSP' by (i) making binary comparisons of all protein sequences and (ii) taking into account the prediction of structural classes of proteins. The second step is to submit each protein of the sub-database to a secondary structure prediction using a predictive algorithm based on sequence similarity. The third step is to iteratively determine the predictive parameters that optimize the prediction quality on the whole sub-database. The last step is to apply the final parameters to the query sequence. This method correctly predicts 69% of amino acids for a three-

state description of the secondary structure (alpha helix, beta sheet and coil) in the whole database (46,011 amino acids).

SOPMA

SOPMA (http://npsa-pbil.ibcp.fr/cgi-bin/npsa_automat.pl?page=npsa_sopma.html) is the improved version of SOPM. SOPMA correctly predicts 69.5% of amino acids for a three-state description of the secondary structure (alpha-helix, beta-sheet and coil) in a whole database containing 126 chains of non-homologous (less than 25% identity) proteins. Joint prediction with SOPMA and a neural networks method correctly predicts 82.2% of residues for 74% of co-predicted amino acids.

SSCP

The program SSCP (http://coot.embl.de/SSCP/) computes the percentage of residues forming (content) helix, strand, and coil for a given protein using the amino acid composition as the only input information. The basic idea of the method consists in the application of analytic vector decomposition methods applied on the composition vector of the query protein. SSCP computes in two methods, the first method relies only on the average amino acid composition of secondary structural segments (helix, sheet, coil) in a submitted set of proteins. The second method relies also on composition fluctuations in the secondary structural segments (helix, sheet, coil) of a learning set of proteins. The output is computer-readable.

SOSUI

SOSUI (http://bp.nuap.nagoya-u.ac.jp/sosui/) predicts a part of the secondary structure of proteins from a given amino acid sequence. The main objective is to classify whether the protein is a soluble or a transmembrane protein. The accuracy of the classification of proteins was 99% and the corresponding value for the transmembrane helix prediction was 97%. There are three basic assumptions in the SOSUI system. First, membrane proteins are characterized by at least one, particularly hydrophobic, primary transmembrane helix. Secondary hydrophilic transmembrane helices may also exist in multispanning membrane proteins even though their hydrophobicity is infact similar to the hydrophobic segments of soluble proteins. The possible role of secondary transmembrane helices is the formation of active sites of proteins. Third, the primary transmembrane helices are stabilized by a combination of amphiphilic side chains at the helix ends as well as high hydrophobicity in the central region.

BLASTP

Basic Local Alignment Search Tool, BLAST (http://www.ncbi.nlm.nih.gov/blast/Blast.cgi) is a set of similarity search programs designed to explore all of the available sequence databases regardless of whether the query is a protein or a DNA. The BLASTP tool compares an amino acid query sequence against a protein sequence database. Standard protein-protein BLAST (blastp) is used for both identifying a query amino acid sequence and for finding similar sequences in protein databases. Blastp finds local regions of similarity also. When sequence similarity spans the whole sequence, blastp will also report a global alignment, which is the preferred result for protein identification purposes.

Blast 2 Sequences

Blast 2 sequences (http://www.ncbi.nlm.nih. gov/blast/bl2seq/wblast2.cgi) tool produces the alignment of two given sequences using BLAST engine for local alignment. Blast 2 Sequences utilizes the BLAST algorithm for pairwise DNA-DNA or protein-protein sequence comparison. The resulting alignments can be obtained in both graphical and text form. The Blast 2 sequences

tool will function in Windows, Mac and several UNIX based platforms.

Scanprosite

The ScanProsite (www.expasy.ch/tools/scan-prosite/) server allows scanning protein sequences (either from UniProt Knowledgebase (SwissProt/TrEMBL) or PDB or provided by the user) for the occurrence of patterns, profiles and rules (motifs) stored in the PROSITE database, or to search protein databases for hits by specific motifs.

Protein Sequence Retrieval and Selection

Exhaustive database searches were performed in the various protein sequence databases to identify the Marfan syndrome associated human proteins. Protein sequence databases such as EMBL (www.ebi.ac.uk/embl/), GenBank (http://www. ncbi.nlm.nih.gov/Genbank/), PRF (http://www. prf.or.jp/en), DBJ (http://www.ddbj.nig.ac.jp/), UniProtKB/Swiss-Prot (http://www.expasy. org/sprot), PDB (http://www.pdb.org/pdb/) and UniProtKB/TrEMBL (http://www.uniprot.org/ database/knowledgebase.shtml) were scanned for the keywords "Marfan syndrome", "Fibrillin 1", "FBN1" using the search interface available in the NCBI Entrez protein (http://www.ncbi. nlm.nih.gov/entrez/) database and the BLASTP (http://www.ncbi.nlm.nih.gov/blast/) similarity search tool. The search results yielded 215 protein sequences from different organisms. From these 215 proteins, the human protein sequences were downloaded in FASTA (Lipman, D.J., 1985) format and submitted to Protein Mutant Database (http://pmd.ddbj.nig.ac.jp/) to identify and confirm the presence of mutations causing Marfan syndrome. Protein sequences without mutations were discarded and totally 23 human protein sequences with Marfan syndrome causing mutations were identified. These 23 protein sequences were matched with each other using the online server "Blast 2 sequences" (http://www.ncbi.nlm.nih.gov/blast/bl2seq/wblast2.cgi) and finally 14 dissimilar protein sequences were selected for analysis. The Fibrillin 1 human protein sequences and other sequences similar to Fibrillin 1 and with Marfan syndrome causing mutations selected for analysis and their details are tabulated in Table 1.

Computational Tools and Servers

The amino acid composition of Marfan syndrome associated human Fibrillin 1 proteins were computed using the tool BioEdit 5.0.9 (Hall, T.A., 1999) and CLC Free Workbench. Percentages of hydrophobic and hydrophilic residues were computed using the program SAPS and primary structural data. The physico-chemical parameters such as theoretical isoelectric point (pI), molecular weight, extinction coefficient (Gill, S.C., 1989), half-life (Bachmair, A., 1986, Gonda, D.K., 1989, Tobias, J.W., 1991, Ciechanover, A.L. 1989, Varshavsky, A., 1997), instability index (Guruprasad,

K., 1990), aliphatic index (Ikai., 1980) and grand average hydrophathy (GRAVY) (Kyte, J., 1982) values were computed using the Expasy's ProtParam (Gasteiger, E., 2005) (http://us.expasy.org/tools/protparam.html) prediction server and Compute pI/Mw tool (http://au.expasy.org/tools/pi_tool.html) and tabulated in Table 2. The tools SOPM (Geourjon, C., 1994) (http://npsa-pbil.ibcp.fr/cgi-bin/npsa_automat.pl?page=/NPSA/npsa_sopm.html) and SOPMA (Geourjon, C., 1995) (http://npsa-pbil.ibcp.fr/cgi-bin/npsa_automat.pl?page=/NPSA/npsa_sopma.html) were used for the secondary structure prediction. Secondary Structural Content Prediction (http://coot.embl.de/SSCP/) server (Eisenhaber, F., 1996) is used for the computation of percentages of α-helical, β-strand and coiled regions and secondary structure class identification (Table 3). The SOSUI (http://bp.nuap.nagoya-u.ac.jp/sosui/) server (Takatsugu Hirokawa, 1998) performed the identification of transmembrane region in Fibrillin 1 proteins. The tool BioEdit is used to compute the Cornette scale (Cornette J.L., 1987) mean hydrophobicity

Table 1. Marfan syndrome associated human protein sequences retrieved from NCBI Entrez protein database

Accession Number	Database	Length	Sequence description
CAA45118.1	embl	3002	Fibrillin 1
EAW77354	genbank	2871	Fibrillin 1
EAW77353	genbank	2869	Fibrillin 1
1713408A	prf	1973	Fibrillin 1
BAD16738.1	dbj	1365	Fibrillin 1
BAD92077	dbj	830	Fibrillin 1
1713407B	prf	754	Fibrillin 1
1713407A	prf	517	Fibrillin 1
Q75N89	swissprot	195	Fibrillin 1
AAB25244.1	genbank	109	Fibrillin 1
1EMO_A	pdb	82	Fibrillin 1
Q504W9	swissprot	55	Fibrillin 1
B4E3I6	tr	1149	highly similar to Fibrillin 1
BAC11489.1	dbj	1497	unnamed protein product

profile of the transmembrane region (Figure 1). Scanprosite (www.expasy.ch/tools/scanprosite/) server was used for the identification of profiles (Figure 2) present in protein sequences.

RESULTS AND DISCUSSION

Primary Structure Analysis

The results of primary structure analysis suggest that all the Marfan syndrome associated human Fibrillin 1 proteins are rich in cysteine and glycine amino acids. All the Fibrillin 1 proteins consist of equal number of hydrophobic and hydrophilic amino acids. The average molecular weight of Marfan syndrome associated proteins calculated is 134086 Dalton.

Physicochemical Analysis

Expasy's ProtParam computes the extinction Coefficient (EC) for a range of (276, 278, 280 and 282nm) wavelength. The EC value at 280nm is favoured because proteins absorb strongly there while other substances commonly in protein solutions do not.

Extinction Coefficient (EC) of Marfan syndrome associated human Fibrillin 1 proteins at 280nm is ranging from 1490 to 259165 M^{-1} cm^{-1} with respect to the concentration of Cys, Trp and Tyr (Table 2). Expasy's ProtParam classifies most of the Marfan syndrome associated human Fibrillin 1 proteins as unstable on the basis of Instability index (II>40) and few proteins (AAB25244.1, 1EMO_A, Q504W9) as stable (II<40) proteins in the room temperature. The aliphatic index (AI) that is defined as the relative volume of a protein occupied by aliphatic side chains (Ala, Val, Ile and Leu) is regarded as a positive factor for increase of thermal stability of globular proteins is very low (39-81) for most of the Marfan syndrome associated human Fibrillin 1 proteins and it infers that the Fibrillin 1 proteins may become unstable at high temperature. The aliphatic index value is comparatively high for the protein Q504W9 and it infers that they may be comparatively stable than other Fibrillin 1 proteins. Grand Average hydropathy (GRAVY) index of all the Marfan syndrome associated human Fibrillin 1 proteins are ranging from (-0.2 to -0.4), this indicates that all these proteins may interact equally and easily with water. Isoelectric point (pI) is the pH at which

Figure 1. Cornette Scale Mean hydrophobicity profile computed for the transmembrane region of 1713407A

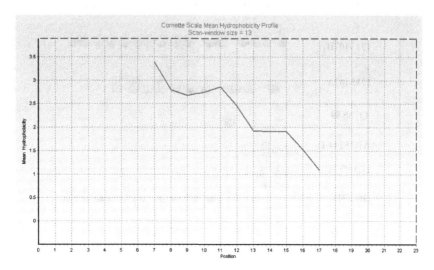

the surface of protein is covered with charge but net charge of the protein is zero. At pI proteins are stable and compact. The computed pI value indicates that most of the proteins are acidic (pI<7) and the proteins Q75N89 and Q504W9 are basic (pI>7) in property (Table 2). The number of basic and acidic amino acids in each Marfan syndrome associated human Fibrillin 1 proteins correlates well with the corresponding pI computed. The computed isoelectric point (pI) will be useful for developing buffer systems for purification by Isoelectric focusing method. The computed protein concentration and extinction coefficients help in the quantitative study of protein-protein and protein-ligand interactions in solution.

Secondary Structure Analysis

The secondary structure predicted with the help of programs SOPM and SOPMA shows that all the Marfan syndrome associated human Fibrillin 1 proteins are found to be with mixed secondary structural content. The computed percentage of residues forming α-helices β-strands and coils are shown in (Table 3).

Figure 2. Organization of profiles identified in the Marfan syndrome associated proteins

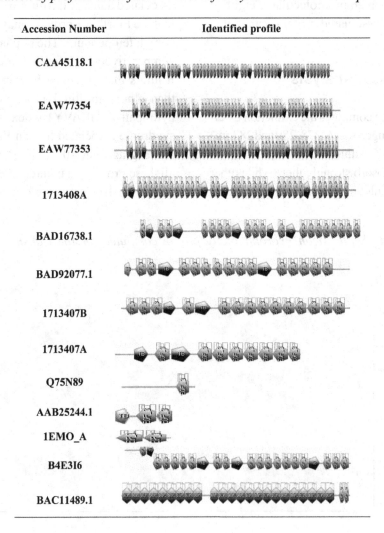

Table 2. Physicochemical parameters computed using Expasy's ProtParam tool

Accession Number	M. Wt	pI	EC	II	AI	GRAVY
CAA45118.1	325687	4.88	259165	44.80	53.33	-0.443
EAW77354	312297	4.81	234060	44.07	53.65	-0.423
EAW77353	312143	4.81	235550	44.14	53.58	-0.425
1713408A	215847	4.63	159205	45.07	51.53	-0.485
BAD16738.1	147449	5.09	104755	45.52	55.50	-0.333
BAD92077.1	90837	4.48	67825	43.72	52.72	-0.440
1713407B	81408	4.38	60570	46.26	53.81	-0.345
1713407A	56529	4.55	38630	51.60	58.45	-0.268
Q75N89	20702	8.74	15940	45.92	52.56	-0.329
AAB25244.1	11583	4.00	10845	31.09	51.83	-0.399
1EMO_A	8745	4.05	3730	34.31	39.15	-0.382
Q504W9	5807	10.87	1490	34.18	81.82	-0.453
B4E3I6	124380	4.66	93745	42.09	52.77	0.385
BAC11489.1	163803	5.50	248325	47.60	61.18	-0.337

M.Wt.-Molecular weight, pI – Isoelectric point, EC – Extinction coefficient at 280nm, II – Instability Index, AI – Aliphatic Index, GRAVY – Grand Average Hydropathy.

Table 3. Percentage of residues forming alpha, beta and coil structures

Accession Number	Alpha	Beta	Coil	Class
CAA45118.1	15.8	55.0	29.2	mixed
EAW77354	5.8	54.8	39.4	beta
EAW77353.1	26.4	49.7	23.9	mixed
1713408A	100.0	0.0	0.0	alpha
BAD16738.1	15.3	22.3	62.4	mixed
BAD92077.1	5.4	61.8	32.7	beta
1713407B	90.2	0.0	9.8	alpha
1713407A	5.9	71.7	22.3	beta
Q75N89	32.7	16.8	50.6	mixed
AAB25244.1	5.9	71.7	22.3	beta
1EMO_A	76.3	20.8	2.9	mixed
Q504W9	24.3	16.2	59.5	mixed
B4E3I6	6.7	62.9	30.3	beta
BAC11489.1	15.9	52.4	31.8	mixed

Transmembrane Region Analysis

The server SOSUI identified one transmembrane region in the 1713407A protein. The transmembrane region (VLVTVVFIFLSYNKMLSSPC-ING) with 23 amino acids found to have more hydrophobic residues and it is also well documented by the Cornette scale mean hydrophobicity profile. (Figure 1) in which all the peaks are above the zero line.

Profile Identification

The Scanprosite server identified two different profiles (EGF-like domain profile and TGF-beta binding domain profile) in all Marfan syndrome associated proteins except the Q504W9 Fibrillin 1 protein. Scanprosite server did not identify any profiles in the Q504W9 Fibrillin 1 protein. In BAC11489.1, extracellular sushi domain that is also known as the complement controle protein module or the short consensus repeat is identified. The organization of all the identified profiles is shown in Figure 2.

CONCLUSION

Marfan syndrome associated human Fibrillin 1 proteins retrieved from the NCBI Entrez database were analyzed and characterized with the aim of providing more detailed description of Marfan syndrome associated proteins. Primary structure analysis reveals that the Marfan syndrome associated proteins are rich in cysteine and glycine residues. Physicochemical analysis give a good idea about the properties such as pI, EC, AI, GRAVY and Instability Index that are essential and vital in providing data about the proteins and their properties.

Secondary structure analysis shows that human Fibrillin 1 proteins are found to be with mixed secondary structural content. Scanprosite server identified EGF-like domain, TGF-beta binding domain and extracellular sushi domain profiles in Marfan syndrome associated proteins.

REFERENCES

Bachmair, A., Finley, D., & Varshavsky, A. (1986). In vivo half-life of a protein is a function of its amino-terminal residue. *Science, 234*(4773), 179–186. doi:10.1126/science.3018930

Ciechanover, A. L., Schwartz. (1989). How are substrates recognized by the ubiquitin-mediated proteolytic system? *Trends in Biochemical Sciences, 14*(12), 483–488. doi:10.1016/0968-0004(89)90180-1

Cornette, J. L., Cease, K. B., Margalit, H., Spouge, J. L., Berzofsky, J. A., & DeLisi, C. (1987). Hydrophobicity scales and computational techniques for detecting amphipathic structures in proteins. *Journal of Molecular Biology, 195*(3), 659–685. doi:10.1016/0022-2836(87)90189-6

Eisenhaber, F., Imperiale, F., Argos, P., & Froemmel, C. (1996). Prediction of Secondary Structural Content of Proteins from Their Amino Acid Composition Alone. *New Analytic Vector Decomposition Methods, Proteins, Struct., Funct. Design, 25*(2), 157–168.

Gasteiger, E., Hoogland, C., Gattiker, A., Duvaud, S., Wilkins, M. R., & Appel, R. D. Bairoch. (2005). Protein Identification and Analysis Tools on the ExPASy Server. In J. M. Walker (Ed.), *The Proteomics Protocols Handbook* (pp. 571-607). New York: Humana Press.

Geourjon, C., & Deleage, G. (1994). SOPM: a self-optimized method for protein secondary structure prediction. *Protein Engineering, 7*(2), 157–164. doi:10.1093/protein/7.2.157

Geourjon, C., & Deleage, G. (1995). SOPMA: significant improvements in protein secondary structure prediction by consensus prediction from multiple alignments. *Computer Applications in the Biosciences, 11*(6), 681–684.

Gill, S. C., & Von Hippel, P. H. (1989). Calculation of protein extinction coefficients from amino acid sequence data. *Analytical Biochemistry, 182*(2), 319–326. doi:10.1016/0003-2697(89)90602-7

Gonda, D. K., Bachmair, A., Wunning, I., Tobias, J. W., Lane, W. S., & Varshavsky, A. (1989). Universality and structure of the N-end rule. *The Journal of Biological Chemistry, 264*(28), 16700–16712.

Guruprasad, K., Reddy, B. V. B., & Pandit, M. W. (1990). Correlation between stability of a protein and its dipeptide composition: a novel approach for predicting in vivo stability of a protein from its primary sequence. *Protein Engineering, 4*(2), 155–161. doi:10.1093/protein/4.2.155

Hall, T. A. (1999). BioEdit: a user-friendly biological sequence alignment editor and analysis program for Windows 95/98/NT. *Nucleic Acids Symposium Series, 41*, 95–98.

Hirokawa, T., Boon-Chieng, S., & Mitaku, S. (1998). SOSUI: classification and secondary structure prediction system for membrane proteins. *Bioinformatics Applications Note, 14*(4), 378–379.

(1895-1898). Ikai. (1980). Thermostability and aliphatic index of globular proteins. *Journal of Biochemistry, 88*(6).

Kyte, J., & Doolittle, R. F. (1982). A simple method for displaying the hydropathic character of a protein. *Journal of Molecular Biology, 157*(1), 105–132. doi:10.1016/0022-2836(82)90515-0

Lipman, D. J., & Pearson, W. R. (1985). Rapid and sensitive protein similarity searches. *Science, 227*(4693), 1435–1441. doi:10.1126/science.2983426

Magenis, R. E., Maslen, C. L., Smith, L., Allen, L., & Sakai, L. Y. (1991). Localization of the fibrillin (FBN) gene to chromosome 15, band q21.1. *Genomics, 11*, 346–351. doi:10.1016/0888-7543(91)90142-2

Online Mendelian Inheritance in Man, Johns Hopkins University, (2000). *Marfan syndrome, Type I; MIM Number 154700: 2/25/00.* Retrieved October 10th, 2007 from http://www.ncbi.nlm.nih.gov/omim/

Online Mendelian Inheritance in Man, Johns Hopkins University, (2008). *Fibrillin 1; MIM Number 134797: 12/11/08.* Retrieved October 11th, 2008 from http://www.ncbi.nlm.nih.gov/sites/entrez

Tobias, J. W., Shrader, T. E., Rocap, G., & Varshavsky, A. (1991). The N-end rule in bacteria. *Science, 254*(5036), 1374–1377. doi:10.1126/science.1962196

Varshavsky, A. (1997). The N-end rule pathway of protein degradation. *Genes to Cells, 2*(1), 13–28. doi:10.1046/j.1365-2443.1997.1020301.x

KEY TERMS AND DEFINITIONS

Accession Number: An identifier supplied by the curators of the major biological databases upon submission of a novel entry that uniquely identifies that sequence (or other) entry.

Amino Acid: One of the 20 chemical building blocks that are joined by peptide linkages to form a polypeptide chain of a protein.

Da (Dalton): Unit of molecular weight of protein; Synonymous with atomic mass units.

FASTA Format: FASTA format starts with >, a unique identifier, a short definition for protein followed by protein sequence.

Hydrophobicity: Hydrophobicity (from the combining form of water in Attic Greek *hydro-* and for fear *phobos*) refers to the physical property of a molecule (known as a hydrophobe) that is repelled from a mass of water. Hydrophobic molecules tend to be non-polar and thus prefer other neutral molecules and nonpolar solvents. Examples of hydrophobic molecules include the alkanes, oils, fats, and greasy substances in general.

Mutation: Mutation is a permanent change in the DNA sequence of a gene. Mutations in a gene's DNA sequence can alter the amino acid sequence of the protein encoded by the gene. Mutations can be caused by copying errors in the genetic material during cell division, by exposure to ultraviolet or ionizing radiation, chemical mutagens, or viruses, or can be induced by the organism, itself, by cellular processes such as hypermutation.

Protein Domains: A protein domain is a part of protein sequence and structure that can evolve, function, and exist independently of the rest of the protein chain. Each domain forms a compact three-dimensional structure and often can be independently stable and folded. Many proteins consist of several structural domains.

Protein Primary Structure: Primary structure refers to the "linear" sequence of amino acids. Proteins are large polypeptides of defined amino acid sequence. The gene that encodes the proteins determines the sequence of amino acids in each protein. The gene is transcribed into a messenger RNA (mRNA) and the mRNA is translated into a protein by the ribosome.

Protein Secondary Structure: Secondary structure is "local" ordered structure brought about via hydrogen bonding mainly within the peptide backbone. The most common secondary structure elements in proteins are the alpha helix and the beta sheet.

Server: A computer that processes requests issued from remote locations by clients machines

ABBREVIATIONS

AI: Aliphatic Index
BioEdit: Biological Sequence Alignment Editor
Cys: Cysteine
Da: Dalton
3D: Three Dimensional
DBJ: DNA Database of Japan
DNA: Deoxyribonucleic acid
EC: Extinction coefficient at 280nm
EMBL: European Molecular Biology Laboratory
FBN1: Fibrillin 1
GRAVY: Grand Average Hydropathy
II: Instability Index
M.Wt.: Molecular weight
NCBI: National Center for Biotechnology Information
NMR: Nuclear Magnetic Resonance
OMIM: Online Mendelian Inheritance in Man
pI: Isoelectric point
PDB: Protein Data Bank (Protein Database)
PIR: Protein Information Resource (Protein Database)
PMD: Protein Mutant Database
PRF: Protein Research Foundation (Protein Database)
RNA: Ribonucleic acid
SAPS: Statistical Analysis of Protein Sequences
SOPM: Self-Optimized Prediction Method
SOPMA: Self-Optimized Prediction Method with Alignment
SRS: Sequence Retrieval Systems
SSCP: Secondary Structural Content Prediction
Trp: Tryptophan
Tyr: Tyrosine

Chapter 10
A Modified High Speed Hopfield Neural Model for Perfect Calculation of Magnetic Resonance Spectroscopy

Hazem El-Bakry
Mansoura University, Egypt

Nikos Mastorakis
Technical University of Sofia, Bulgaria

ABSTRACT

In this chapter, an automatic determination algorithm for nuclear magnetic resonance (NMR) spectra of the metabolites in the living body by magnetic resonance spectroscopy (MRS) without human intervention or complicated calculations is presented. In such method, the problem of NMR spectrum determination is transformed into the determination of the parameters of a mathematical model of the NMR signal. To calculate these parameters efficiently, a new model called modified Hopfield neural network is designed. The main achievement of this chapter over the work in literature (Morita, N. and Konishi, O., 2004) is that the speed of the modified Hopfield neural network is accelerated. This is done by applying cross correlation in the frequency domain between the input values and the input weights. The modified Hopfield neural network can accomplish complex dignals perfectly with out any additinal computation steps. This is a valuable advantage as NMR signals are complex-valued. In addition, a technique called "modified sequential extension of section (MSES)" that takes into account the damping rate of the NMR signal is developed to be faster than that presented in (Morita, N. and Konishi, O., 2004). Simulation results show that the calculation precision of the spectrum improves when MSES is used along with the neural network. Furthermore, MSES is found to reduce the local minimum problem in Hopfield neural networks. Moreover, the performance of the proposed method is evaluated and there is no effect on the performance of calculations when using the modified Hopfield neural networks.

DOI: 10.4018/978-1-60566-768-3.ch010

INTRODUCTION

Applications of magnetic resonance imaging were started in magnetic resonance imaging (MRI) which is a technique imaging the human anatomy, and they include various specialized technique such as diffusion-weighted imaging (DWI), perfusion-weighted imaging (PWI), magnetic resonance angiography (MRA) and magnetic resonance cholangio-pancreatography (MRCP). Functional MRI (fMRI) that is an innovative tool for functional measurement of human brain and that is a technique imaging brain functions, also became practical and has been widely used in recent years. In contrast with MRI and fMRI, magnetic resonance spectroscopy (MRS) is a technique that measures the spectra of the metabolites in a single region, and magnetic resonance spectroscopic imaging (MRSI), which obtains the spectra from many regions by applying imaging techniques to MRS, has also been developed. Although 31P-MRS was widely performed in MRS before, proton MRS is primarily performed recently. 13C-MRS using heteronuclear single-quantum coherence (HSQC) method has also been developed recently. MRS and MRSI, however, have remained underutilized together due to their technical complexities compared with MRI.

At present, MRS is technically evolved and its operation has remarkably improved. The measurement of MRS also has started to be automatically analyzed and indicated, and there are some representative analysis software introduced in the Internet, LCModel: an automatic software packages for in-vivo proton MR spectra including the curve-fitting procedure (Provencher, S.W. 2001), and MRUI: Magnetic Resonance User Interface including the time-domain analysis of in-vivo MR data (Naressi, A., Couturier, C., Castang, I., de. Beer, R. and Graveron-Demilly, D., 2001). The technique proposed in this chapter is also used for in the time-domain. It probably a better result of the analysis is obtained by combining the algorithms

of MRUI with our technique, because both of them are performed in the time-domain.

MRSI has the big feature that is not in MRI and fMRI, that is, it can detect internal metabolite non-invasively, track the metabolic process and perform the imaging. Thus the importance of it is huge. Furthermore, MRSI is also expected as an imaging technique realizing the molecular imaging. I believe that MRSI has the value beyond fMRI, because of its potential.

For commonly performing the MRSI, it is an indispensable technique to quantify NMR spectra automatically, and it is also expected to progress the automatic analysis techniques. Therefore, it is necessary to develop a novel method introducing neural network techniques including our proposing method, as well as existing analysis software. Consequently, it is important to proceed with the research of this territory.

MRS is used to determine the quantity of metabolites, such as creatine phosphate (PCr) and adenocine triphosphate (ATP), in the living body by collecting their nuclear magnetic resonance (NMR) spectra. In the field of MRS, the frequency spectrum of metabolites is usually obtained by applying the algorithm of fast Fourier transform (FFT) (Cooley, J.W. and Tukey, J.W., 1965) to the NMR signal obtained from the living body. Then, quantification of the metabolites is carried out by estimating the area under each spectral peak using a curve fitting procedure (Maddams 1980, Sijens et al., 1998, Mierisová et al., 2001). However, this method is not suitable for processing large quantities of data because human intervention is necessary. The purpose of this chapter is to present an efficient automatic spectral determination method to process large quantities of data without human intervention.

This chapter is organized as follows: first conventional determination methods of NMR spectra are described and a brief outlines of the proposed algorithm is given. Second, an over efficient view of NMR signal theory; a mathematical model of the NMR signal are discussed. The proposed approach to spectral determination is presented. Third, design of complex-valued Hopfield neural networks for fast and efficient spectral estimation is introduced. Fourth, Modified Sequential Extension of Section (MSES) explains the concept of MSES. For performance evaluation of the proposed method, simulations were carried out using sample signals that imitate an actual NMR signal, and the results of those simulations are given. The results are evaluated and discussed. Finally, conclusions and future work are given.

MATHEMATICAL MODEL OF THE NMR SIGNAL AND DETERMINATION OF SPECTRA

Magnetic resonance imaging (MRI) systems, which produce medical images using the nuclear magnetic resonance (NMR) phenomenon, have recently become popular. Additional technological innovations, such as high-speed imaging technologies (Mansfield 1977, Henning et al., 1986, Melki et al., 1991, Feinberg et al., 1991, Meyer et al., 1992) and imaging of brain function using functional MRI (Ogawa et al., 1990, Belliveau et al., 1991, Kwong et al., 1992) are also rapidly progressing. Currently, the above-mentioned imaging technologies mainly take advantage of the NMR phenomena of protons. The atomic nuclei used for analyzing metabolism in the living body include proton, phosphorus-31, carbon-13, fluorine-19 and sodium-22. Phosphorus-31 NMR spectroscopy has been widely used for measurement of the living body, because it is able to track the metabolism of energy.

NMR was originally developed and used in the field of analytical chemistry. In that field, NMR spectra are used to analyze the chemical structure of various materials. This is called NMR spectroscopy. In medical imaging, it is also possible to obtain NMR spectra. In this case, the technique is called magnetic resonance spectroscopy

(MRS), and it can be used to collect the spectra of metabolites in organs such as the brain, heart, lung and muscle. The difference between NMR spectroscopy and MRS is that in MRS, spectra is collected from the living body in a relatively low magnetic field (usually, about 1.5 Tesla); in NMR spectroscopy, small chemical samples are measured in a high magnetic field.

In MRI systems, Fourier transform is widely used as a standard tool to produce an image from the measured data and to obtain NMR spectra. In NMR spectroscopy, a frequency spectrum can be obtained by applying the fast Fourier transform (FFT) to the free induction decay (FID) that is observed as a result of the magnetic relaxation phenomenon (Derome, A.E., 1987). Here the FID is an NMR signal in the time domain and it is a time series, that is, it can be modeled as a set of sinusoids exponentially damping with time. When FFT is applied to such a signal, the spectral peaks obtained are of the form called a Lorentz curve (Derome, A.E., 1987). If the signal is damped rapidly, the height of the spectral peaks will be decreased and the width of the peaks will increase. This is an inevitable result of applying FFT to FIDs. In addition, the resolution of the spectrum collected in a low magnetic field is much lower than a typical spectrum obtained by NMR spectroscopy. Therefore, the problems of spectral analysis in MRS and NMR spectroscopy are quite different. The spectral peaks obtained in MRS are spread out and the spectral distribution obtained is very different from the original distribution. Therefore, peak height to quantify metabolites cannot be used. Instead, the area under each peak is estimated by using curve-fitting procedures (non-linear least square methods) (Maddams 1980, Sijens et al. 1998, Mierisová et al. 2001). However, existing curve-fitting procedures are inadequate for processing large quantities of data because they require human intervention. The aim of our research is to devise a method that does not require such human intervention.

Two approaches can be considered to solve this problem: (1) automating the description of spectral peaks and the determination of the peak areas, and (2) using methods of determination and quantification other than the Fourier transform. In the first approach, attempts at automatic quantification of NMR spectra using hierarchical neural networks have been reported (Ala-Korpela et al. 1997, Kaartinen et al. 1998). In this research, a three-layered network based on back propagation (Rumelhart et al. 1986) was employed and the spectra in the frequency domain were used as the training data of the network. The fully-trained network had the ability to quantify unknown spectra automatically, and curve fitting procedures were not necessary. However, large amounts of training data were necessary to increase the precision of quantification. These methods quantify the spectra instead of performing the curve fitting procedures. In the second approach, the maximum entropy method (MEM), derived from the autoregressive (AR) model and the linear prediction (LP) method, and other similar methods have been studied widely (Haselgrove et al., 1988, van Huffel et al., 1994). These are parametric methods, that is, in these methods, a mathematical model of the signal is assumed and the parameters of that model are estimated from observed data. The spectrum can then be estimated from the model parameters. However, methods based on AR modeling require large amounts of calculation.

The main objective of this research is to develop a method to estimate NMR spectra without human intervention or complicated calculations. Therefore, a parametric approach, in which a neural network is used (Han 1997), is considered. Fixed weights Hopfield neural networks (Hopfield 1982 and 19841984) are used. It is possible to estimate the parameters using the ability of these neural networks to find a local minimum solution or a minimum solution. In addition, it was noted that NMR signals are complex-valued and a method to estimate the spectrum using complex-valued Hopfield networks (Hirose 1992a, 1992b, Zhou

et al. 1993), in which the weights and thresholds of conventional networks are expanded to accommodate complex numbers, was developed. Both a hierarchical type (Nitta 1991 and 1997,1997 Benvenuto 1992, Georgiou 1992) and a recurrent type (Nemoto 1991 and Jankowski 1996) have been proposed. The operation of these networks was accelerated as described by (El-Bakry H.M., and Zhao Q., 2005, El-Bakry H.M., 2006) and this is main achievement of this chapter. Furthermore, a technique that takes into account the damping of the NMR signal, which we call "sequential extension of section" (MSES) has been devised, and used with the above-mentioned network.

MATHEMATICAL MODEL OF NMR SIGNAL

If an atomic nucleus possessing a spin is placed in a static magnetic field, it begins a rotation called "precession" around the direction of the static magnetic field. It is assumed that the direction of the static magnetic field is the z-direction, and the orthogonal plane for the z-direction is the x-y plane. Considering an atomic ensemble, a macroscopic magnetization M resulting from the sum of the spin of each nucleus appears in the z-direction. When the ensemble is exposed to an external rotating magnetic field at the resonance frequency of the precession, each nucleus in the ensemble resonates. As a result, a component of magnetization in the x-y plane appears, and the component in the z-direction decreases. It is assumed that the magnetization M has rotated and is now operating around the z-axis. The resonant magnetic field pulse that tilts M 90 degrees to the x-y plane is called a 90-degree pulse.

After a 90-degree pulse, the magnetization M returns to its original orientation in the z-direction. During that time, the component in the x-y plane is exponentially damped with time t and time constant T_2, so the signal is represented by an equation in the form $\exp(-t/T_2)$. The component in the z-direction recovers with time t and time constant T_1; this process is represented by an equation in the form $1 - \exp(-t/T_2)$. This phenomenon is called the magnetic relaxation. The change in the component in the x-y plane is called the transverse relaxation, and the change in the component in the z-direction is called the longitudinal relaxation. T_2 and T_1 are called the transverse relaxation time and the longitudinal relaxation time, respectively. Because of inhomogeneity in the static magnetic field, the transverse relaxation time is actually shortened. Thus, we usually observe this shortened transverse relaxation, called T_2^* ($T_2^* < T_2$), unless we use a technique such as the spin echo method (Derome 1987).

In NMR, the component in the x-y plane is called an "NMR signal" or "free-induction decay" (FID), and it is expressed in a complex form because it is in essence a rotation.

An NMR signal (FID) with m components is modeled as follows:

$$\hat{x}_n = \sum_{k=1}^{m} A_k \exp(-b_k n) \exp[j(2\pi f_k n + \varphi_k)],$$
$$n = 0, 1, \cdots, N-1 \qquad (1)$$

where $\hat{x}_n$ ($n = 0, 1, \cdots, N-1$) denotes the observed signal, which is complex-valued, and n denotes the sample point on the time axis. A_k, b_k, f_k, and φ_k denote the spectral composition, damping factor, rotation frequency, and phase in the rotation, respectively, of each metabolite, and m is the number of the metabolites composing a spectrum (each of these is a real number, and $j = \sqrt{-1}$).

NMR SPECTRA

The position of each peak appearing in a NMR spectrum depends on its offset frequencies (chemical shifts) from the resonance frequency of a target

nucleus under a specified static magnetic field (Derome 1987). These offset frequencies are f_k ($k=1,\ldots,m$).

In a common pulse method, each peak possesses the offset phase expressed by a linear function of its offset frequencies f, as follows (Derome 1987):

$$\theta(f) = \alpha + \beta \cdot f \qquad (2)$$

where, α is called the zero-dimensional term of phase correction, and is a common phase error influencing each peak. β is called the one-dimensional term of phase correction, and is a phase error that is dependent on the offset frequencies, or more specifically, the positions of each peak.

Thus, in NMR spectra, the position of each peak and the scale of their offset phase are decided by the measurement condition used. Because of this fact, it is possible to make a rough prediction of the position of each peak of a NMR spectrum under specified measurement conditions. This positional can be used as a constrained condition when estimating unknown parameters using neural networks. In addition, because the relationship between a specified static magnetic field and the apparent transverse relaxation time T_2^* of a target nucleus are known in MRS (Haselgrove et al. 1988), it is possible to determine the rough scale of T_2^* for a target nucleus when the strength of the static magnetic field is known. This information regarding T_2^* can also be used as a constrained condition. That is, it can be used for the determination of b_k.

DETERMINATION OF NMR SPECTRA

The following approach is used in our method of parametric spectral determination.

1. A mathematical model of the NMR signal is given, as described above.

2. Adequate values are supplied as initial values of the parameters A_k, b_k, f_k, and φ_k, and an NMR signal is simulated.

3. The sum of the squares of the difference, at each sample point, between the simulated signal and the actual observed signal is calculated.

4. The parameters are changed to give optimum estimates for the observed signal by minimizing the sum-squared error.

DESIGN OF MODIFIED HOPFIELD NEURAL NETWORK FOR SPECTRAL ESTIMATION

Conventional Hopfield neural networks accept input signal with fixed size (n). Therefore, the number of neurons equals to (n). Instead of treating (n) inputs, the idea is to collect all the input data together in a long signal (for example 100xn). Then the input signal is processed by Hopfield neural networks as a single pattern with length L (L=100xn). Such a process is performed in the frequency domain.

Given any two functions f and d, their cross correlation can be obtained by:

$$d(x) \otimes f(x) = \left(\sum_{n=-\infty}^{\infty} f(x+n)d(n) \right) \tag{3}$$

Therefore, the output of each neuron can be written as follows [8,9]:

$$O_i = g\left(W_i \otimes Z \right) \tag{4}$$

where Z is the long input signal, W is the weight matrix, O_i is the output of each neuron and g is the activation function.

Now, the above cross correlation can be expressed in terms of one dimensional Fast Fourier Transform as follows:

$$W_i \otimes Z = F^{-1}\left(F(Z) \bullet F^*\left(W_i \right) \right) \tag{5}$$

It is clear that the operation of the modified Hopfield neural network depends on computing the Fast Fourier Transform for both the input and weight matrices and obtaining the resulting two matrices. After performing dot multiplication for the resulting two matrices in the frequency domain, the Inverse Fast Fourier Transform is calculated for the final matrix. Here, there is an excellent advantage with the modified Hopfield neural network that should be mentioned. The Fast Fourier Transform is already dealing with complex numbers, so there is no change in the number of computation steps required for the modified Hopfield neural network. Hence, by evaluating this cross correlation, a speed up ratio can be obtained comparable to conventional Hopfield neural networks.

For the determination of NMR spectra, the sum-squared error of the parameter determination problem is defined as the energy function of a Hopfield network. This converts the parameter determination problem to an optimization problem for the Hopfield network. The energy function is defined as:

$$E = -\frac{1}{2} \sum_{n=0}^{N-1} \left| \hat{x}_n - \sum_{k=1}^{m} A_k \exp(-b_k n) \exp\{j(2\pi f_k n + \varphi_k)\} \right|^2 \tag{6}$$

where, as in Eq.(1), n denotes the sample point on the time axis and $\hat{x}_n$ denotes the complex-valued observed signal at n.

The energy function E of complex-valued neural networks should have the following properties (Hashimoto et al. 1999, Kuroe et al. 2002):

1) A function that relates the state $\hat{x}$ denoted by a complex number to a real-valued number.

2) To converge on the optimum solution, it is always necessary to satisfy the following condition in the dynamic updating of the Hopfield network:

$$\frac{dE(\cdot)}{dt} \leq 0 \tag{7}$$

The energy function defined by Eq.(6) satisfies property 1. In Eq.(6), if

$$\hat{d}_n = \hat{x}_n - \sum_{k=1}^{m} A_k \exp(-b_k n) \exp\{j(2\pi f_k n + \varphi_k)\} \tag{8}$$

then the energy function can be expressed as:

$$E = -\frac{1}{2} \sum_{n=0}^{N-1} \left| \hat{d}_n \right|^2 = -\frac{1}{2} \sum_{n=0}^{N-1} \hat{d}_n \hat{d}_n^{*}$$
(* : denotes the complex conjugate) $\tag{9}$

From Eq.(6), when the parameters A_k, b_k, f_k, and φ_k in Eq.(1) are replaced by P_k, the time variation of the above energy function can be expressed as

$$\frac{dE}{dt} = \sum_{k=1}^{m} \sum_{P_k} \frac{\partial E}{\partial P_k} \frac{dP_k}{dt} \ , \ (P_k; \ A_k, \ b_k, \ f_k, \ \varphi_k, \ k=1,\cdots,m) \tag{10}$$

Here, suppose that

$$\frac{dP_k}{dt} = -\left(\frac{\partial E}{\partial P_k}\right)^{*} \tag{11}$$

Then,

$$\frac{dE}{dt} = -\sum_{k=1}^{m} \sum_{P_k} \left| \frac{\partial E}{\partial P_k} \right|^2 \leq 0 \tag{12}$$

will hold, and property 2 is satisfied, so convergence in the dynamic updating of the modified

complex-valued Hopfield neural network is guaranteed.

From Eq.(9), the variation of the energy function with the variation of the parameters is as follows:

$$\frac{\partial E}{\partial P_k} = -\frac{1}{2} \sum_{n=0}^{N-1} \left(\frac{\partial \hat{d}_n}{\partial P_k} \hat{d}_n^{*} + \frac{\partial \hat{d}_n^{*}}{\partial P_k} \hat{d}_n \right) \tag{13}$$

From the form of the right-hand side of Eq.(13),

$$\left(\frac{\partial E}{\partial P_k} \right)^{*} = \frac{\partial E}{\partial P_k} \tag{14}$$

Then, by Eqs.(11) and (14), we have

$$\frac{dP_k}{dt} = -\frac{\partial E}{\partial P_k} = \frac{1}{2} \sum_{n=0}^{N-1} \left(\frac{\partial \hat{d}_n}{\partial P_k} \hat{d}_n^{*} + \frac{\partial \hat{d}_n^{*}}{\partial P_k} \hat{d}_n \right) \tag{15}$$

Equation (15) expresses the time variation of the parameters P_k, that is, the updating of the parameters. Suppose that $\hat{d}_n^{*}$ and $\hat{d}_n$ on the right-hand side of the equation are the inputs to the modified Hopfield neural network and $\partial \hat{d}_n / \partial P_k$ and $\partial \hat{d}_n^{*} / \partial P_k$ are the input weights in the network, then a Hopfield complex-valued network can be designed. The inputs and the input weights are then calculated by Eq.(8). In this network, two complex-valued input systems conjugated to each other are input to the network. The updating of the parameters is then carried out by complex calculation. However, because the two terms on the right-hand side of Eq.(15) are complex conjugates of each other, the left-hand side is a real number. The structure of the complex-valued network is depicted in Figure 1, where the coefficient 1/2 in Eq.(15) is omitted. Two complex-valued input systems conjugated to each other are input to one

Figure 1. Structure of the modified Hopfield neural network

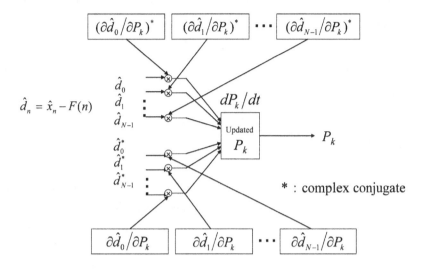

unit, and the updating of the parameters is carried out by complex-valued calculation.

Eq.(8) can be decomposed into a real part $d_{re}(n)$ and an imaginary part $d_{im}(n)$:

$$\hat{d}_n = d_{re}(n) + jd_{im}(n) \tag{16}$$

Suppose that the real and imaginary parts of $\hat{x}_n$ are denoted as $x_r(n)$ and $x_i(n)$, respectively. Then we have:

$$d_{re}(n) = x_{re}(n) \\ - \sum_{k=1}^{m} A_k \exp(-b_k n) \cos\{j(2\pi f_k n + \varphi_k)\} \tag{17}$$

$$d_{im}(n) = x_{im}(n) \\ - \sum_{k=1}^{m} A_k \exp(-b_k n) \sin\{j(2\pi f_k n + \varphi_k)\} \tag{18}$$

From these,

$$\begin{aligned} \left|\hat{d}_n\right|^2 &= \hat{d}_n \hat{d}_n^* \qquad (*: \text{complex conjugate}) \\ &= \left\{d_{re}(n) + jd_{im}(n)\right\}\left\{d_{re}(n) - jd_{im}(n)\right\} \\ &= \left|d_{re}(n)\right|^2 + \left|d_{im}(n)\right|^2 \end{aligned} \tag{19}$$

Then, Eq.(9) can be developed as follows:

$$E = -\frac{1}{2}\sum_{n=0}^{N-1}\left|\hat{d}_n\right|^2 - E_{re} + E_{im} \tag{20}$$

$$E_{re} = -\frac{1}{2}\sum_{n=0}^{N-1}\left|d_{re}(n)\right|^2 \tag{21}$$

$$E_{im} = -\frac{1}{2}\sum_{n=0}^{N-1}\left|d_{im}(n)\right|^2 \tag{22}$$

From Eqs.(10) and (20), we obtain

$$\begin{aligned} \frac{dE}{dt} &= \sum_{k=1}^{m}\sum_{P_k}\frac{\partial E}{\partial P_k}\frac{dP_k}{dt} \\ &= \frac{dE_{re}}{dt} + \frac{dE_{im}}{dt} \\ &= \sum_{k=1}^{m}\sum_{P_k}\left(\frac{\partial E_{re}}{\partial P_k} + \frac{\partial E_{im}}{\partial P_k}\right)\frac{dP_k}{dt} \end{aligned} \tag{23}$$

where,

$$\frac{dE_{re}}{dt} = \sum_{k=1}^{m}\frac{\partial E_{re}}{\partial P_k}\frac{dP_k}{dt} \tag{24}$$

$$\frac{dE_{im}}{dt} = \sum_{k=1}^{m} \frac{\partial E_{im}}{\partial P_k} \frac{dP_k}{dt}$$
$$(P_k ;\ A_k,\ b_k,\ f_k,\ \varphi_k)$$

(25)

Assume that

$$\frac{dP_k}{dt} = - \left(\frac{\partial E_{re}}{\partial P_k} + \frac{\partial E_{im}}{\partial P_k} \right)$$
$$(P_k ;\ A_k,\ b_k,\ f_k,\ \varphi_k, \qquad k = 1, \cdots, m)$$

(26)

We can get the following:

$$\frac{dE}{dt} = \sum_{k=1}^{m} \sum_{P_k} \left| \frac{\partial E_{re}}{\partial P_k} + \frac{\partial E_{im}}{\partial P_k} \right|^2 \leq 0$$

(27)

From Eqs.(21) and (22), we obtain

$$\frac{\partial E_{re}}{\partial P_k} = - \sum_{n=0}^{N-1} \frac{\partial d_{re}(n)}{\partial P_k} d_{re}(n)$$

(28)

$$\frac{\partial E_{im}}{\partial P_k} = - \sum_{n=0}^{N-1} \frac{\partial d_{im}(n)}{\partial P_k} d_{im}(n)$$

(29)

Hence, Eq.(26) can be expressed as follows:

$$\frac{dP_k}{dt} = - \left(\frac{\partial E_{re}}{\partial P_k} + \frac{\partial E_{im}}{\partial P_k} \right)$$
$$= \sum_{n=0}^{N-1} \left(\frac{\partial d_{re}(n)}{\partial P_k} d_{re}(n) + \frac{\partial d_{im}(n)}{\partial P_k} d_{im}(n) \right)$$
$$(P_k ;\ A_k,\ b_k,\ f_k,\ \varphi_k, \qquad k = 1, \cdots, m)$$

(30)

From Eqs.(15) and (30), the complex-valued network can be expressed as an equivalent real-valued network which has two real-valued input systems. That is, let the parameters change with time as shown in Eq.(30). Then, the energy function E satisfies property 2, above. Thus, convergence in the updating of the complex-valued network can be guaranteed. The equivalent network is depicted in Figure 2. The parameters are updated by the steepest descent method as follows:

$$P_k = P_k + \varepsilon \frac{dP_k}{dt} , \qquad (\varepsilon > 0)$$
$$(P_k ;\ A_k,\ b_k,\ f_k,\ \varphi_k, \qquad k = 1, \cdots, m)$$

(31)

Figure 2. The structure of real-valued network equivalent to complex-valued network shown in Figure 1

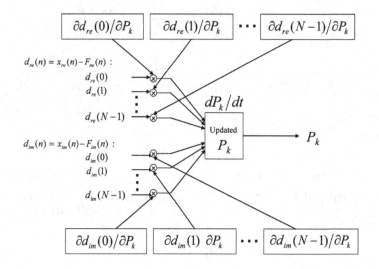

166

For every parameter P_k, this equivalent network forms a unit which has two input systems, corresponding to the real and imaginary parts of the NMR signal. Each input system has an input d and an input weight $\partial \hat{d}_n / \partial P_k$ corresponding to the number of sample points. d is calculated by Eqs.(17) and (18), which means that the inputs and input weights are calculated using the previous values of the parameters and the observed signal. By means of this input, the state of the unit is changed and each parameter is updated. The input and the input weights are recalculated with the updated parameters. This is the equivalent network implemented in this chapter.

Incidentally, k in Eqs.(30) and (31) represents one of the components of an NMR signal. Because Eq.(30) is applied to each k, the updating of parameters A_k, b_k, f_k, and φ_k is simultaneously carried out on k. As described, in the network used in this chapter, the sequential updating of each unit, which is a feature of the Hopfield network, is transformed to sequential updating of every unit group A_k, b_k, f_k, and φ_k on k.

As expressed in Eq.(1), the NMR signal is a set of sinusoidal waves in which the spectral components A_k are exponentially damped with time n. The operation shown in Figure 3 is introduced so that the proposed network would recognize the decay state more accurately. In the figure, the horizontal axis shows the sample points at time n, and the vertical axis shows the NMR signal. In this operation, first, appropriate values are assigned as initial values for each of the parameters. Then, our network operates on section A from time 0 ($n = 0$) to an adequate time k_1 ($n = k_1$). The parameter estimates are obtained when the network has equilibrated. Next, the network operates on section B from time 0 to an adequate time k_2 ($k_1 < k_2$). The equilibrium values in section A are used as the initial values in section B. Thereafter, we extend the section in the same way, and finally, the network operates on the entire time interval corresponding to all sample points. This operation is equivalent to recognizing the shape of the signal by gradually extending the observation section while taking into account the detailed aspects of the signal during its most rapid change.

Figure 3. Illustration of the modified sequential extension of section (SES) method

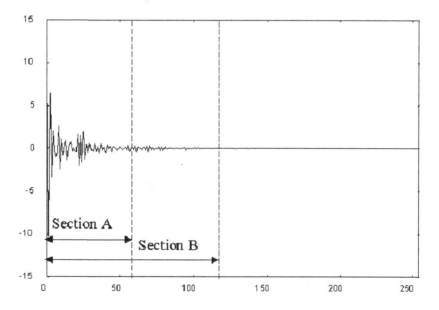

SIMULATION RESULTS AND DISCUSSION

Sample signals, equivalent to NMR signals that consisted of 1024 data points on a spectrum with a bandwidth of 2000 Hz for the atomic nucleus of phosphorus-31 in a static magnetic field of 2 Tesla, are simulated. The three signals shown in Table 1 and Figure 4, Figure 5 and Figure 6 were used.

In Table 1, peaks 1 through 7 represent phosphomonoesters (PME), inorganic phosphate (Pi), phosphodiesters (PDE), creatine phosphate (PCr), γ-adenocine triphosphate (γ-ATP), α-ATP, and β-ATP, respectively.

Signal 1 and 3 are equivalent to the spectra of healthy cells with a normal energy metabolism. Pi is relatively small in the spectral components in their signals. Signal 2 is equivalent to the spectrum of a cell that is approaching necrosis. In such cells, the metabolism of energy is decreased and Pi is large in comparison with other components, as shown in Table 1. This signal is analogous to a single-component spectral signal (a monotonic damped signal) compared with signal 1 and 3. Among these signals, only the spectral component A_k is different.

Table 1. Parameters of sample signals

Peak	$f_k{}^*$	b_k	$\Phi_k{}^{**}$	A_k (1)	A_k (2)	A_k (3)
1	0.368	0.05395	0.4774	0.726	0.7	0.996
2	0.397	0.03379	0.3699	1.02	6.246	0.5
3	0.435	0.05918	0.2296	2.1	1.8	2.1
4	0.485	0.03785	0.051	2.37	1.2	3.6
5	0.526	0.04858	-0.1002	1.89	0.5	1.15
6	0.616	0.05744	-0.4264	2.04	0.5	2.2
7	0.763	0.04035	-0.9657	1.1	0.3	0.7

Figure 4. Signal 1

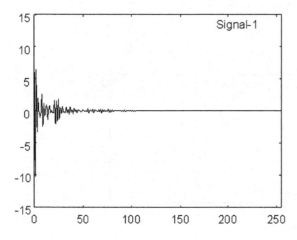

Figure 5. Signal 2

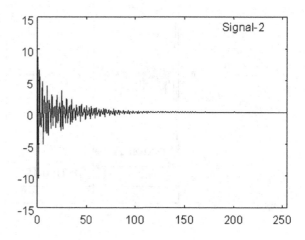

Figure 6. Signal 3

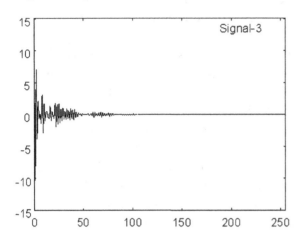

IMPLEMENTATION OF THE NETWORK

We next introduce some auxiliary operations that are necessary for stable implementation of the proposed network. The settings of the initial values of the parameters are shown in Table 2. For the amplitude A_k, the amplitudes of the real part and the imaginary part are compared; the larger is divided by 7, the number of signal components; and the result is used as the initial value for all seven components. For the initial values of the frequency f_k and the damping coefficient b_k of each metabolite, rough values are known for f_k and b_k under observation conditions, as described in "NMR spectra" above. Therefore, the initial values were set close to their rough values. All of the initial phases are set to zero.

In the steepest descent method in Eq.(15), two values, 10^{-5} and 10^{-6} are used as ε. By setting the upper limit of the number of the parameter updates to 50,000, we ensure that the units continue to be renewed until the energy function decreases. Then, the parameters can be updated while the energy function is decreasing and the number of renewals does not exceed the upper limit. By using these procedures, it is possible to operate the network in a stable condition. In addition, the following two conditions for stopping the network are set.

1. The updates of all parameters are terminated.
2. The energy function reaches an equilibrium point.

The criterion for condition 1 is a limit on the time variation of the parameter P_k: if $dP_k/dt \leq 0.01$, we set $dP_k/dt = 0$ and terminate the updating of the parameter.

Regarding condition 2, we judge that the energy function has reached an equilibrium point when the energy function increases, or when the number of updates exceeds the upper limit mentioned above. Theoretically, the network stops and an optimum solution is obtained when the above two conditions are satisfied simultaneously. However, because a monotonic decrease of the energy function is

Table 2. Initial values of parameters

Peak	f_k	b_k	Φ_k	A_k (1)	A_k (2)	A_k (3)
1	0.35	0.1	0.0	1.481595	1.502836	1.505209
2	0.4	0.1	0.0	1.481595	1.502836	1.505209
3	0.45	0.1	0.0	1.481595	1.502836	1.505209
4	0.5	0.1	0.0	1.481595	1.502836	1.505209
5	0.55	0.1	0.0	1.481595	1.502836	1.505209
6	0.6	0.1	0.0	1.481595	1.502836	1.505209
7	0.75	0.1	0.0	1.481595	1.502836	1.505209

produced by the above-mentioned operations, in practice, we force the network to stop when either of the two conditions occurs.

In each renewal of the unit, we also adjust the network so that the update values do not depart greatly from the actual values by using the prior knowledge of the spectrum outlined in "Initial values of the parameters" above. For the frequency f_k, we adopt only values within a range of 0.05 around the values in Table 1. A similar procedure is also carried out for the phase φ_k: the range is ±1.0. For the damping coefficient b_k, we adopt only values below 0.1.

As shown in Figures 4-6, the sample signals have decayed to near-zero amplitude after 255 points on the time axis (each full data set has 1024 points). Therefore, we performed the MSES method for the following three sets of sections:

1. Four sections: [0-63], [0-127], [0-255], [0-1023]
2. Five sections: [0-31], [0-63], [0-127], [0-255], [0-1023]
3. Six sections: [0-15], [0-31], [0-63], [0-127], [0-255], [0-1023]

DETERMINATION BY USING MODIFIED HOPFIELD NEURAL NETWORK

First parameter estimations of the sample signals are performed using the modified Hopfield neural network without the MSES technique. For signal 1, the result of the determination using $\varepsilon = 10^{-6}$ was better than for $\varepsilon = 10^{-5}$. In the case using $\varepsilon = 10^{-6}$, the spectral composition A_k and the frequency f_k were more accurately estimated than the damping coefficient b_k and the phase φ_k. Except for peaks 1 and 2, the errors in A_k were less than 20% in relative terms, and all of the errors in f_k were less than 10%.

Although the effect of the difference in ε became quite small for signals 2 and 3, the same tendency was also shown. However, even using $\varepsilon = 10^{-6}$, the estimation of signals 2 and 3 was not as good as the estimation of signal 1. In these results, all of the errors in f_k are less than 10%, but the only peaks with errors of less than 20% of A_k were peak 3 (about 11%) in signal 2, and peaks 3 (19.7%) and 6 (4.38%) in signal 3.

In summary, the estimation for signal 1 was the best of the three signals. The estimation errors for signals 1 and 2 using $\varepsilon = 10^{-6}$ are shown in Tables 3 and 4.

DETERMINATION COMBINED WITH MODIFIED SEQUENTIAL EXTENSION OF SECTION (MSES)

Using the MSES technique, the determination results were improved. For signal 1, which was best estimated using modified Hopfield neural network alone, when we applied the four-section extension method using $\varepsilon = 10^{-6}$, we were able to

Table 3. The estimated error (%) of signal 1 using the complex-valued neural network

Peak	f_k	b_k	Φ_k	A_k
1	7.82	-20.9	-83.4	58.9
2	9.65	195.9	-25.4	37.5
3	0.16	57.5	22.9	-14.6
4	0.47	-32.9	-100.4	-19.9
5	0.08	-29.3	-13.1	-2.1
6	-0.28	-2.92	49.3	-3.97
7	0.05	0.77	0.58	0.45

Table 4. The estimated error (%) of signal 2 using the complex-valued neural network

Peak	f_k	b_k	Φ_k	A_k
1	7.80	-61.4	-109.5	258.4
2	0.29	3.0	47.0	-66.7
3	-8.05	-53.2	-421.0	11.1
4	-0.78	164.2	-625.1	38.0
5	-4.94	105.8	180.5	206.4
6	-0.75	74.0	-55.9	97.4
7	-1.68	147.8	-111.1	98.2

obtain the best result. For signal 1, the result of the estimation using $\varepsilon = 10^{-6}$ and four sections is shown in Table 5. Compared to Table 3, the accuracy of estimation of the damping coefficient b_k and the phase φ_k are improved. However, the accuracy of estimation of the frequency f_k is only slightly improved overall, and the resolution of the spectrum is also only slightly improved. The accuracy of the spectral composition A_k is improved at peaks 2 and 4, but degraded at peaks 1 and 3.

For signal 2, when we applied the MSES technique using six sections with $\varepsilon = 10^{-5}$, the errors were improved overall, compared to using the complex-valued network alone, but we still did not obtain an estimation as accurate as that for signal 1 (Table 6). For signal 3, we could not obtain accurate estimation using any combination of the choices for ε and the number of sections, especially for the spectral composition A_k.

ESTIMATION FOR THE SIGNAL WITH NOISE

Real NMR signals always include noise. Therefore, we need to verify the ability of the proposed method, that is, the modified Hopfield neural network combined with MSES, to estimate parameters for NMR signals that include noise. For that purpose, we used sample signals in which three levels of white Gaussian noise with signal to noise ratios (SNRs) of 10, 5, or 2 were added to signal 1, which was well-estimated compared to other two signals. The SNR is defined as follows:

$$SNR = \sum_{k=0}^{n-1} | F(t_k) |^2 \Big/ n\sigma^2 \qquad (32)$$

where, $F(t_k)$ is the signal composition at time t_k, σ^2 is the variance of the noise, and n is the total

Table 5. The estimated error (%) of signal 1 with modified sequential extension of section using 4 sections ($\varepsilon = 10^{-6}$)

Peak	f_k	b_k	Φ_k	A_k
1	8.67	-20.8	-163.2	78.9
2	5.26	195.9	6.84	28.4
3	-0.11	-15.1	123.4	-18.8
4	0.04	-3.9	55.9	-4.3
5	0.08	-0.5	3.4	-1.2
6	0.02	-1.83	-2.1	-1.96
7	0.09	-2.03	-0.99	-2.1

number of sample points (in this case, 1024). The sample signal with SNR = 2 is shown in Figure 7, and the results of the estimation of signals with each SNR are shown in Table 7, Table 8, and Table 9. For these results, we used $\varepsilon = 10^{-6}$ and the MSES method with four sections.

Comparing these results to those obtained from the sample data with no noise reported in Table 5, there is almost no change in estimation error for the frequency f_k, and the estimation error exceeds 10% only at peak 2 (11.9%) for SNR = 2. For the spectral composition A_k, peaks 4, 5, 6, and 7 had less than 10% error in Table 5. In the case where noise was added with SNR = 2, peaks 4 and 6 are estimated with better than 10% error, but -16.3% is obtained at peak 5 and -21.9% is obtained at peak 7. For the damping coefficient b_k, the peaks with small estimation errors in Table 5 maintain the same error level in the presence of noise.

Thus, we conclude that the proposed estimation method is not significantly influenced by noise for the estimation of f_k, A_k, and b_k. However, the phase φ_k had greater variation than in Table 5,

Figure 7. Noisy NMR signal with SNR = 2

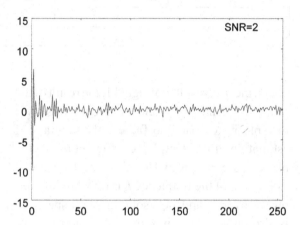

Table 6. The estimated error (%) of signal 2 with modified sequential extension of section using 6 sections ($\varepsilon = 10^{-5}$)

Peak	f_k	b_k	Φ_k	A_k
1	-0.33	85.4	-220.8	205.7
2	-2.14	96.2	170.3	29.1
3	3.59	68.8	-535.5	26.7
4	0.39	43.7	55.2	38.2
5	0.87	39.9	365.1	50.2
6	-0.10	15.7	-21.3	19.3
7	-0.30	5.84	-45.5	2.40

Table 7. The estimated error (%) of signal 1 with noise, SNR = 10

Peak	f_k	b_k	Φ_k	A_k
1	8.69	-13.7	-135.6	91.7
2	6.78	195.9	-0.53	15.3
3	0.14	-26.7	82.7	-31.4
4	-0.04	2.01	144.7	0.97
5	0.13	-1.59	-24.1	0.26
6	0.03	4.14	4.92	-0.34
7	0.16	-4.83	-3.54	-5.45

Table 8. The estimated error (%) of signal 1 with noise, SNR = 5

Peak	f_k	b_k	Φ_k	A_k
1	8.67	-31.5	-111.1	81.8
2	7.68	195.9	-16.40	15.9
3	0.16	-31.2	49.2	-34.6
4	0.04	-11.0	38.2	-8.31
5	0.27	-1.34	-115.5	0.11
6	-0.16	1.83	18.6	4.75
7	0.08	11.4	-3.55	5.0

Table 9. The estimated error (%) of signal 1 with noise, SNR = 2

Peak	f_k	b_k	Φ_k	A_k
1	8.39	-60.4	-121.8	26.7
2	11.9	195.9	-60.4	72.6
3	0.29	-18.9	-216.4	-24.8
4	-0.29	9.33	420.4	7.97
5	0.17	-22.2	-53.7	-16.30
6	-0.52	-1.49	49.4	-5.54
7	0.07	-3.84	-1.44	-21.9

revealing that the estimation of phase is easily influenced by noise.

DISCUSSION

The results of the simulations indicate that the modified Hopfield neural network has the ability to estimate four different parameters of the NMR signal. The simulation results show that the frequency composition f_k and the spectral composition A_k can be estimated with less error than the damping coefficient b_k and the phase φ_k. When MSES was applied to this neural network method, it was found that the estimation precisions of b_k and φ_k were improved. In addition, it was shown that this combined method experiences no rapid decline in accuracy when applied to signals to which noise was added. However, the optimal sections on which to apply MSES and the optimal step size of ε are different for every simulated NMR

signal, and it was verified that they do influence the estimation accuracy.

In the proposed estimation method, preliminary knowledge about the targeted spectrum is indispensable when determining the initial value of the parameters and updating them during the estimation process. If there is no preliminary knowledge, the network must search for the solution in an unlimited solution space, and the probability of reaching an optimum solution in a reasonable time period becomes very small. In addition, because the steepest-descent method is used to update the parameters, it is difficult for the network to reach the optimum solution if it starts from inappropriate initial values.

MSES uses the equilibrium values of the parameters calculated using one section as the initial values for the following section in a sequence. In other words, every time a section is extended, the neural network is used to minimize a new energy function with new initial values and a new group of data. Therefore, when calculation on the new section begins, the direction in which a minimum solution has previously been sought is reset, and the network is free to search in another direction. This may reduce the danger of falling into a local minimum solution. However, when the damping of the target signal is monotonic (depending on the determination of the initial section), it appears that the search direction may no longer be effectively reset and the network cannot escape from a local solution.

The signal in Figure 7, which contains noise, maintains the characteristics of the initial damping for the noise-free version of the same signal in Figure 4. It seems that this fact was advantageous in MSES. Therefore, comparably stable estimation accuracy in the presence of noise is obtained using preliminary knowledge of the parameters and the MSES method.

Usually, a Hopfield network cannot reach the optimum solution from a local solution without restarting from different initial values (Dayhoff, J. E., 1989). SES carries out this operation automatically. A Boltzmann machine (Geman et al., 1984, David et al., 1985, Hinton et al., 1986) might be used to avoid local solutions and approach the optimum solution. However, in that method, the state of the network is not indeterminate and it is changed stochastically. Thus, stability of the decrease in energy with state transitions is not guaranteed. Compared with the avoidance of the local solution by the Boltzmann machine, MSES seems to be more elegant because it is free of the uncertainty associated with the stochastic operation. However, the stability of convergence to the optimal solution is influenced by the damping state of the targeted signal, and we must overcome this problem.

CONCLUSION

An efficient modified Hopfield neural network for NMR spectrum estimation has been presented.

The main valuable achievement of this chapter is that the estimation operation is accelerated by performing cross correlation in the frequency domain between the input data and the input weights of neural networks. Unlike the conventional quantitative methods of NMR spectrum estimation using hierarchical neural networks, the proposed algorithm does not need a learning process. In addition, the MSES method has been devised and used in combination with Hopfield neural network in order to take into account the damping state of the NMR signal. For performance evaluation of the proposed estimation method, simulations have been carried out using sample signals composed of seven different metabolites to simulate in vivo ^{31}P-NMR spectra, with and without added noise.

Simulations results have shown that the proposed method has the ability to estimate the modeling parameters of the NMR signal. However, it was also shown that its ability differs according to the damping state of the signals.

The investigation here has indicated that MSES reduces the danger of falling into a local minimum in the search for the optimum solution using a Hopfield neural network. Although there another technique such as a Boltzmann machine might be used to avoid local solutions, it is stochastic and requires much futile searching before it reaches the optimum solution. On the other hand, it has been observed that the proposed method could find the optimum solution stably if the variation in the targeted signal could be identified accurately.

REFERENCES

Ala-Korpela, M., Changani, K.K., Hiltunen, Y., Bell, J.D., & Fuller, B.J., Bryant, et al. (1997). Assessment of quantitative artificial neural network analysis in a metabolically dynamic ex vivo 31P NMR pig liver study. *Magnetic Resonance in Medicine, 38*, 840–844. doi:10.1002/mrm.1910380522

Belliveau, J. W., Kennedy, D. N. Jr, McKinstry, R. C., Buchbinder, B. R., Weisskoff, R. M., & Cohen, M. S. (1991). Functional mapping of the human visual cortex by magnetic resonance imaging. *Science, 254*(5032), 716–719. doi:10.1126/science.1948051

Benvenuto, N., & Piazza, F. (1992). On the complex backpropagation algorithm. *IEEE Transactions on Signal Processing, 40*(4), 967–969. doi:10.1109/78.127967

Cooley, J. W., & Tukey, J. W. (1965). An algorithm for machine calculation of complex Fourier series. *Mathematics of Computation, 19*(90), 297–301. doi:10.2307/2003354

David, H.A., Hinton, G.E. & Sejnowski, T.J. (1985). A Learning Algorithm for Boltzmann Machines. *Cognitive Science: A Multidisciplinary Journal, 9*(1), 149-169.

Dayhoff, J. E. (1989). *Neural Network Architectures: An Introduction.* New York: Van Nostrand Reinhold.

Derome, A. E. (1987). *Modern NMR Techniques for Chemistry Research (Organic Chemistry Series, Vol 6).* Oxford, UK: Pergamon Press.

El-Bakry, H. M. (2006). New Fast Time Delay Neural Networks Using Cross Correlation Performed in the Frequency Domain. *Neurocomputing Journal, 69*, 2360–2363. doi:10.1016/j.neucom.2006.03.005

El-Bakry, H. M., & Zhao, Q. (2005). Fast Time Delay Neural Networks. *International Journal of Neural Systems, 15*(6), 445–455. doi:10.1142/S0129065705000414

Feinberg, D. A., & Oshio, K. (1991). GRASE (gradient and spin echo) MR imaging: A new fast clinical imaging technique. *Radiology, 181*, 597–602.

Geman, S., & Geman, D. (1984). Stochastic Relaxation, Gibbs Distribution and the Bayesian Restoration of Images. *IEEE Transactions on Pattern Analysis and Machine Intelligence, 6*, 721–741. doi:10.1109/TPAMI.1984.4767596

Georgiou, G. M., & Koutsougeras, C. (1992). Complex domain backpropagation. *IEEE Transactions on Circuits and Systems. 2, Analog and Digital Signal Processing, 39*(5), 330–334. doi:10.1109/82.142037

Han, L., & Biswas, S. K. (1997). Neural networks for sinusoidal frequency estimation. *Journal of the Franklin Institute, 334B*(1), 1–18. doi:10.1016/S0016-0032(96)00079-8

Haselgrove, J. C., Subramanian, V. H., Christen, R., & Leigh, J. S. (1988). Analysis of in-vivo NMR spectra. *Reviews of Magnetic Resonance in Medicine, 2,* 167–222.

Henning, J., Nauerth, A., & Fnedburg, H. (1986). RARE imaging: A first imaging method for clinical MR. *Magnetic Resonance in Medicine, 3*(6), 823–833. doi:10.1002/mrm.1910030602

Hinton, G. E., & Sejnowski, T. J. (1986). Learning and Relearning in Boltzmann Machine. *Parallel distributed processing: explorations in the microstructure of cognition, vol. 1: foundations* (pp. 282-317). Cambridge, MA: MIT press.

Hirose, A. (1992a). Dynamics of fully complex-valued neural networks. *Electronics Letters, 28*(16), 1492–1494. doi:10.1049/el:19920948

Hirose, A. (1992b). Proposal of fully complex-valued neural networks. [), Baltimore, MD.]. *Proceedings of International Joint Conference on Neural Networks, 4*, 152–157.

Hopfield, J. J. (1982). Neural networks and physical systems with emergent collective computational abilities. *Proceedings of the National Academy of Sciences of the United States of America, 79*, 2554–2558. doi:10.1073/pnas.79.8.2554

Hopfield, J. J. (1984). Neurons with graded response have collective computational properties like those of two-state neurons. *Proceedings of the National Academy of Sciences of the United States of America, 81*, 3088–3092. doi:10.1073/pnas.81.10.3088

Jankowski, S., Lozowski, A., & Zurada, J. M. (1996). Complex-valued multistate neural associative memory. *Proceedings of IEEE Transactions on Neural Networks, 7*(6), 1491–1496. doi:10.1109/72.548176

Kaartinen, J., Mierisova, S., Oja, J. M. E., Usenius, J. P., Kauppinen, R. A., & Hiltunen, Y. (1998). Automated quantification of human brain metabolites by artificial neural network analysis from in vivo single-voxel 1H NMR spectra. *Journal of Magnetic Resonance (San Diego, Calif.), 134*, 176–179. doi:10.1006/jmre.1998.1477

Kuroe, Y. Hashimoto. N. & Mori, T. (2002). On energy function for complex-valued neural networks and its applications, Neural information proceeding. *Proceedings of the 9th International Conference on Neural Information Processing Computational Intelligence for the E-Age* (Vol.3, pp. 1079-1083).

Kwong, K., Belliveau, J. W., Chesler, D. A., Goldberg, I. E., Weisskoff, R. M., & Poncelet, B. P. (1992). Dynamic magnetic resonance imaging of human brain activity during primary sensory stimulation. *Proceedings of the National Academy of Sciences of the United States of America, 89*(12), 5675–5679. doi:10.1073/pnas.89.12.5675

Maddams, W. F. (1980). The scope and Limitations of Curve Fitting. *Applied Spectroscopy, 34*(3), 245–267. doi:10.1366/0003702804730312

Mansfield, P. (1977). Multi-planar image formation using NMR spin echoes. *Journal of Physics. C. Solid State Physics, 10*, 55–58. doi:10.1088/0022-3719/10/3/004

Melki, P. S., Mulkern, R. V., Panych, L. S., & Jolesz, F. A. (1991). Comparing the FAISE method with conventional dual-echo sequences. *Journal of Magnetic Resonance Imaging, 1*, 319–326. doi:10.1002/jmri.1880010310

Meyer, C. H., Hu, B. S., Nishimura, D. G., & Macovski, A. (1992). Fast Spiral Coronary Artery Imaging. *Magnetic Resonance in Medicine, 28*(2), 202–213. doi:10.1002/mrm.1910280204

Miersová, S., & Ala-Korpela, M. (2001). MR spectroscopy quantification: a review of frequency domain methods. *NMR in Biomedicine, 14*, 247–259. doi:10.1002/nbm.697

Morita, N., & Konishi, O. (2004). A Method of Estimation of Magnetic Resonance Spectroscopy Using Complex-Valued Neural Networks. *Systems and Computers in Japan, 35*(10), 14–22. doi:10.1002/scj.10705

Naressi, A., Couturier, C., Castang, I., de. Beer, R., & Graveron-Demilly, D. (2001). Java-based graphical user interface for MRUI, a software package for quantitation of in vivo medical magnetic resonance spectroscopy signals. *Computers in Biology and Medicine, 31*, 269–286. doi:10.1016/S0010-4825(01)00006-3

Nitta, T. (1997). An Extension of the Back-Propagation Algorithm to Complex Numbers. *Neural Networks, 10*(8), 1392–1415. doi:10.1016/S0893-6080(97)00036-1

Nitta, T. (2000). An Analysis of the Fundamental Structure of Complex-Valued Neurons. *Neural Processing Letters, 12*(3), 239–246. doi:10.1023/A:1026582217675

Nitta, T. (2002). Redundancy of the Parameters of the Complex-valued Neural Networks. *Neurocomputing, 49*(1-4), 423–428. doi:10.1016/S0925-2312(02)00669-0

Nitta, T. (2003). On the Inherent Property of the Decision Boundary in Complex-valued Neural Networks. *Neurocomputing, 50*(c), 291–303. doi:10.1016/S0925-2312(02)00568-4

Nitta, T. (2003). Solving the XOR Problem and the Detection of Symmetry Using a Single Complex-valued Neuron. *Neural Networks, 16*(8), 1101–1105. doi:10.1016/S0893-6080(03)00168-0

Nitta, T. (2003). The Uniqueness Theorem for Complex-valued Neural Networks and the Redundancy of the Parameters. *Systems and Computers in Japan, 34*(14), 54–62. doi:10.1002/scj.10363

Nitta, T. (2003). Orthogonality of Decision Boundaries in Complex-Valued Neural Networks. *Neural Computation, 16*(1), 73–97. doi:10.1162/08997660460734001

Nitta, T. (2004). Reducibility of the Complex-valued Neural Network. *Neural Information Processing - Letters and Reviews, 2*(3), 53-56.

Ogawa, S., Lee, T. M., Nayak, A. S., & Glynn, P. (1990). Oxygenation-sensitive contrast in magnetic resonance image of rodent brain at high magnetic fields. *Magnetic Resonance in Medicine, 14*(1), 68–78. doi:10.1002/mrm.1910140108

Provencher, S., W. (2001). Automatic quantification of localized in vivo ^{1}H spectra with LC-Model. *NMR in Biomedicine, 14*(4), 260–264. doi:10.1002/nbm.698

Rumelhart, D. E., Hinton, G. E., & Williams, R. J. (1986). Learning internal representations by error propagation. In Rumelhart, D,E and McClelland, J.L.(Eds.), *Parallel Distributed Processing,* (Vol. 1: *Foundations*, pp. 318-362). Cambridge, MA: MIT press.

Sijens, P. E., Dagnelie, P. C., Halfwrk, S., van Dijk, P., Wicklow, K., & Oudkerk, M. (1998). Understanding the discrepancies between 31P MR spectroscopy assessed liver metabolite concentrations from different institutions. *Magnetic Resonance Imaging, 16*(2), 205–211. doi:10.1016/S0730-725X(97)00246-4

van den Boogaart, A., Van Hecke, P., Van Hulfel, S., Graveron-Dermilly, D., van Ormondt, D., & de Beer, R. (1996). MRUI: a graphical user interface for accurate routine MRS data analysis. *Proceeding of the European Society for Magnetic Resonance in Medicine and Biology 13th Annual Meeting*, Prague, (p. 318).

van Huffel, S., Chen, H., Decanniere, C., & Hecke, P. V. (1994). Algorithm for time-domain NMR data fitting based on total least squares. *Journal of Magnetic Resonance. Series A., 110*, 228–237. doi:10.1006/jmra.1994.1209

Zhou, C., & Liu, L. (1993). Complex Hopfield model. *Optics Communications, 103*(1-2), 29–32. doi:10.1016/0030-4018(93)90637-K

Chapter 11
Advances in Biosensors
for In Vitro Diagnostics

David P. Klemer
University of Wisconsin-Milwaukee, USA

ABSTRACT

The array of tools available to the medical practitioner for diagnosis of disease has experienced extremely rapid expansion over the past decades. Traditional "blood chemistries" and hematological testing have been augmented with immunoassays for serological testing and PCR-based assays for genomic screening. Rapid, inexpensive point-of-care assays with enhanced sensitivity and specificity have the potential for altering the manner in which medicine is practiced; pharmacogenomics and the advent of "personalized medicine" permit the tailoring of therapeutic pharmacologic regimens to the genetic makeup of an individual. Facilitating this are novel biosensing approaches for in vitro diagnostics, developed at the interface of engineering, physics, chemistry and biology. New discoveries promise to sustain the high rate of growth of this important field of research and development. This chapter examines recent advances in techniques for biosensing and in vitro biomedical diagnostics, building on progress in materials science, nanotechnology, semiconductor devices, and biotechnology. The importance of this topic is motivated through the presentation of case studies of biosensing applications within various medical specialties.

INTRODUCTION

It is difficult to craft a precise definition of a "biosensor," given that the concept of biosensing is very broad in scope, involving a wide range of implementations. At the most fundamental level, biosensors are devices which are capable of pro-

viding qualitative and/or quantitative information about biologically-relevant entities or components. These entities can span the dimensional scale from ions to complete organisms. For example, an optical probe which measures hydrogen ion concentration (pH) using colorimetric dyes can be considered a biosensor, used to detect and quantify urea or glucose. For that matter, an even simpler biosen-

DOI: 10.4018/978-1-60566-768-3.ch011

sor might consist of a strip of pH paper! At the other extreme, sentinel animals such as chickens may also be considered biosensors, frequently used to detect the presence of viruses and larger microorganisms responsible for epidemics of infectious disease over a large geographical region (Rizzoli, 2007). Indeed, the proverbial "canary in a coal mine" served as a sensor for toxic chemical species which could endanger life—a living "biosensor" used in centuries past.

In this chapter, we will limit the scope of our discussion of biosensing, considering only sensors that are designed to identify and quantify the presence of biomolecular species of interest. In addition, we will focus our attention on sensors which perform measurements of an analyte outside the human body, so-called *in vitro* sensing or *in vitro* diagnostics. This is the type of sensing that a general practitioner may be interested in when drawing a blood sample from a patient in a family medicine clinic. In the discussion which follows, we will not address the type of sensing accomplished, for example, using a radioactively labeled "molecular probe" injected into the body and used to detect the presence of cancer using a PET (positron emission tomography) scanner. Although this *in vivo* approach to diagnostics based on biomedical imaging is of great interest, such techniques fall outside the scope of this chapter.

BIOSENSING TECHNIQUES AND EXAMPLES

It is helpful to develop a framework to organize a discussion of biosensors. The basic schematic block diagram of a biosensor is shown in Figure 1. The sensing platform typically consists of a molecule—often termed a "biorecognition element"—which has a particular affinity for another biomolecule to be detected (Prasad, 2003). Of the four principal families of organic biomolecules which contribute to life (proteins, nucleic acids, saccharides and lipids), proteins and nucleic acids have received a great deal of attention as biorecognition elements, given their high specificity of intermolecular interactions. Binding of the recognition element to a target biomolecule is transformed into a measurable signal via a transduction process; this process may be facilitated by an external excitation or stimulus. As an example, consider a target biomolecule which is tagged with a fluorescent molecule and bound to a biorecognition probe molecule immobilized onto a biosensor. In this case, an ultraviolet excitation signal might be used to excite a fluorescence emission which is sensed using a photodetector and transduced into an electrical current. The current is then converted to a detected voltage, processed and used for further analysis or display.

Biosensor implementations can be organized and classified using various schemas, based on the

Figure 1. Conceptual framework for biosensor implementation

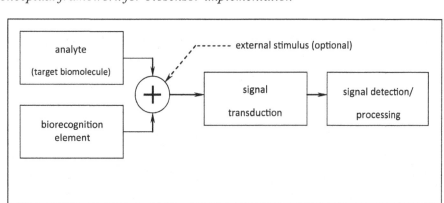

form of the biorecognition element, the physico-chemical or electrical property which is converted into a measurable signal, or the type of external stimulus used to obtain that transduced signal. This is illustrated in Table 1, which describes a number of factors which may be involved in the implementation of a particular form of biosensor.

Table 1 is by no means comprehensive; researchers are continually investigating interactions between biomolecules and electromagnetic waves across the spectrum, from radio waves to gamma rays – from Raman spectroscopy in the infrared region (Lyandres, 2008) to gamma ray emission for *in vivo* nuclear imaging modalities such as positron emission tomography (PET) imaging. Biomolecular recognition techniques also continue to increase in complexity, with the advent of nanomaterials used as fluorescent labels (Su, 2008) and man-made peptide nucleic acids with high binding affinity to oligonucleotides (Joung, 2008). The rapidly declining cost of techniques

used in biomolecular synthesis guarantees continued interest in novel biosensor implementations based on new biotechnologies.

For purposes of illustration, we now examine several biosensor implementations using the framework given in Figure 1 and Table 1.

Colorimetric devices. As a prototypical example, the simple colorimetric immunoassay is one of the most widely used biosensors, ranging from the implementation as a sensitive assay in biomedical laboratories (the enzyme-linked immunosorbent assay, ELISA (Crowther, 1995)) to inexpensive tests such as the home pregnancy test sold in retail pharmacies. In this biosensor, an immobilized probe antibody serves as a biorecognition element for detection of the associated antigen, and a secondary enzyme-conjugated antibody is used to confirm binding through enzymatic activation of a substrate material, resulting in a visible color change. The 'lock and key' nature of

Table 1. Examples of physicochemical properties, biomolecular interactions and transduction signals used in the implementation of practical biosensors

Physical/chemical/electrical property	Biorecognition/analyte interaction	Transduction signal
Mass	**Protein**	**Photonic**
	antibody ↔ antigen	fluorescence
Optical interactions	receptor ↔ ligand	absorbance
wavelength/frequency	enzyme ↔ substrate	reflectance (refractive index/SPR angle)
amplitude/absorption/scattering		polarization state
phase	**Protein lectin**	
polarization	lectin ↔ carbohydrate	**Electronic**
	lectin ↔ glycoprotein	impedance (resistance, reactance)
Electrical characteristics		resonance frequency
conductivity	**Nucleic acids**	piezoelectric properties
electronic charge	ssDNA ↔ ssDNA	
dielectric constant	PNA ↔ ssDNA	**Chemical**
		pH
Biochemical characteristics		voltammetric signal
pH		

(ssDNA = single-stranded deoxyribonucleic acid, PNA = peptide nucleic acid, SPR = surface plasmon resonance) antigen-antibody binding provides high specificity for this type of sensor. Figure 2 illustrates one of several possible implementations of a colorimetric immunoassay.

Biophotonic devices. Another widely-used sensor for biomolecular detection based on an optical signal is the surface plasmon resonance (SPR) sensor (Liedberg, 1995). SPR sensors have been fabricated using various biorecognition elements, such as protein-based (antigen-antibody) immunosensors or oligonucleotide sensors for DNA detection. In this sensor, binding of the target molecule in an analyte solution with an immobilized probe molecule results in a change in angle of minimum optical reflectivity at a reflecting surface (the so-called 'surface plasmon resonance angle'), and this change can be readily detected using an optical photodetector. Advantages of this type of biosensor include the ability to perform label-free DNA detection, as well as the ability to observe intermolecular binding kinetics in real time.

Piezoelectric sensors. Also termed quartz-crystal microbalances, these devices detect a mass change when a target molecule in analyte solution binds to a probe recognition molecule which is immobilized onto a quartz crystal. The ensuing change in mass results in a change in the resonance frequency of the quartz crystal, which can readily be detected electronically.

Figure 3 illustrates a typical quartz crystal microbalance used for biosensing applications, alongside a flow cell which permits an analyte to be introduced to the quartz crystal. The crystal (with probe biomolecules immobilized onto the surface) is clamped in place within the cell, and target-probe biomolecular binding results in a mass change. This variation in mass induces a corresponding change in the resonance frequency of an electronic oscillator (Shen, 2008) which can be easily detected. Piezoelectric immunosensors have found wide use across many subspecialties of medicine, from infectious diseases (Pohanka, 2008) to oncology (Zeng, 2006).

MEMS sensors. Using standard techniques developed for the silicon semiconductor industry, microelectromechanical systems (i.e., MEMS devices) can be designed to realize complex electromechanical functions on a dimensional scale measured in micrometers. Researchers have fabricated DNA biosensors using miniaturized cantilevers etched into a silicon substrate, and have used these devices as sensitive DNA sensors (Ivanov, 2006). Changes in MEMS cantilever resonance frequency—associated with the mass change occurring from hybridization of DNA target and probe strands—are used to derive a measurable sensor signal.

Nanomaterial-based sensors. A great deal of recent attention in the research community

Figure 2. A typical approach to immunosensing

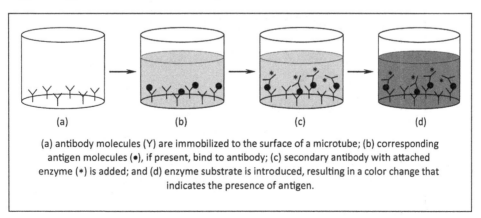

(a) antibody molecules (Y) are immobilized to the surface of a microtube; (b) corresponding antigen molecules (•), if present, bind to antibody; (c) secondary antibody with attached enzyme (*) is added; and (d) enzyme substrate is introduced, resulting in a color change that indicates the presence of antigen.

has focused on materials and structures with features having very small dimensions, from sub-nanometers to tens of nanometers. Such nanomaterials exhibit physical, chemical and electrical properties which may be quite different than their macroscopic counterparts. The design of biosensors incorporating nanomaterials as part of the sensing platform is a rich area for research possibilities.

As one example of a biosensing implementation, DNA oligonucleotides have been immobilized onto carbon nanotubes as a biorecognition molecule, and changes in the electrical properties of carbon nanotube structures have been used to detect probe-target biomolecular binding events (Rivas, 2007).

Hybrid technologies which couple the novel properties of nanomaterials with established techniques for biosensing are under intense investigation. Nanomaterials have been incorporated into piezoelectric immunosensors in order to enhance sensitivity and specificity (Zhang, 2008). Another example of a hybrid approach involves the incorporation of quantum dot nanocrystals into fluorescence imaging techniques widely used

in molecular and cellular biology. Quantum dots are a type of inorganic semiconductor nanocrystal with unique fluorescence properties that can be exploited in the implementation of optical-based fluorescence biosensors (Gill, 2008). Quantum dot nanocrystals resist photobleaching and can be easily incorporated into multiplexed sensors, in which a single excitation source is used to excite quantum dots of various colors, each color used to identify a specific target biomolecule. The emission wavelengths of quantum dots can be extended into the near-infrared range, a region of the spectrum in which relatively few fluorophores are available (Sun, 2007).

Microelectronic sensors. Solid-state devices have long been used in biosensing applications. Ion-sensitive field-effect transistors (ISFETs) can be used to measure physiologic parameters (McKinley, 2008) such as pH, thus can qualitatively and quantitatively indicate the presence of biomolecules such as urea. Microelectronic impedimetric sensors have been developed which measure the impedance change associated with biomolecular interactions taking place between an interdigitated electrode pair.

Figure 3. Piezoelectric ('quartz crystal microbalance') immunosensor

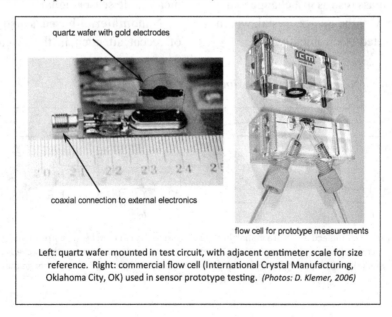

Left: quartz wafer mounted in test circuit, with adjacent centimeter scale for size reference. Right: commercial flow cell (International Crystal Manufacturing, Oklahoma City, OK) used in sensor prototype testing. *(Photos: D. Klemer, 2006)*

These are just a few examples of novel biosensor implementations which have received recent attention. The possibilities of yet-undiscovered biosensor implementations which combine the technologies given in Table 1 are vast. For example, optical detection of the fluorescence of a single-stranded DNA oligonucleotide tagged with a quantum-dot fluorophore might be used to implement a DNA biosensor based on nanocrystal labeling.

APPLICATIONS AND CASE EXAMPLES IN BIOMEDICINE

There is a wide array of applications for biosensing devices within medicine and life science research. In addition, biosensors have great relevance to numerous other fields, such as environmental science, veterinary medicine, food safety and pharmaceutical research, to name a few. In this section, we will restrict attention to medical applications of biosensors, considering several case examples in various specialties of medicine.

Pharmacogenomics. New diagnostic applications for biosensors are motivated by intense research in this area. Knowledge of an individual's genotype can potentially allow for the prescription of optimal pharmacologic treatment regimens, so-called "personalized medicine." Indeed, the economic benefit of screening for certain genotypes is the subject of discussion and debate for such medications as the blood thinning medication warfarin (Lackner, 2008). Sensors which can rapidly and inexpensively discriminate between specific genotypes will facilitate advances in this growing area of research.

Infectious Diseases. New approaches to in vitro point-of-care diagnostics will be necessary as the diagnosis and management of old and new infectious diseases increases in complexity. The great concern over emerging infectious diseases such as coronavirus-induced SARS, West Nile encephalitis, and Bovine Spongiform Encephal-

opathy (BSE) (i.e., Mad-Cow Disease), provides a strong motivation for the development of rapid diagnostic tests which are highly sensitive and specific. Immunoassays or DNA-based assays implemented in the form of electronic biosensors will augment the armamentarium available for diagnosis and management of infectious diseases.

Oncology. With our increasing understanding of the genetic factors which predispose to neoplastic disease and cancer, there is an associated demand for low-cost diagnostic tests which can detect and discriminate between certain disease-specific genotypes. Genetic screening for breast cancer genes (e.g., BRCA-1 and BRCA-2) and DNA-based testing for colon cancer are two examples of genetic testing which have received a great deal of attention in recent years. Given the increasing number of biomarkers for cancer which have been documented, accurate biosensors for these molecular markers can facilitate early, low-cost screening programs, prevent significant morbidity and mortality, and potentially reduce overall healthcare costs.

Neurology. Neurologists are presently attempting to define a set of specific molecular biomarkers which can permit an objective diagnosis of diseases such as Alzheimer's Dementia, replacing more subjective test instruments such as the Mini Mental Status Exam questionnaire. Low-cost biosensors which can detect and quantify these biomarkers will allow for a simple, objective assessment in the screening, diagnosis, management and treatment of this and other neurological diseases.

Biomedical Research. New, inexpensive methods for sensitive and specific detection and quantification of biomolecules in an analyte solution can have a profound impact on biomedical research. Application of microelectronic techniques for in situ fabrication of DNA microarrays allows for the simultaneous quantification of an extremely large number of genes and genetic variants. Application of new technologies can alter the way research is conducted, from a very limited 'hypothesis-based' approach in which a single gene is studied, to a 'data-mining' approach in which the behaviors of thousands of genes are simultaneously examined, with sophisticated statistical techniques used to test for patterns of significance.

CONCLUSION: THE FUTURE OF BIOSENSING

As our understanding of human physiology increases—along with the sophistication of approaches for diagnosis and treatment of illness—there will be a concomitant increase in the need for biosensing devices capable of sensitive and specific detection and quantification of biomolecules such as proteins and nucleic acids. One present limitation to the adoption of complex highly-technical sensing schemes is the cost; a DNA-based test used for screening of colon cancer may have a cost which is orders of magnitudes higher than inexpensive but nonspecific tests which are traditionally used. Application of high-technology approaches in the field of *in vitro* diagnostics and biosensing has a strong potential for major changes in the way medicine is currently being practiced, with major benefits in cost and healthcare outcomes in both the developed and developing world.

REFERENCES

Crowther, J. R. (Ed.). (1995). *ELISA: Theory and Practice*. Totowa, NJ: Humana Press.

Gill, R., Zayats, M., & Willner, I. (2008). Semiconductor quantum dots for bioanalysis. *Angewandte Chemie International Edition in English, 47*(40), 7602–7625. doi:10.1002/anie.200800169

Ivanov, D. (2006). BioMEMS sensor systems for bacterial infection detection: progress and potential. *BioDrugs, 20*(6), 351–356. doi:10.2165/00063030-200620060-00005

Joung, H. A., Lee, N. R., Lee, S. K., Ahn, J., Shin, Y. B., & Choi, H. S. (2008). High sensitivity detection of 16s rRNA using peptide nucleic acid probes and a surface plasmon resonance biosensor. *Analytica Chimica Acta, 630*(2), 168–173. doi:10.1016/j.aca.2008.10.001

Lackner, T. E. (2008). Pharmacogenomic dosing of warfarin: ready or not? *The Consultant Pharmacist, 23*(8), 614–619.

Liedberg, B., Nylander, C., & Lundstrom, I. (1995). Biosensing with surface plasmon resonance, how it all started. *Biosensors & Bioelectronics, 10*, i–ix. doi:10.1016/0956-5663(95)96965-2

Lyandres, O., Yuen, J. M., Shah, N. C., VanDuyne, R. P., Walsh, J. T., & Glucksberg, M. R. (2008). Progress toward an in vivo surface-enhanced Raman spectroscopy glucose sensor. *Diabetes Technology & Therapeutics, 10*(4), 257–265. doi:10.1089/dia.2007.0288

McKinley, B. A. (2008). ISFET and fiber optic sensor technologies: in vivo experience for critical care monitoring. *Chemical Reviews, 108*(2), 826–844. doi:10.1021/cr068120y

Pohanka, M., Skládal, P., & Pavlis, O. (2008). Label-free piezoelectric immunosensor for rapid assay of Escherichia coli. *Journal of Immunoassay & Immunochemistry, 29*(1), 70–79. doi:10.1080/15321810701735120

Prasad, P. N. (2003). *Introduction to Biophotonics* (pp. 312-314). New York: John Wiley & Sons.

Rivas, G. A., Rubianes, M. D., Rodríguez, M. C., Ferreyra, N. F., Luque, G. L., & Pedano, M. L. (2007). Carbon nanotubes for electrochemical biosensing. *Talanta, 74*(3), 291–307. doi:10.1016/j.talanta.2007.10.013

Rizzoli, A., Rosa, R., Rosso, F., Buckley, A., & Gould, E. (2007). West Nile virus circulation detected in northern Italy in sentinel chickens. *Vector Borne and Zoonotic Diseases (Larchmont, N.Y.)*, *7*(3), 411–417. doi:10.1089/vbz.2006.0626

Shen, G., Liu, M., Cai, X., & Lu, J. (2008). A novel piezoelectric quartz crystal immnuosensor based on hyperbranched polymer films for the detection of alpha-Fetoprotein. *Analytica Chimica Acta*, *630*(1), 75–81.

Su, J., Zhang, J., Liu, L., Huang, Y., & Mason, R. P. (2008). Exploring feasibility of multicolored CdTe quantum dots for in vitro and in vivo fluorescent imaging. *Journal of Nanoscience and Nanotechnology*, *8*(3), 1174–1177.

Sun, J., Zhu, M. Q., Fu, K., Lewinski, N., & Drezek, R. A. (2007). Lead sulfide near-infrared quantum dot bioconjugates for targeted molecular imaging. *International Journal of Nanomedicine*, *2*(2), 235–240.

Zeng, H., Wang, H., Chen, F., Xin, H., Wang, G., & Xiao, L. (2006). Development of quartz-crystal-microbalance-based immunosensor array for clinical immunophenotyping of acute leukemias. *Analytical Biochemistry*, *351*(1), 69–76. doi:10.1016/j.ab.2005.12.006

Zhang, Y., Wang, H., Yan, B., Zhang, Y., Li, J., Shen, G., & Yu, R. (2008). A reusable piezoelectric immunosensor using antibody-adsorbed magnetic nanocomposite. *Journal of Immunological Methods*, *332*(1-2), 103–111. doi:10.1016/j.jim.2007.12.019

KEY TERMS AND DEFINITIONS

Analyte: A biological component which is to be detected by a biosensor.

Biorecognition Element: A molecule in a biosensor which serves to uniquely identify a target analyte through a specific biomolecular interaction. Binding between the biorecognition element and the target biomolecule is used to generate a signal indicating the intermolecular interaction.

Biosensor: A device which is capable of detecting a biological component or analyte (e.g., molecule or microorganism), providing a reproducible and measurable signal upon recognition of that component.

Immunosensor (Immunoassay): A biosensor which employs a protein antibody as a biorecognition element, for detection of the corresponding antigen. Alternatively, the roles of antigen and antibody may be reversed, with an antigen serving as a biorecognition element for detection of the corresponding antibody.

In Vitro **Diagnostics:** Qualitative and/or quantitative detection of biological substances (e.g. biomolecules such as proteins or nucleic acids), through measurements external to the human body, for the purposes of healthcare diagnosis or management of disease. Compare *in vivo* diagnostics.

In Vivo **Diagnostics:** Measurements made on or internal to the human body, for the purposes of healthcare diagnosis or management of disease. Typical implementation of *in vivo* diagnostics involves medical imaging modalities, such as ultrasound or magnetic resonance imaging.

Nanomaterials: Materials with features having dimensions measured in the range of nanometers, exhibiting physical, chemical and/or electrical properties which differ markedly from their macroscopic counterparts.

Oligonucleotide: A short polymer of nucleotides, typically 25-30 nucleotides in length, often referred to as an "oligo." In biosensing applications, an oligonucleotide is a short, single-stranded DNA segment which serves as a biorecognition probe for the complementary target DNA.

Pharmacogenomics: A field of study within pharmacology dealing with the influence of an organism's genetic makeup on the response to treatment with a given pharmacologic agent.

Advances in pharmacogenomics may be used to develop an individualized approach to medical treatment that is tailored to a person's genetic background, so-called 'personalized medicine.'

Quantum Dot: A nanocrystal having a size which is typically tens of nanometers or less in all three dimensions, usually synthesized as a crystal of compound semiconductor materials (e.g., the II-VI material cadmium telluride). Quantum dots have unique optical and electrical properties; most notable are their fluorescence properties which have been exploited in various biomolecular imaging applications.

Chapter 12
Image Processing Tools for Biomedical Infrared Imaging

Gerald Schaefer
Loughborough University, UK

Arcangelo Merla
University G. D'Annunzio – Chieti-Pescara, Italy

ABSTRACT

Medical infrared imaging captures the temperature distribution of the human skin and is employed in various medical applications. Unfortunately, many of the conventional and commercial suites for image processing provide only very basic tools for the processing of medical thermal images which represent a challenging combination of both functional and morpho-structural imaging. In this chapter, several more advanced approaches are discussed which in turn provide tremendous help to the clinician. As an example, it is often useful to cross-reference thermograms with visual images of the patient, either to see which part of the anatomy is affected by a certain disease or to judge the efficacy of the treatment. It is shown that image registration techniques can be effectively used to generate an overlay of visual and thermal images to provide a useful diagnostic visualisation. Image registration can also be performed based on two thermograms and a warping-based method for this is presented. Segmenting the background from the foreground (i.e., the patient) is a crucial task and it is highlighted how this can be accomplished. Finally, it is shown how descriptors, extracted from medical infrared images, can be usefully employed to search through a large database of cases as well as to aid in diagnosis.

INTRODUCTION

Advances in camera technologies and reduced equipment costs have lead to an increased interest in the application of thermography in the medical field (Jones, 1998). Thermal medical imaging (or medical infrared imaging) uses a camera with sensitivities in the infrared to provide a picture of the temperature distribution of the human body or parts thereof. It is a non-invasive, non-contact, passive, radiation-free technique that can also be used in combination with anatomical investigations based on x-rays and three-dimensional scanning techniques such as CT and MRI and often reveals problems when the

DOI: 10.4018/978-1-60566-768-3.ch012

anatomy is otherwise normal. It is well known that the radiance from human skin is an exponential function of the surface temperature which in turn is influenced by the level of blood perfusion in the skin. Thermal imaging is hence well suited to pick up changes in blood perfusion which might occur due to inflammation, angiogenesis or other causes. Asymmetrical temperature distributions as well as the presence of hot and cold spots are known to be strong indicators of an underlying dysfunction (Uematsu, 1985).

While image analysis and pattern recognition techniques have been applied to medical infrared images for many years in astronomy and military applications, relatively little work has been conducted on the automatic processing of thermal medical images. Computerised image processing techniques have been used in acquiring and evaluating thermal images and proved to be important tools for clinical diagnostics (Plassmann & Ring, 1997; Wiecek, Zwolenik, Jung, & Zuber, 1999) but the available tools are rather limited and provide only basic functionality such as identifying isotherms, analysing image histograms or manually marking regions of interest for further investigation.

In this chapter we discuss several more advanced approaches to analysing medical thermograms. We start by showing how image registration techniques can be employed to generate an overlay of thermal and visual images of a patient as well as to align two thermograms. This proves useful for relating the patient's anatomy to any hot or cold areas that might show up on the thermal image and can be useful for the following of a disease and its treatment. We then present a simple technique that can be used to segment the patient from the background and can be used for region of interest identification. Finally, we we show how image features, extracted directly from the images, can be used to aid diagnosis and to search through a large database of cases.

THERMAL-VISUAL IMAGE OVERLAYS

Often visual and infrared images of the patient are taken to relate inflamed skin areas to the human anatomy which is useful for medial diagnosis as well as for assessing the efficacy of any treatment. Currently this process requires great expertise and is subject to the individual clinician's ability to mentally map the two distinctly different images. Therefore, an overlay of the two image types resulting in a composite image which makes it possible to cross-reference regions with unusual temperature distributions to the human anatomy will provide a useful tool for improved medical diagnosis.

Such an overlay can be achieved through application of an image registration technique (Tait et al., 2006). Registration is a method used to geometrically align or overlay two images taken from different sensors, viewpoints or instances in time (Zitova & Flusser, 2003). A reference (fixed) and a sensed (moving) image are aligned through a combination of scaling, translation and rotation, i.e. through an affine transform which is also the type of transform that we employ in our approach. Registration techniques can typically be classified as either intensity or landmark-based. Both techniques have advantages and disadvantages in their own unique approach. The main difficulty of landmark-based algorithms is the need to identify a set of corresponding control points in both images based upon which the best matching transform is sought. Landmarks can be found either manually or automatically. While manual selection of control points can be fairly time consuming and requires user interaction, automatic identification of landmarks constitutes a challenging problem and often requires a priori knowledge of the image features involved.

In order to provide a fully automated approach and require only little input and time of the clinician, application of an intensity-based technique is hence preferred. Here, all image information is utilised and the best alignment is derived as that which optimises a pre-defined similarity metric between the registered images. The steps involved are transform optimisation, image re-sampling and similarity computation which are applied in an iterative manner until the process has converged. We employ a gradient-decent optimiser, B-spline interpolation for the re-sampling and a mutual information measure as similarity metric (Tait et al., 2006). For scenes where the visual image contains unwanted background objects, a skin detection algorithm can be applied to discard background regions (Schaefer, Tait, & Zhu, 2006). If required, the workload can also be distributed through a parallel processing approach (Tait, Schaefer, & Hopgood, 2008).

Once an appropriate transform has been found and image registration performed (typically the visual image is selected as reference and the thermal one as sensed image), a composite image is created. This is simply performed by computing a weighted sum of the respective pixel values of the original visual image and the thresholded thermogram. Equal weights will generate an average of the two images whereas different weight factors will put more emphasis on one of the two modalities.

An example overlay is given in Figure 1 which shows the original visual image, the thermogram, the patient segmented from the visual image based on the skin detection step, and the final, overlaid image. The final image was weighted as 80% visual and 20% thermal. As can be seen, an accurate overlay of the two image types is achieved. This can prove useful for the assessment of morphea (localised scleroderma) patients (Howell et al., 2004). In Figure 1 the warmer area of the chest overlay indicates the distribution of a morphea lesion.

WARPING-BASED REGISTRATION OF THERMAL IMAGE SEQUENCES

Image registration techniques can also be employed to align two thermograms which can be

Figure 1. Example of thermal-visual overlay, original visual image (top left), thermogram (top right), segmented visual image (bottom left), composite image (bottom right)

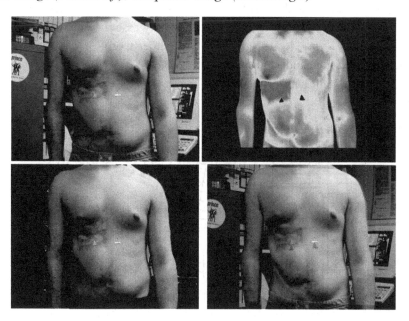

useful for images of the same patient taken at different times or for images of different patients. In this case one has not only to consider the difference in geometry and pose but also possible changes in the body structure of the patient(s). The approach presented in the previous section is therefore not applicable as it is restricted to affine transformations which are unable to account for these local adjustments. However, a registration algorithm based on warping can be used in this case (Tangherlini, Merla, & Roman, 2006) as it is able to achieve deformation both in scale and orientation as depicted in Figure 2.

This warping transforms a source image (the image that has to be registered) into a destination image (e.g., a previously taken image of a patient or a template image) through morphing which is performed by:

1. Defining one or more mapping oriented segments on the destination image.
2. Identifying the corresponding mapping segments on the source image.
3. Re-mapping the source image computing the actual position of each pixel with respect to the source image mapping segments.
4. Creating a new warped image using the mapping segments defined in 1. and the actual co-ordinates obtained in 3.

This technique was used in a study to assess whether a controlled change of the natural posture may effect the skin temperature distribution in healthy volunteers (Tangherlini, Merla, & Romani, 2006). Here, image warping is required for objective assessment of variations in skin temperature distribution because postural changes may modify apparent size and shapes of the body (Merla et al., 2002), and the skin temperature itself. The change of posture was obtained asking subjects to wear special shoes, where one heel was 2 centimeters higher than the other one. Two total body images sequences were acquired from the same subject: the first with the subject having bare feet and the second while the subject was wearing the special shoes. Paper markers were put on a few anatomical landmarks (reference points) to provide objective features to define the extremities of the mapping segments.

For each series, the first image was assumed as destination image. The remaining images of the series were used as source images and consequently warped as shown in Figure 3. On the series of warped images, image subtraction was performed for images corresponding to the same region of interest in order to evaluate the temperature differences and obtain objective image comparison. An example of this is shown in Figure 4. The shown image difference allows identification of local skin temperature variations which are possibly associated with postural changes due to the shoes worn.

Figure 2. Example of warping transformed image

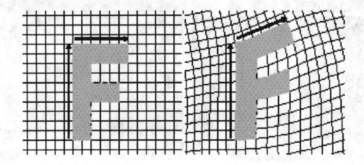

Figure 3. Warping of the posterior view for one of the studied subjects a) destination image with the user-defined mapping segments, b) source image to be warped with actual position of the corresponding mapping segments, c) warped image

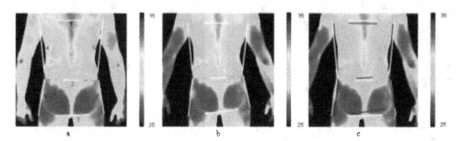

REGION OF INTEREST SEGMENTATION IN THERMOGRAMS

In many thermal imaging applications the patient's body or parts thereof need to be extracted so that analysis of the temperature profiles of these extracted regions of interest can be performed. An example is the extraction of regions corresponding to the arms of renal fistula patients. Arterio-venous fistulae (AVF) enable the long-term treatment of end stage renal disease using haemodialysis. Poor AVF blood flow can result from fistula narrowing or occlusion while at the other extreme vascular steal from the hand can result from very high fistula blood flow rates. Accurate prediction of blood flow rate is therefore of crucial importance.

It was shown in (Allen et al., 2006) that thermal images of both arms of a fistula patient and subsequent comparison of bilateral maximum temperatures in the arm regions can be used to estimate fistula blood flow. However, in tools such as (Plassmann & Ring, 1997) regions of interest from which temperature values can be extracted, need to be defined manually by the clinician which is a time-consuming and tiring task. An automated method that segments, without user interaction, the regions of interest in question can therefore be of great help and can be achieved through relatively simple image processing operations (Allen et al., 2005).

Firstly, the background, i.e. all non-body parts in the thermogram, were removed. In the image processing literature thresholding algorithms are commonly employed to divide the intensities in an image into two ranges, below and above a certain threshold value which is ideally adaptively

Figure 4. Warping of the posterior view for one of the studied subjects a) destination image with the user-defined mapping segments, b) warped image, c) difference image between destination and warped images

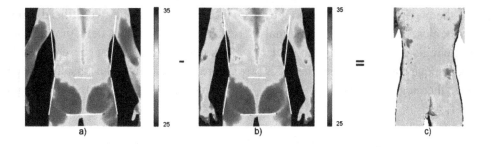

Figure 5. Arm thermogram of a patient (left) and region of interest automatically extracted (right)

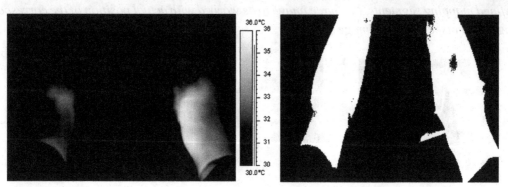

selected based on the image so as to account for variation in the intensity distribution due to image capture, contrast etc. The threshold is derived based on the algorithm by Otsu (1979) which maximises the between-class variance. Areas with intensities (i.e. temperatures) below the threshold are deemed as background and hence discarded.

Secondly, morphological image processing (Soille, 1999) is applied to fill small holes and merge them with larger corresponding areas. The two largest regions left after this should then be those of the arms of the patients. In cases where those areas are merged into one, e.g. if some other parts of the body are visible in the thermogram, the regions are split at the point that produces the shortest boundary between the areas.

Finally, extracted arm regions are labelled according to their side (i.e. "fistula arm" and "non-fistula arm") and the maximum regional temperatures and their bilateral differences (fistula side - non-fistula side) can now be calculated.

An example is given in Figure 5 which shows a clearly defined fistula together with the automatic segmentation of the two arm regions. Bilateral differences of the maximum temperature values were shown to be close to those based on a manual definition of the regions of interest (Allen et al., 2006). This technique can hence form the basis for automated, thermography-based estimation of AVF blood flow.

IMAGE FEATURES FOR AIDING DIAGNOSIS

One area where thermography has shown promising potential is the diagnosis of breast cancer, especially when the tumor is in its early stages or in dense tissue (Anbar et al., 2001; Head, Wang, Lipari, & Elliott, 2000). Image processing techniques, in particular algorithms that extract certain visual features from defined regions of interest, can be useful in this context. The idea is to calculate these features for both left and right breast regions and to utilise bilateral differences as a cue for diagnosis. Various features can be employed; in the following we give some details of those used in (Schaefer, Zavisek, & Nakashima, 2009).

Basic Statistical Features

Clearly the simplest feature to describe a temperature distribution such as those encountered in thermograms is to calculate its statistical mean. To express symmetry features, the mean for both breasts is calculated and the absolute value of the difference of the two utilised. Similarly, the standard temperature deviation and its bilateral absolute difference can be employed. Furthermore, one can use the absolute differences of the median temperature and the 90-percentile as further descriptors.

Moments

Image moments are defined as

$$m_{pq} = \sum_{y=0}^{M-1} \sum_{x=0}^{N-1} x^p y^q g(x,y) \tag{1}$$

where x and y define the pixel location and N and M the image size. Useful features are moments m_{01} and m_{10} which essentially describe the centre of gravity of the breast regions, as well as the distance (both in x and y direction) of the centre of gravity from the geometrical centre of the breast. For all four features, the absolute differences of the values between left and right breast are taken.

Histogram Features

Histograms record the frequencies of certain temperature ranges of the thermograms. Normalised histograms for both regions of interest (i.e. left and right breast) are constructed and the cross-correlation between the two histograms used as a feature. From the difference histogram (i.e. the difference between the two histograms), the absolute value of its maximum, the number of bins exceeding a certain threshold (empirically set to 0.01 in our experiments), the number of zero crossings, energy and the difference of the positive and negative parts of the histogram are calculated.

Cross Co-Occurrence Matrix

Co-occurrence matrices have been widely used in texture recognition tasks (Haralick, 1979) and can be defined as

$$\gamma_{T_i,T_j}^{(k)}(I) = \underset{p_1 \in I_{T_i}, p_2 \in I}{PR}[p_2 \in I_{T_j}, |\, p_1 - p_2 \,|= k] \tag{2}$$

with

$$|\, p_1 - p_2 \,| = \max |\, x_1 - x_2 \,|, |\, y_1 - y_2 \,| \tag{3}$$

where T_i and T_j denote two temperature values and (x_k, y_k) denote pixel locations. In other words, given any temperature T_i in the thermogram, γ gives the probability that a pixel at distance k away is of temperature T_j. In order to arrive at an indication of asymmetry between the two sides a cross co-occurrence matrix defined as (Schaefer, Zavisek, & Nakashima, 2009)

$$\gamma_{T_i,T_j}^{(k)}(I(1), I(2)) = \underset{p_1 \in I(1)_{T_i}, p_2 \in I(2)}{PR}[p_2 \in I(2)_{T_j}, |\, p_1 - p_2 \,|= k] \tag{4}$$

is used from which several features, namely homogeneity, energy, contrast and symmetry (Haralick, 1979) are extracted. Furthermore, the first four moments m_1 to m_4 of the matrix

$$m_p = \sum_k \sum_l (k-l)^p \gamma_{k,l} \tag{5}$$

are used.

Mutual Information

The mutual information MI between two distributions can be calculated from the joint entropy H of the distributions and is defined as

$$MI = H_L + H_R + H \tag{6}$$

with

$$H_L = -\sum_k P_L(k) \log_2 p_L(k) \tag{7}$$

$$H_R = -\sum_l P_R(l) \log_2 p_R(l)$$

$$H = \sum_k \sum_l P_{LR}(k,l) \log_2 p_{LR}(k,l)$$

and

$$p_{LR}(k,l) = x_{k,l} / \sum_{k,l} x(k,l) \tag{8}$$

$$p_L(l) = \sum_l p_{LR}(k,l)$$

$$p_R(k) = \sum_k p_{LR}(k,l)$$

and is employed as a further descriptor.

Fourier Analysis

One can also calculate the Fourier spectrum and use the difference of absolute values of the ROI spectra. Useful features are the difference maximum and the distance of this maximum from the centre.

IMAGE FEATURES FOR DATABASE RETRIEVAL

Image features can also be used for searching through large datasets of thermograms. A sample application could be to look for cases that are similar to a current one, which in turn can be approached by looking for thermograms with similar characteristics. Again, various visual features can be employed for this. For example, one can apply a wavelet transform to the thermal images and utilise the resulting wavelet coefficients, which represent image details of various sizes and ori-

entations, for establishing similarity between two images (Schaefer & Starosolski, 2008).

In particular, in the Fast Multiresolution Image Quering method (FMIQ) (Jacobs, Finkelstein, & Salesin, 1995), following the wavelet decomposition, the coefficients are thresholded and only a small number (100) of wavelet coefficients of largest absolute values (i.e., representing the most important details of image) are used as a fingerprint of the image. It is also possible to design a special metric that takes into account only these N nonzero wavelet coefficients, thanks to which efficient comparison of fingerprints is possible. Also, depending on the position of a given coefficient it is weighted using one of 6 weights (for 6 ranges of distances from top left corner to coefficient's position). Weights permit adjusting the algorithm to given image types.

An example query is given in Figure 6 which shows a sample thermogram together with those 10 thermal images out of a dataset of 400 heterogeneous infrared images, that have been found to be the closest matches. As can be seen, all retrieved thermograms are of the same anatomical areas.

Image features cannot only be used for direct retrieval of relevant images or cases, but they can also be employed to browse a whole dataset of

Figure 6. Example retrieval of thermograms based on wavelet features

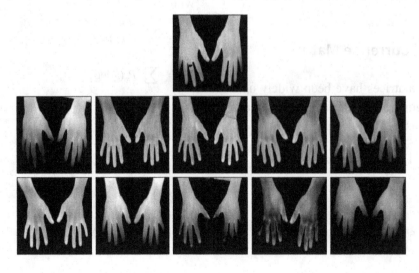

thermograms as was demonstrated in (Schaefer, Zhu, & Ruszala, 2005). Here, features based on moments as in Equation (1) were used. Image moments were first normalised in order to provide invariance with respect to translation and scale, and then combinations of the resulting normalised moments (Maitra, 1979) extracted which are also invariant with respect to rotation and contrast. These moment invariants can then be projected into a 2-dimensional visualisation space through application of multi-dimensional scaling (Kruskal & Wish, 1978) and in this visualisation plane the

user can navigate through a large collection of thermograms as shown in Figure 7.

CONCLUSION

Medical infrared imaging has proven to have various useful applications. In this chapter we have shown, through various examples, how advanced image processing methods can be successfully employed in this context. We have shown that overlays of thermal and visual images of a patient

Figure 7. Image feature based browsing of a database of thermograms

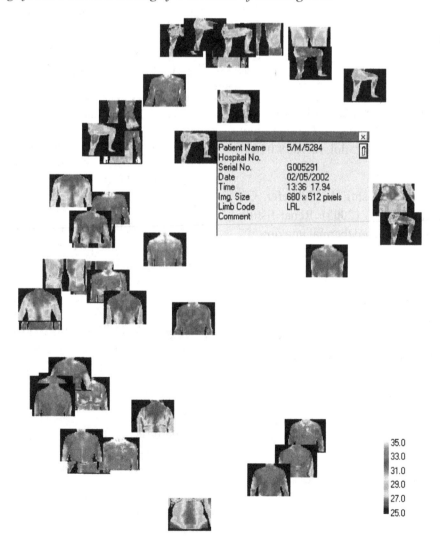

can be useful in relating hot or cold areas to ana-tomical structures. Similarly, image registration can be employed to align different thermograms from which local temperature changes can be esti-mated. Automatic extraction of regions of interest is a common and difficult task in many medical imaging applications and we have discussed how a relatively simple method can be used to achieve reasonable results that allow estimation of blood flow for renal fistula patients. Finally, it was demonstrated that visual features of medical infrared images can be useful for several tasks such as aiding diagnosis and browsing large databases of thermograms.

REFERENCES

Allen, J., Oates, C., Chishti, A., Ahmed, I., Talbot, D., & Murray, A. (2006). Thermography and colour duplex ultrasound assessments of arterio-venous fistula function in renal patients. *Physiological Measurement*, *27*, 51–60. doi:10.1088/0967-3334/27/1/005

Allen, J., Oates, C., Chishti, A., Schaefer, G., Zhu, S., Ahmed, I., et al. (2005). Renal fistula assessment using automated thermal imaging. *3rd European Medical and Biological Engineering Conference.*

Anbar, N., Milescu, L., Naumov, A., Brown, C., Button, T., Carly, C., & AlDulaimi, K. (2001). Detection of cancerous breasts by dynamic area telethermometry. *IEEE Engineering in Medicine and Biology Magazine*, *20*(5), 80–91. doi:10.1109/51.956823

Haralick, R. (1979, May). Statistical and structural approaches to texture. *Proceedings of the IEEE*, *67*(5), 786–804. doi:10.1109/PROC.1979.11328

Head, J., Wang, F., Lipari, C., & Elliott, R. (2000). The important role of infrared imaging in breast cancer. *IEEE Engineering in Medicine and Biology Magazine*, *19*, 52–57. doi:10.1109/51.844380

Howell, K., Visentin, M., Lavorato, A., Jones, C., Martini, G., & Smith, R. (2004). Thermography, photography, laser doppler flowmetry and 20 Mhz b-scan ultrasound for the assessment of localised scleroderma activity: A pilot protocol. *Thermology International*, *14*(4), 144–145.

Jacobs, C., Finkelstein, A., & Salesin, D. (1995). Fast multiresolution image querying. *Siggraph*, *95*, 277–286.

Jones, B. (1998). A reappraisal of infrared thermal image analysis for medicine. *IEEE Transactions on Medical Imaging*, *17*(6), 1019–1027. doi:10.1109/42.746635

Kruskal, J., & Wish, M. (1978). *Multidimensional scaling*. London: Sage Publications.

Maitra, S. (1979). Moment invariants. *Proceedings of the IEEE*, *67*, 697–699. doi:10.1109/PROC.1979.11309

Merla, A., Romano, V., Zulli, F., Saggini, R., Di Donato, L., & Romani, G. (2002). Total body infrared imaging and postural disorders. *24th IEEE Int. Conference on Engineering in Medicine and Biology.*

Otsu, N. (1979). A threshold selection method from grey-level histograms. *IEEE Transactions on Systems, Man, and Cybernetics*, *9*(1), 62–66. doi:10.1109/TSMC.1979.4310076

Plassmann, P., & Ring, E. (1997). An open system for the acquisition and evaluation of medical thermological images. *European Journal of Thermology*, *7*, 216–220.

Schaefer, G., & Starosolski, R. (2008). A comparison of two methods for retrieval of medical images in the compressed domain. *30ᵗʰ IEEE Int. Conference Engineering in Medicine and Biology*, (pp. 402–405).

Schaefer, G., Tait, R., & Zhu, S. (2006). Overlay of thermal and visual medical images using skin detection and image registration. *28ᵗʰ IEEE Int. Conference Engineering in Medicine and Biology*, (pp. 965–967).

Schaefer, G., Zavisek, M., & Nakashima, T. (2009). Thermography based breast cancer analysis using statistical features and fuzzy classification. *Pattern Recognition*.

Schaefer, G., Zhu, S., & Ruszala, S. (2005). Visualisation of medical infrared image databases. *27ᵗʰ IEEE Int. Conference Engineering in Medicine and Biology*, (pp. 1139–1142).

Soille, P. (1999). *Morphological image analysis: Principles and applications*. Berlin: Springer-Verlag.

Tait, R., Schaefer, G., & Hopgood, A. (2008). Intensity-based image registration using multiple distributed agents. *Knowledge-Based Systems*, *21*(3), 256–264. doi:10.1016/j.knosys.2007.11.013

Tait, R., Schaefer, G., Howell, K., Hopgood, A., Woo, P., & Harper, J. (2006). Automated overlay of visual and thermal medical images. *Int. Biosignal Conference*.

Tangherlini, A., Merla, A., & Romani, G. (2006). Field-warp registration for biomedical high-resolution thermal infrared images. *28ᵗʰ IEEE Int. Conference on Engineering in Medicine and Biology*, (pp. 961–964).

Uematsu, S. (1985). Symmetry of skin temperature comparing one side of the body to the other. *Thermology*, *1*(1), 4–7.

Wiecek, B., Zwolenik, S., Jung, A., & Zuber, J. (1999). Advanced thermal, visual and radiological image processing for clinical diagnostics. *21ˢᵗ IEEE Int. Conference on Engineering in Medicine and Biology*.

Zitova, B., & Flusser, J. (2003). Image registration methods: a survey. *Image and Vision Computing*, *21*, 977–1000. doi:10.1016/S0262-8856(03)00137-9

KEY TERMS AND DEFINITIONS

Image Features: Descriptors extracted directly from the raw image data.

Image Registration: Geometrically aligning two images taken under different conditions or at different times.

Image Retrieval: Identification of similar images based on image features.

Image Segmentation: Extraction of certain parts of an image.

Thermal Medical Imaging: Taking of images using infrared sensors for medical purposes.

Chapter 13
Image Analysis for Exudate Detection in Retinal Images

Gerald Schaefer
Loughborough University, UK

Albert Clos
Aston University, UK

ABSTRACT

Diabetic retinopathy is recognised as one of the most common causes of blindness. Early diagnosis is important and is based on detection of features such as exudates during eye fundus image screening. In this chapter it is shown how areas corresponding to exudates can be automatically detected using a neural network that, following contrast enhancement and vessel and optic disc extraction steps, classifies each image pixel as exudate or non-exudate. Experimental results on an image set with known ground truth verify the usefulness of the presented approach.

INTRODUCTION

Diabetic retinopathy is recognised as one of the most common causes of blindness (Aiello et al., 1998). It can however be diagnosed early based on various features that can be detected in eye fundus images. One of these indicators is exudates which are typically formed in groups or rings surrounding leakage of plasma and often appear yellowish in the image. An example of a retinal image with exudates is given in Figure 1.

Various approaches for detecting exudates have been presented in the literature. In (Gardner et al.,

1996) a neural network for retinal image analysis was developed. Their algorithm was able to identify vessels, exudates and haemorrhages. A retinal image was divided into disjoint 20×20 pixel regions and each region assigned by an expert as either exudate or non-exudate. Each pixel of the window corresponds to an input of a backpropagation network giving a total of 400 inputs. A sensitivity of 93.1% in detecting exudates was reported.

Osareh et al. (2003) used histogram specification as a preprocessing step to eliminate colour variations and then segmented the images based on a fuzzy c-means clustering technique. Each segmented region is classified as either exudate or non-exudate and characterised by 18 visual features.

DOI: 10.4018/978-1-60566-768-3.ch013

Figure 1. Sample retinal image with exudates

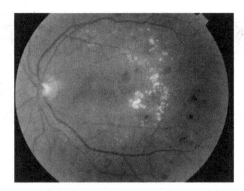

A two-layer perceptron network was trained with these and a sensitivity of 93% and specificity of 94.1% were achieved.

Walter et al. (2002) utilise morphological image processing to isolate exudate regions. First, candidate regions are found based on high local contrast variations while the exact contours of exudate regions are then extracted using morphological operators. They report a sensitivity of 92.8%.

In (Sinthanayothin et al., 2002) a recursive region growing technique is employed which groups similar pixels together. Following a thresholding step, the resulting binary image shows the extracted exudate areas. Using this approach a sensitivity of 88.5% and specificity of 99.7% were achieved.

The approach that we present in this chapter is also centred around the use of neural networks for exudate classification, but in contrast to earlier work it proceeds on a pixel-by-pixel basis and utilises only the colour information at each pixel location (Clos, Schaefer & Nolle, 2007). We first pre-process the images in order emphasise the differences between exudate and non-exudate regions, and to reduce the variation of colours in the images. We then perform vessel tracking and optic disc detection and discard the associated areas. The remaining regions are passed to a backpropagation neural network based on a sliding window data extraction mechanism. Principal component analysis (PCA) is applied

in order to reduce the dimensionality of the data and speed up the training of the network. The network is then trained to differentiate exudate from non-exudate regions and hence to detect the locations of exudates in the images. Experimental results based on a ground truth dataset with known exudate locations confirm the efficacy of our technique.

IMAGE PRE-PROCESSING

As it is difficult to control lighting conditions but also due to variations in ethnic background and iris pigmentation, retinal images usually exhibit large colour and contrast variations, both on global and local scales. Figure 2 shows four different fundus images and it is apparent that the colour variations between them need to be considered in order to successfully analyse the structures in the images.

Goatman et al. (2003) evaluated three pre-processing techniques on retinal images including greyworld normalisation, histogram equalisation and histogram specification in order to reduce the variation in background colour among different images. They found that histogram specification performed best followed by histogram equalisation.

In our experiments we therefore employ histogram specification as a pre-processing step. Histogram specification modifies the histogram

Figure 2. Colour variations in different retinal images

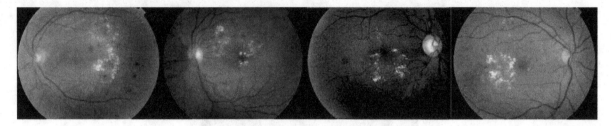

of a target image to match that of a source image (which is the same for all images and is the image shown with a red border in Figure 2). We use only the green channel of the image as it typically provides the best contrast for exudates.

VESSEL AND OPTIC DISC DETECTION

In order to achieve the best possible performance for the following classification it is advisable to remove all potentially confusing training data. In terms of retina images this means discarding those parts of the image which definitely do not correspond to exudate regions. In particular, we are interested in finding three different categories of areas/pixels in the image: non-retinal pixels, retinal vessels, and the optical disc.

With "non-retinal pixels" we mean those pixels outside the retinal area in an image. This essentially evaluates to the darkest pixels in the image but still requires some enhancement to be able to completely eliminate it without affecting the retinal area. Problems associated with this come from the noise of image compression and the position of the retinal area which differs from image to image. We apply a Gaussian filter to equalise non-retinal pixel values and extract the highest value of a big area in the filtered dark pixels as the threshold to separate the retina area from the background.

Retinal vessels are one of the striking parts of the retina and important in many studies. As exudates and vessels won't coincide at the same location, we are interested in automatically detecting and removing vessel pixels. Blood vessels are typically of a more red and darker colour compared to the background retinal tissue colour. In our approach we apply the method introduced by Walter (2003) which is based on mathematical morphology operators to detect the vessels followed by a post-filtering stage. An example of a retina image and the corresponding extracted vessel tree is provided in Figure 3.

Even more important than detecting blood vessels is the identification of the image area that corresponds to the optic disc. The reason for this is

Figure 3. A retinal image (left) and its segmented vessel tree (right)

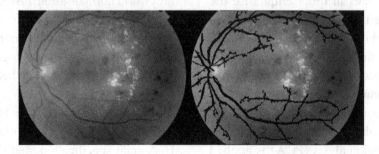

that in many cases the optic disc has very similar colour characteristics to those of exudates and is hence likely to be misclassified as a lesion region. We adopt the optic disc detection approach from Walter (2003). First, the location of the disc needs to be identified, and then based on this location the actual boundary of the disc is established.

Locating the disc is the easier part as the optic disc is typically the brightest part of the image. We utilise the vessel segmentation discussed above, dilate it and multiply the result with the red channel of the original image to obtain an intensity image containing the optic disc, the vessels and some pixels around them as they converge on the optical nerve head. We then segment out the middle part of the image in which the optical disc will never appear to eliminate possible bright lesions and then compare the sum of intensities in both sides. The brightest part will contain the optic disc location at its maximum intensity value. An example of the technique is illustrated in Figure 4.

In order to extract the exact boundaries of the disc, we apply a Gaussian filter followed by a morphological closing of the image in order to eliminate vessels and other artefacts. After that, we use an edge detector to detect the boundaries between the background and the disc area, dilate the result to obtain a closed "circle" and then fill

it to define the desired segmentation mask. Figure 5 shows an example of these stages.

PRINCIPAL COMPONENT ANALYSIS

We employ principal component analysis (PCA) in order to reduce the high dimensionality of the input data as well as to be able to build a simpler network (i.e., with fewer neurons) which in turn should prove easier and faster to train. We use a sliding window approach to extract image areas of 9×9 pixels corresponding to 81 network inputs. Following PCA, we can compress this data into a vector of 40 inputs corresponding to the first 40 principal components which capture 97% of the variance in the original data.

NEURAL NETWORK

Our system is basically a 2-class classifier. Figure 6 provides an overview of our classification approach.

We employ a backpropagation one-output neural network (Bishop, 2002) with the output representing the likelihood of the input representing an exudate.

Figure 4. Different stages of optic disc localisation

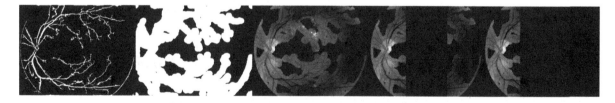

Figure 5. Optic disc boundary extraction

Figure 6. Overview of classification system

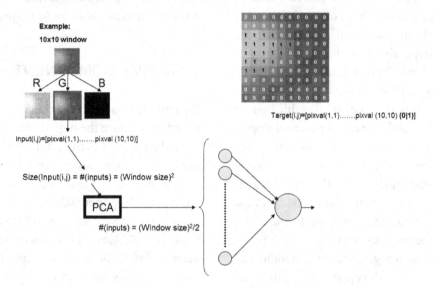

Testing various network configurations we found the best results were obtained with a network of 40 inputs (stemming from the result of the PCA on 9×9 windows reduced to 40 values as discussed above), 1 output, and no hidden neurons. As output function we chose a linear function, which provided the best classification performance.

Training is a crucial stage in employing neural networks. Training data was derived from 17 images with exudate regions marked by an expert. This data was divided into training, test, and validation sets in order to arrive at a neural network configuration that provides good classification performance without overfitting deficiencies. As the exudate data is fairly sparse, we created datasets that contained the same number of exudate and non-exudate samples. Once successfully trained, the neural network can be used to detect the locations of exudates in unknown input images.

RESULTS

Experiments were performed on a dataset of 17 retina images with ground truth exudate regions manually marked by experts. We tested various combinations of inputs (corresponding to the

Figure 7. Example retinal image with exudate areas by neural network (left) and manually marked ground truth (right)

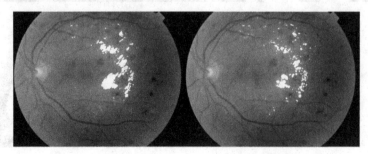

various stages of the processing pipeline as discussed above). As expected, the best results were obtained by using histogram specification, followed by excluding image areas corresponding to the background, blood vessels, and the optic disc, as well as applying PCA and taking the 40 components that capture most of the variance in the data. Using this configuration we obtained exudate detection performance that compares well with other results reported in the literature (Clos, Schaefer, & Nolle, 2007) which demonstrates the usefulness of our presented approach. An example image together with exudate regions marked by the neural network and compared to the ground truth is shown in Figure 7.

CONCLUSION

Detection of exudates in retinal fundus images is an important step in the early detection of diabetic retinopathy but is also a time-consuming task for the expert. Automated approaches are therefore highly sought after and in this chapter we have shown that such an approach is indeed feasible and can be incorporated in a computer-aided diagnosis system. In our approach, retinal images are contrast enhanced using histogram specification which also eliminates colour variations between images. Areas corresponding to retinal vessels and the optic disc are then automatically extracted while the remaining pixel data is fed into a neural network which, for each pixel, provides a likelihood that the pixel constitutes part of an exudate area. Experiments on a set of fundus images with ground truth have demonstrated the usefulness of our approach.

REFERENCES

Aiello, L., Callerano, J., Gardner, T., King, D., Blankenship, G., Ferris, F., & Klein, R. (1998). Diabetic retinopathy. *Diabetes Care, 21*, 143–156.

Bishop, C. M. (2002). *Neural networks for pattern recognition*. Oxford, UK: Oxford University Press.

Clos, A., Schaefer, G., & Nolle, L. (2007). Exudate detection in eye digital fundus images using neural networks. In *13th Int. MENDEL Conference on Soft Computing*, (pp. 121-127).

Gardner, G. G., Keating, D., Williamson, T. H., & Elliott, A. T. (1996). Automatic detection of diabetic retinopathy using an artificial neural network: a screening tool. *The British Journal of Ophthalmology, 80*(11), 940–944. doi:10.1136/bjo.80.11.940

Goatman, K. A., Whitwam, A. D., Manivannan, A., Olson, J. A., & Sharp, P. F. (2003). Colour normalisation of retinal images. In *Medical Image Understanding and Analysis*.

Osareh, A., Mirmehdi, M., Thomas, B., & Markham, R. (2003). Automated identification of diabetic retinal exudates in digital colour images. *The British Journal of Ophthalmology, 87*(10), 1220–1223. doi:10.1136/bjo.87.10.1220

Sinthanayothin, C., Boyce, J. F., Williamson, T. H., Cook, H. L., Mensah, E., Lal, S., & Usher, D. (2002). Automated detection of diabetic retinopathy on digital fundus images. *Diabetic Medicine, 19*(2), 105–112. doi:10.1046/j.1464-5491.2002.00613.x

Walter, T. (2003). *Application de la morphologie mathématique au diagnostic de la rétinopathie diabétique à partir d'images couleur*. PhD Thesis, Ecole Nationale Superieure des Mines de Paris.

Walter, T., Klein, J.-C., Massin, P., & Erginay, A. (2002). A contribution of image processing to the diagnosis of diabetic retinopathy-detection of exudates in color fundus images of the human retina. *IEEE Transactions on Medical Imaging, 21*(10), 1236–1243. doi:10.1109/TMI.2002.806290

Chapter 14
Computer Aided Risk Estimation of Breast Cancer:
The "Hippocrates–mst" Project

George M. Spyrou
Academy of Athens, Greece

Panos A. Ligomenides
Academy of Athens, Greece

ABSTRACT

In this chapter the authors report about their experiences in designing, implementing, prototyping and evaluating a system for computer aided risk estimation of breast cancer. The strategy and architecture of "Hippocrates-mst" along with its functionalities are going to be presented. Also, the evaluation results in the clinical practice concerning the performance of "Hippocrates-mst" in the "Ippokrateio" University Hospital of Athens will be presented. The feedback from medical experts along with the new features of the system that are under development will be discussed.

INTRODUCTION

Breast cancer is worldwide the most frequent cancer in women. Years of experience have revealed that mammography constitutes the most efficient method in the early diagnosis of this type of cancer (Elmore et al., 2005). Clustered microcalcifications belong to the worthy mammographic findings since they have been considered as important indicators of the presence of breast cancer (Fondrinier et al., 2002, Gulsun et al., 2003, Buchbinder et al., 2002, Bassett, 1992). Several efforts have been made to classify the microcalcifications as benign or ma-

lignant according to their characteristics (Lanyi, 1977, 1985, Le Gal et al., 1984, 1976, Timins, 2005). However, problems still appear in mammographic imaging methods due to subtle differences in contrast between benign and malignant lesions, mammographic noise, the insufficient resolution and local low contrast. These inherent difficulties of mammographic image reading prevent the medical expert from quantifying the findings and from making a correct diagnosis (Wright et al., 2003, Shah et al., 2003, Yankaskas et al., 2001, Elmore et al., 2003). Therefore, a non-negligible percentage of biopsies following mammographic examination are classified as false positive (Esserman et al., 2002, Roque & Andre 2002).

DOI: 10.4018/978-1-60566-768-3.ch014

Improvements towards the early diagnosis of breast cancer, as far as the computer science is concerned, belong to the field of systems for computer-aided mammography (Doi et al., 1997, Wu et al., 1992). Such systems have mainly dealt with detection and classification of microcalcifications and use image processing and analysis techniques as well as artificial intelligence methods. The most well known methods of computer aided detection or diagnosis in mammography are artificial neural networks (ANN) with different architectures and variations, the segmentation methods, multiscale analysis - wavelets and morphologic analysis to distinguish between malignant and benign cases (Wu et al., 1993, Zhang et al., 1994, Chan et al., 1995, Zhang et al., 1996, Gurcan et al., 2001, Gurcan et al., 2002, Cooley & Micheli-Tzanakou, 1998, Bocchi et al., 2004, Dengler et al., 1993, Gavrielides et al., 2002, Li et al., 1997, Zhang et al., 1998, Lado et al., 2001, Mata Campos et al., 2000, Shen et al., 1994, Chang, et al., 1998, Spyrou et al., 1999).

The project that we present here, describes a system that deals with the digital or the digitized mammographic image, offering computer aided risk assessment starting from a selected region with microcalcifications (Spyrou et al., 1999, Spyrou et al., 2002a, 2002b). The proposed system is based on methods of quantifying the critical features of microcalcifications and classifying them as well as their clusters according to their probability of being cancerous. A risk-index calculation model has been developed and is included in the system. Furthermore, the calculated risk-index can be refined through other information such as the position and the direction of the cluster along with information from the patient record such as the age and the medical history of the patient (Berg et al., 2002, Cancer Facts and Figures, 2004). The design of the interface follows the clinical routine and allows for interaction with the physician at any time, providing information about the procedures and parameters used in the model of diagnosis.

Apart from the very encouraging laboratory testing, the prototype system is under a long-term evaluation phase in "Ippokrateio" University Hospital of Athens. Up to now the results from the current evaluation procedures indicate that the system estimates the risk of breast cancer towards the right direction. As far as the malignant cases are concerned, the system has a very high sensitivity. For the benign cases, the system is able to achieve a significant reduction on the unnecessary biopsies. Nevertheless, the sensitivity and specificity levels are not balanced. Actually, there is an overestimation of risk driven from the attempt to minimize the false negative results. Therefore, we further investigate for a better compromise which will emerge after a fine tuning in the calculation of parameters and the setup of thresholds concerning the selected features of the microcalcifications and their clusters.

SYSTEM AND PROGRAM DESCRIPTION

The method used in this computer aided risk assessment scheme is outlined in Figure 1. According to this scheme, the doctor may select from the patient's archive the set of the four mammographic images (Right and Left CranioCaudal-CC and MedioLateral-ML) that corresponds to a specific date of mammographic examination. Every selected image from this set can be displayed inside a form with various digital tools. These tools can be applied either to the whole image or to a Region Of Interest (ROI) with the help of a digital lens (a rectangular region where several digital tools can be applied). With the digital lens the user can select a ROI and subsequently apply several techniques such as histogram equalization, zoom, edge detection, differentiation of contrast and brightness. Some of them are very effective visualization tools, especially in breast periphery where the tissue is overexposed and thus very dark.

Figure 1. The computer aided risk assessment scheme

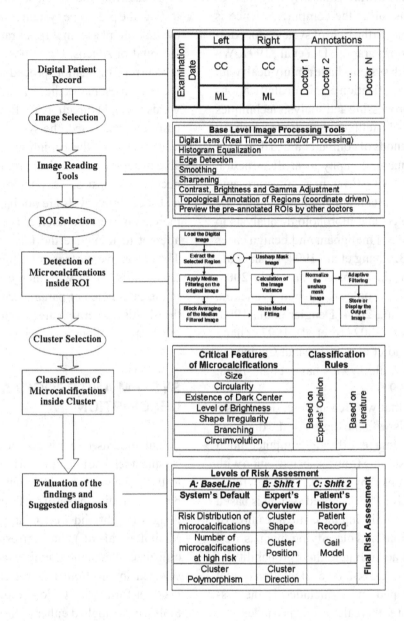

The module for the detection of the microcalcifications implements a very specific function of the digital lens tool described with a flowchart on the corresponding part of Figure 1 and it is based on the application of High Pass filtering, Variance Normalization and Adaptive filtering in order to reveal the microcalcifications from a low contrast image environment (Karssemeijer, 1993, Lorenz, et al., 1993).

After the detection of the microcalcifications, image analysis is needed for the microcalcification feature extraction and quantification. The considered microcalcification features are: (i) size, (ii) circularity, (iii) existence of dark center, (iv) level of brightness, (v) irregularity regarding the shape, (vi) level of branching and (vii) circumvolution.

As far as the shape features are concerned, we apply a four projections method for recognition, based on the principle of viewing each microcalcification from four different points of view (Horizontal, Vertical and two Diagonals) as shown in Figure 2. These calculations take place on every microcalcification of the cluster inside a sub region of the initial ROI selected by the doctor. Subsequently, the risk of each microcalcification is estimated and a classification is taking place, according to an empirical model comprising of rules either from the related literature or from selected medical experts.

The final risk assessment may be synthesized from the default base line risk estimation (A), and two possible shifts of risk (B) and (C) as described below:

(A) The default base line risk estimation is based on the evaluation of the findings and specifically on the following parameters: (A1) the risk distribution of microcalcifications, (A2) the number of microcalcifications at high risk, (A3) the cluster polymorphism.

(B) A first possible shift of risk estimation which is based on the doctor's expertise regarding the cluster location and direction inside breast.

Figure 2. The four-projections method for morphology recognition

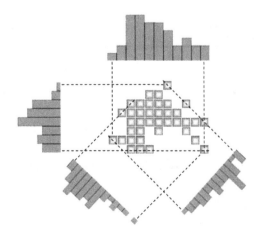

(C) A second possible shift of risk estimation which is based on the patient's record. It can either be provided by the doctor or calculated by a well known algorithm (Gail Model) that has been implemented to the system (Gail et al., 1989, Gail & Costantino, 2001).

Finally, the risk estimation can be classified in the four virtual zones of risk:

(i) Zone 1, with risk between 0% and 35% (definitely benign case: discouraging biopsy),

(ii) Zone 2, with risk between 35% and 55% (benign case with doubts: can not avoid biopsy),

(iii) Zone 3, with risk between 55% and 75% (malignant case with doubts: encouraging biopsy),

(iv) (Zone 4, with risk between 75% and 100% (definitely malignant case: strong prompt for biopsy).

INTERFACE

The above risk assessment scheme has been implemented and integrated into a software system with graphical interface that is adapted to the physician's needs. The user (physician) may log in with the corresponding username / password and be subsequently driven with forms mainly concerning (i) the patient's archive management, (ii) general image analysis and visualization tools, (iii) microcalcifications analysis tools, and (iv) risk assessment. A characteristic form is shown in Figure 3, with the digital tools that can be applied to the image. In this form, microcalcifications have been revealed inside a region of interest. There is also a sub region where each of the revealed microcalcifications has been also classified and coloured according to the risk estimation model.

A second form is shown in Figure 4, where the physician can be informed about the base line risk assessment (A) as it is estimated from the

Figure 3. A characteristic form of the interface concerning the image analysis tools

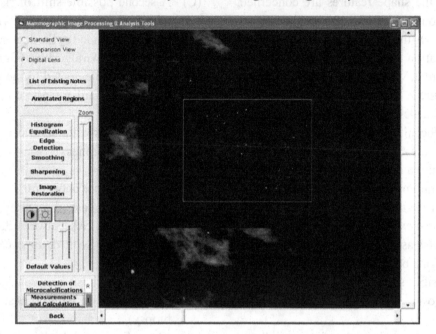

Figure 4. A characteristic form of the interface concerning risk estimation

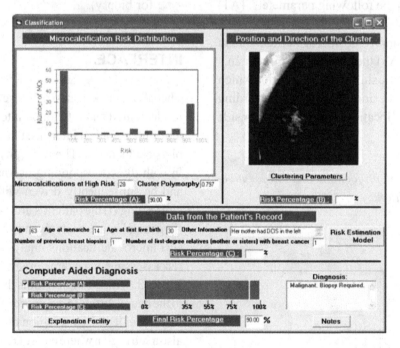

computer model. There is also specific information concerning the topology of the sub region as well as the patient's record, providing the user

with the possibility of two possible shifts of risk estimation (B and C).

HARDWARE

The system consists of a modern personal computer with sufficient memory, processing power and storage capability for high-resolution image handling (see Figure 5). The computer must be connected with a high-resolution film scanner or a digital mammography apparatus. The film scanner must have optical resolution of at least 300 dpi, wide optical density range (0.0 to at least 3.5 O.D.), at least 12-bit output, quick scan rate and scanning area greater than 18 cm x 24 cm, namely greater than the mammographic film dimensions.

EVALUATION

We have done until now three evaluation studies: The first one used an images set that was containing 260 mammograms, of both craniocaudal and mediolateral views, with microcalcifications. Each case was accompanied by a histopathological examination of resected speciment and with the patient's demographic data and medical history as well. The results from the Hippocrates-mst system are presented in a risk percentage scale from 0 to 100. Percentage lesser than or equal to 35% means there is not enough evidence to support sending the patient for a biopsy test. Percentage from 35%

to 55% declares benign state and a biopsy test is suggested, while percentage greater than 55 indicates malignancy and a biopsy test has to be made as soon as possible. At this point it must be stated that the system only makes suggestions to the doctor and it is the doctor that makes the final decision whether a case has to be sent for a biopsy test or not. From these women 73 have malignant histopathological test results (biopsy: mal) and the rest 187 women have benign histopathological test results (biopsy: ben). The average of the results for malignant cases was 81.21, giving the proposed system a high sensitivity in malignant microcalcifications detection. The average of the results for the benign cases was 49.46. According to those results, the system shows a high sensitivity (98.63%) by classifying correctly all malignant cases except one, which had high density -bad quality mammography- and was detected as a false negative. However, the system suffers from low specificity. Nevertheless its specificity is much better at 29.95% (56 cases correctly identified as benign with a percentage equal or smaller of 35% out of 187 referred cases which were benign according to the biopsy results) compared to the doctors' decisions for biopsy referrals.

The other two studies were performed during the evaluation period of the system in the "Ippokrateio" University Hospital of Athens. The first one included 186 cases with BI-RADS 3,4,5

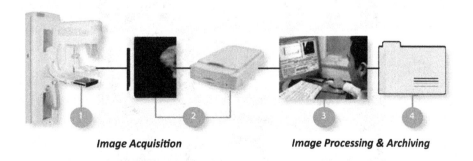

Figure 5. Image acquisition, processing and archiving is achieved with the system

Image Acquisition   *Image Processing & Archiving*

accompanied with biopsy. From them, 34 were malignant. The system showed high levels of sensitivity identifying correctly the 97.06% of the malignant cases and at the same time succeeded in 28.79% reduction of the unnecessary biopsies. In the other study, 110 cases (6 malignant) were used, exclusively diagnosed as BI-RADS 3. The system showed high levels of sensitivity identifying correctly the 100% of the malignant cases and at the same time succeeded in 31.81% reduction of the unnecessary biopsies.

DISCUSSION

Screening mammography with computer-aided diagnosis systems has been shown to improve radiologists performance in the detection of clinically occult cancers that otherwise would be overestimated. "Hippocrates-mst" presents high sensitivity but it appears to have low specificity

(see Figure 6). This indicates that there is an over-estimation of the risk, driven from the attempt to minimize the false negative results, which is the most important factor in these systems. However, in any case its specificity is much better than the specificity obtained only with the doctor's visual examination of the mammogram. What is also very important in the use of CAD systems is their appropriate use. The users or the radiologists should firstly be trained and informed about the systems software and about what these systems can do and how. The present system aids the radiologists as a second reviewer in evaluating mammograms and these systems are rapidly gaining acceptance in the breast imaging community. However, the radiologists should not become overly reliant on the system for detection of malignant lesions and should continue to search the films diligently.

The system estimates the risk of breast cancer towards the right direction. However, although it presents high sensitivity, it appears to have low

Figure 6. The system "Hippocrates-mst" in clinical evaluation

specificity. This indicates that there is an overestimation of the risk, driven from the attempt to minimize the false negative results. Thus, there is the need for a better compromise which will emerge after a fine tuning in the calculation of parameters and the setup of thresholds concerning the selected features of the microcalcifications and their clusters. For this reason we are planning to perform a more accurate evaluation procedure that is going to take place in several hospitals of Greece. Some of the issues that must be taken into account are the following: (i) all the cases have to contain microcalcifications, related to the existence (or not) of the lesion and (ii) the biopsies have to be taken from the region that contains the microcalcifications. The feedback from such an accurate procedure will lead to a promising methodology refinement.

REFERENCES

Bassett, L. W. (1992). Mammographic analysis of calcifications. *Radiologic Clinics of North America*, *30*(1), 93–105.

Berg, W. A., D'Orsi, C. J., Jackson, V. P., Bassett, L. W., Beam, C. A., Lewis, R. S., & Crewson, P. E. (2002). Does training in the Breast Imaging Reporting and Data System (BI-RADS) improve biopsy recommendations or feature analysis agreement with experienced breast imagers at mammography? *Radiology*, *224*(3), 871–880. doi:10.1148/radiol.2243011626

Bocchi, L., Coppini, G., Nori, J., & Valli, G. (2004). Detection of single and clustered microcalcifications in mammograms using fractals models and neural networks. *Medical Engineering & Physics*, *26*(4), 303–312. doi:10.1016/j. medengphy.2003.11.009

Buchbinder, S. S., Leichter, I. S., Lederman, R. B., Novak, B., Bamberger, P. N., Coopersmith, H., & Fields, S. I. (2002). Can the size of microcalcifications predict malignancy of clusters at mammography? *Academic Radiology*, *9*(1), 18–25. doi:10.1016/S1076-6332(03)80293-3

Cancer Facts and Figures. (2004). American Cancer Society. Retrieved from http://www.cancer.org/downloads/STT/CAFF_finalPWSecured.pdf

Chan, H. P., Lo, S.-C., Sahiner, B., Lam, K. L., & Helvie, M. A. (1995). Computer-aided detection of mammographic microcalcifications: pattern recognition with an artificial neural network. *Medical Physics*, *22*, 1555–1567. doi:10.1118/1.597428

Chang, Y. H., Zheng, B., Good, W. F., & Gur, D. (1998). Identification of clustered microcalcifications on digitized mammograms using morphology and topography-based computer-aided detection schemes. A preliminary experiment. *Investigative Radiology*, *33*(10), 746–751. doi:10.1097/00004424-199810000-00006

Cooley, T., & Micheli-Tzanakou, E. (1998). Classification of Mammograms Using an Intelligent Computer System. *Journal of Intelligent Systems*, *8*(1/2), 1–54.

Dengler, J., Behrens, J., & Desaga, J. F. (1993). Segmentation of microcalcifications in mammograms. *IEEE Transactions on Medical Imaging*, *12*, 634–642. doi:10.1109/42.251111

Doi, K., Giger, M. L., Nishikawa, R. M., & Schmidt, R. A. (1997). Computer-Aided Diagnosis of Breast Cancer on Mammograms. *Breast Cancer (Tokyo, Japan)*, *4*(4), 228–233. doi:10.1007/BF02966511

Elmore, J. G., Armstrong, K., Lehman, C. D., & Fletcher, S. W. (2005). Screening for breast cancer. *Journal of the American Medical Association*, *293*(10), 1245–1256. doi:10.1001/jama.293.10.1245

Elmore, J. G., Nakano, C. Y., Koepsell, T. D., Desnick, L. M., D'Orsi, C. J., & Ransohoff, D. F. (2003). International variation in screening mammography interpretations in community-based programs. *Journal of the National Cancer Institute, 95*(18), 1384–1393.

Esserman, L., Cowley, H., Eberle, C., Kirkpatrick, A., Chang, S., Berbaum, K., & Gale, A. (2002). Improving the accuracy of mammography: volume and outcome relationships. *Journal of the National Cancer Institute, 94*(5), 369–375.

Fondrinier, E., Lorimier, G., Guerin-Boblet, V., Bertrand, A.F., Mayras, C., Dauver, N. (2002). Breast microcalcifications: multivariate analysis of radiologic and clinical factors for carcinoma. *World J Surg. Mar., 26*(3), 290-6.

Gail, M. H., Brinton, L. A., Byar, D. P., Corle, D. K., Green, S. B., Schairer, C., & Mulvihill, J. J. (1989). Projecting individualized probabilities of developing breast cancer for white females who are being examined annually. *Journal of the National Cancer Institute, 81*(24), 1879–1886. doi:10.1093/jnci/81.24.1879

Gail, M. H., & Costantino, J. P. (2001). Validating and improving models for projecting the absolute risk of breast cancer. *Journal of the National Cancer Institute, 93*(5), 334–335. doi:10.1093/jnci/93.5.334

Gavrielides, M. A., Lo, J. Y., & Floyd, C. E. Jr. (2002). Parameter optimization of a computer-aided diagnosis scheme for the segmentation of microcalcification clusters in mammograms. *Medical Physics, 29*(4), 475–483. doi:10.1118/1.1460874

Gulsun, M., Demirkazik, F. B., & Ariyurek, M. (2003). Evaluation of breast microcalcifications according to Breast Imaging Reporting and Data System criteria and Le Gal's classification. *European Journal of Radiology, 47*(3), 227–231. doi:10.1016/S0720-048X(02)00181-X

Gurcan, M. N., Chan, H. P., Sahiner, B., Hadjiiski, L., Petrick, N., & Helvie, M. A. (2002). Optimal neural network architecture selection: improvement in computerized detection of microcalcifications. *Academic Radiology, 9*(4), 420–429. doi:10.1016/S1076-6332(03)80187-3

Gurcan, M. N., Sahiner, B., Chan, H. P., Hadjiiski, L., & Petrick, N. (2001). Selection of an optimal neural network architecture for computer-aided detection of microcalcifications--comparison of automated optimization techniques. *Medical Physics, 28*(9), 1937–1948. doi:10.1118/1.1395036

Karssemeijer, N. (1993). Adaptive noise equalization and recognition of microcalcification clusters in mammograms. *Int.J.Patt.Rec.&Im. Analysis., 7*.

Lado, M., Tahoces, P. G., Mendez, A. J., Souto, M., & Vidal, J. J. (2001). Evaluation of an automated wavelet-based system dedicated to the detection of clustered microcalcifications in digital mammograms. *Medical Informatics and the Internet in Medicine, 26*(3), 149–163. doi:10.1080/14639230110062480

Lanyi, M. (1977). Differential diagnosis of micro-calcifications, X-ray film analysis of 60 intraductal carcinoma, the triangle principle. *Der Radiologe, 17*(5), 213–216.

Lanyi, M. (1985). Microcalcifications in the breast--a blessing or a curse? A critical review. *Diagnostic Imaging in Clinical Medicine, 54*(3-4), 126–145.

Le Gal, M., Chavanne, G., & Pellier, D. (1984). Diagnostic value of clustered microcalcifications discovered by mammography (apropos of 227 cases with histological verification and without a palpable breast tumor). *Bulletin du Cancer, 71*(1), 57–64.

Le Gal, M., Durand, J. C., Laurent, M., & Pellier, D. (1976). Management following mammography revealing grouped microcalcifications without palpable tumor. *La Nouvelle Presse Medicale, 5*(26), 1623–1627.

Li, H., Liu, K. J., & Lo, S. C. (1997). Fractal modeling and segmentation for the enhancement of microcalcifications in digital mammograms. *IEEE Transactions on Medical Imaging, 16*(6), 785–798. doi:10.1109/42.650875

Lorenz, H., Richter, G. M., Capaccioli, M., & Longo, G. (1993). Adaptive filtering in astronomical image processing. I. Basic considerations and examples. *Astronomy & Astrophysics, 277*, 321.

Mata Campos, R., Vidal, E. M., Nava, E., Martinez-Morillo, M., & Sendra, F. (2000). Detection of microcalcifications by means of multiscale methods and statistical techniques. *Journal of Digital Imaging, 13*(2Suppl 1), 221–225.

Roque, A. C., & Andre, T. C. (2002). Mammography and computerized decision systems: a review. *Annals of the New York Academy of Sciences, 980*, 83–94. doi:10.1111/j.1749-6632.2002.tb04890.x

Shah, A. J., Wang, J., Yamada, T., & Fajardo, L. L. (2003). Digital mammography: a review of technical development and clinical applications. *Clinical Breast Cancer, 4*(1), 63–70. doi:10.3816/CBC.2003.n.013

Shen, L., Rangayyan, R. M., & Desautels, J. E. L. (1994). Application of shape analysis to mammographic calcifications. *IEEE Transactions on Medical Imaging, 13*, 263–274. doi:10.1109/42.293919

Spyrou, G., Nikolaou, M., Koufopoulos, K., & Ligomenides, P. (2002). A computer based model to assist in improving early diagnosis of breast cancer. *In Proceedings of the 7th World Congress on Advances in Oncology and 5th International Symposium on Molecular Medicine*, October 10-12, 2002, Creta Maris Hotel, Hersonissos, Crete, Greece.

Spyrou, G., Nikolaou, M., Koussaris, M., Tsibanis, A., Vassilaros, S., & Ligomenides, P. (2002). *A System for Computer Aided Early Diagnosis of Breast Cancer based on Microcalcifications Analysis*. 5th European Conference on Systems Science, 16-19 October 2002, Creta Maris Hotel, Crete, Greece.

Spyrou, G., Pavlou, P., Harissis, A., Bellas, I., & Ligomenides, P. (1999). Detection of Microcalcifications for Early Diagnosis of Breast Cancer. In *Proceedings of the 7th Hellenic Conference on Informatics, University of Ioannina Press, August 26-28, 1999*, Ioannina, Greece, (p. V104).

Timins, J. K. (2005). Controversies in mammography. *New Jersey Medicine, 102*(1-2), 45–49.

Wright, T., & McGechan, A. (2003). Breast cancer: new technologies for risk assessment and diagnosis. *Molecular Diagnosis, 7*(1), 49–55. doi:10.2165/00066982-200307010-00009

Wu, Y., Doi, K., Giger, M. L., & Nishikawa, R. M. (1992). Computerized detection of clustered microcalcifications in digital mammograms: applications of artificial neural networks. *Medical Physics, 19*, 555–560. doi:10.1118/1.596845

Wu, Y., Giger, M. L., Doi, K., Schmidt, R. A., & Metz, C. E. (1993). Artificial neural networks in mammography: application to decision making in the diagnosis of breast cancer. *Radiology, 187*, 81–87.

Yankaskas, B. C., Schell, M. J., Bird, R. E., & Desrochers, D. A. (2001). Reassessment of breast cancers missed during routine screening mammography: a community-based study. *AJR. American Journal of Roentgenology, 177*(3), 535–541.

Zhang, W., Doi, K., Giger, M. L., Wu, Y., Nishikawa, R. M., & Schmidt, R. A. (1994). Computerized detection of clustered microcalcifications in digital mammograms using a shift-invariant artificial neural networks. *Medical Physics, 21*, 517–524. doi:10.1118/1.597177

Zhang, W., Doi, K., Giger, M. L., Wu, Y., Nishikawa, R. M., & Schmidt, R. A. (1996). An improved shift-invariant artificial neural networks for computerized detection of clustered microcalcifications in digital mammograms. *Medical Physics*, *23*, 595–601. doi:10.1118/1.597891

Zhang, W., Yoshida, H., Nishikawa, R. M., & Doi, K. (1998). Optimally weighted wavelet transform based on supervised training for detection of microcalcifications in digital mammograms. *Medical Physics*, *25*(6), 949–956. doi:10.1118/1.598273

KEY TERMS AND DEFINITIONS

BI-RADS: Breast Imaging Reporting and Data System.

Chapter 15
e–OpenDay:
Open Virtual Environment for Biomedical Related Research, Business and Public Resources

Vasileios G. Stamatopoulos
Biomedical Research Foundation of the Academy of Athens, Greece

George E. Karagiannis
Royal Brompton and Harefield NHS Trust, UK

Michael A. Gatzoulis
Royal Brompton and Harefield NHS Trust, UK

Anastasia N. Kastania
Biomedical Research Foundation of the Academy of Athens, Greece

Sophia Kossida
Biomedical Research Foundation of the Academy of Athens, Greece

ABSTRACT

This chapter presents the feasibility study of a virtual platform for medical related technology transfer, continuing medical education and e-conference. The concept extends the idea of live events (e.g. conferences, open day events) in one physical location. It exploits the creation of a virtual platform where the research world in the area of biomedicine, can showcase their success, interact and co-operate with the business community and collaborate on potentially valuable outcomes and learn without time or place restrictions. The main objective of the project was to offer a pilot service that can showcase the e-OpenDay market potential and technical feasibility. By developing a prototype and through user feedback and evaluation processes, a set of services was identified, developed and validated. The e-OpenDay project made clear that health information services are facing rapid development and expansion to wider markets and user groups. Based on the project results, a business plan was developed that showcased potential in commercial exploitation.

DOI: 10.4018/978-1-60566-768-3.ch015

INTRODUCTION

This chapter presents the feasibility study of a virtual platform for medical related technology transfer, continuing medical education and e-conference, as a result of the work carried out in the framework of the e-OpenDay project. The e-OpenDay project was co-funded by European Commission, Directorate-General Information Society, under the eTEN scheme. The main objectives were to evaluate the technical feasibility and validate the market potential of the proposed virtual platform and similar services.

e-OpenDay originated from real research open day events organised by the National Heart and Lung Institute (Imperial College, School of Medicine) and the Royal Brompton & Harefield NHS Trust (RBHT) in the UK which aimed at bringing closer together the fast moving worlds of medical research and of business. The rate of growth of these events both in terms of numbers of presentations and numbers of attendees was very significant and consequently the idea to transfer, re-design and expand the whole concept across the Internet gained many supporters.

The e-OpenDay concept extends on the idea of hosting live events (e.g. conferences, open day events) in a single physical location. It exploits the creation of a virtual platform, where the research community in the area of healthcare, can showcase their success, interact and co-operate with the business community and collaborate on potentially valuable outcomes without time or place restrictions. The proposed service is envisaged as a combination of knowledge management, e-commerce and e-education aiming to establish an alternative, virtual meeting place for the busy medical research, business and healthcare professional communities, where they can exchange ideas and develop collaborative relationships. It is also envisaged as a mean to offer Continuing Medical Education (CME) accreditation courses without having to physically attend the organised events.

Three major objectives were set prior to the project's commencement:

- to conduct a trans-European market survey
- to provide pilot service and validate the product under real operating conditions
- to set out a detailed business plan

E-OPENDAY SERVICES OFFERED

One of the main objectives of the project was to offer a pilot service that can showcase the "proof of concept" potential. The e-OpenDay service (Figure 1) that was developed provided two main sets of applications as a virtual platform to the research and business communities in the area of healthcare, to facilitate interaction and co-operation.

E-OpenDay InfoExchange Application

This application consisted essentially of a set of interfaces, through which the members of the e-OpenDay service could interact with the system. This is the web-space, where the medical researchers could log-on to submit their own work, while at the same time, several business partners had the opportunity to browse/search for interesting articles to read.

LiveWeb Application

This e-OpenDay service covered Open Days, Conferences and CME courses, allowing users to follow these events virtually. The e-OpenDay platform was able to webcast live events and record them and to make them available for playback, at a later time, for the registered users (Figure 2).

The information content of the virtual platform consisted of:

Figure 1. Schematic representation of the e-OpeDay Service Model

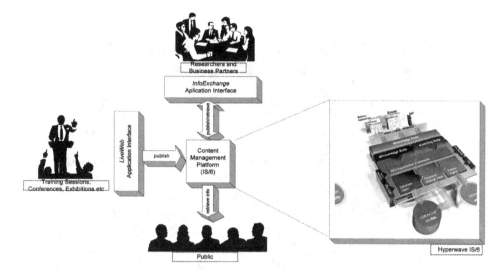

- Research Open Days (virtual posters)
- Conferences (video presentations and PowerPoint presentations)
- Research Community – Promotion (patents, papers, abstracts, experimental protocols and novel techniques)
- Continuing Medical Education (video presentations, PowerPoint presentations and other learning material)

PROJECT RESULTS AND MAIN ACHIEVEMENTS

Definition of User Requirements

The user requirements were defined according to the user needs, the offered system services, the content categories, the user interfaces and the navigation scenarios as well as the technologies used. Emphasis was put into identifying the detailed user profiles of each user group. Based on the user profiles, the specific user needs were then recognised. The analysis and description of the overall services to be offered to the target groups facilitated the definition of the business model. A series of tasks was carried out:

- Interviews with people who have had organised or were organising open days, conference events, CME courses, key people in Research and Development (R&D) in the healthcare industry, junior researchers and industry people.
- Questionnaires were developed and distributed to the business and researchers communities.
- Results were analysed and user requirements were updated accordingly.
- The updated user requirements were presented in a structured form.
- The detailed pilot application protocols for each target group were defined.
- The service was developed based upon the updated user requirements.

Technical Implementation

The two platforms (Content Management server and Media Server) were customised according to the user requirements and integrated to develop the e-OpenDay service prototype.

e-OpenDay was based mainly on existing and mature technologies, thus reducing the risk and

Figure 2. Interface of the LiveWeb application. Live web-casting of the presenter's video and the Power-Point presentation. The user can also navigate through the slides when watching a recorded session.

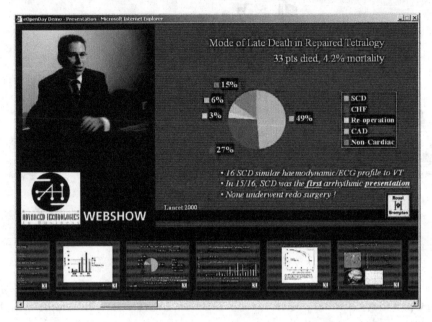

time associated with the development of a new system. The main challenge was in the extensive customisation and integration of the used platforms that was necessary in order to fulfil the user requirements. A usable, attractive and effective environment for the users had to be designed and operated, which would also incorporate content management, billing and administrative functionalities.

The Server or Content Management Platform was the foundation of the overall system hosted by an Internet Service Provider (ISP) with a dedicated line connection to the Internet. The ISP was providing all the data centre facilities, like daily back up, crash recovery support, antivirus protection, firewalls etc. A video-streaming server was also used to produce the streaming of the video to the end user (through the Content Management Platform) in real time.

The e-Open Day InfoExchange application consisted essentially of a set of interfaces, through which the members of the e-OpenDay service could interact with the system. This was the web-

space, where the medical researchers could log-on to submit their work, and at the same time, business partners had the opportunities to browse for interesting articles to read. The content available through InfoExchange was classified over several categories in order to help end-users to access information efficiently with minimum time lost for queries and navigation.

A LiveWeb service offered by e-OpenDay covered Open Days, CME events and Conferences, allowing users to follow these events virtually. The e-OpenDay platform was able to web cast live these events, and record them in order to make them available for playback at a later time for registered users. Live static digital video cameras were connected to terminal computers. These terminals were connected over a Local Area Network (LAN) to a central Server situated at the conference/exhibition site transmitting the Data to the Content Management Platform via Internet. Streaming Server Software was installed as part of the LiveWeb and was producing the streaming of the video to the end user (through the Content

Management Platform) via a Real Time Protocol for real time delivery of data.

e-OpenDay was a user-friendly platform because of the personalisation of the user interfaces. Following an initial and brief registration, the system was aware of the characteristics of each user. Thus, when logging-on, each user faced a dynamically built interface, where information was prioritised according to his/her choices, and search filters/agents could also be personalised. This offered a much more efficient matching of requested and posted material, promoting further the collaboration between the Science and Business Communities.

A set of interfaces was implemented to cover all different type of user groups. Essentially, four (4) user groups were identified to cover the needs of the end users.

- The first group was for all e-OpenDay Guest Users. It referred to the users who wished to retrieve free information and upload documents, abstracts, papers and so on. The purpose of providing free access was to advertise and disseminate the e-OpenDay services to the wider scientific community by searching and retrieving valuable content while promoting the e-OpenDay events, and medical conferences
- The second group of users were the medical researchers
- The third group of users were subscribers from the industry
- The fourth group referred to the administrators that had access to simple back office applications, such as the editing of user data, management of the new registration requests and so on

The services provided by the system to the end users were:

- The Research Open Days
- Conferences

- Research Community – Promotion
- Continuing Medical Education
- Expert search (agents)

The system administration services provided by the system were:

- Subscription Management and
- Back Office support

The main technical achievement of e-OpenDay was the delivery of an integrated platform meeting all the above challenges. Information delivery and supporting transactions were offered over the Internet via a personal computer. The telecommunication infrastructure was based on digital data networks. Internet-based communications ensured universal user access and high expandability of the services.

The basic access to e-OpenDay services was through a web interface and any common Internet connection using a common PC trough a standard browser. Through this interface, the users could register, create their profile and access all the services. Any network medium giving access to the Internet could be used such as PSTN, ISDN, xDLS, GPRS. The infoexchange services in general did not require high-speed connection. Access to the liveweb application however needed higher speed and ideally a DSL connection. The service however is functional even on 64k ISDN or good quality PSTN connections.

For testing, the target-of-test functionality of the e-OpenDay prototype was examined. The functionality delivered by the development team was also examined as to if it was in line with the functional specifications and the user requirements. The testing exercised each test case, use-case scenario, individual use-cases flows or functions and features, using valid and simulated data, to verify that the system is functional and without serious flaws. The system and services were thoroughly tested for defects, inconsistencies, missing functionalities and areas of improvement. Testing experts executed the trial using simulated test data by testing experts and its operational behaviour was observed by filling reporting forms with all problems and inconsistencies observed.

Service Validation

The validation process focused on the updated user requirements and tested the system to ensure that it can successfully meet these requirements. For this reason an e-OpenDay event was organised by the Royal Brompton & Harefield NHS Trust in London, UK. Users that performed the validation were selected and trained at the predefined sites. Following training, the participants accessed the service and filled in questionnaires. The findings and results from validation process were used in the preparation of the final exploitation and business plan in order to update the business model.

Methodology for Service Validation

The service validation was the assessment of the system's efficiency, usability, reliability and scientific soundness by the three main user groups of the system, namely the Researchers, the Healthcare Industry and by the Administrator group. Specialised validation questionnaires for each of the three different user groups were also prepared.

The validation was conducted with the help of users from the research community and the healthcare industry.

The validation trials were conducted during and after the validation event organised in London. Computers with Internet connections were available at the validation sites. After a small introduction, the user read a number of scenarios and followed the instructions. Each user validation session lasted approximately one (1) hour.

Results of Service Validation

Responses to open questions on general system features / system effectiveness suggested the following areas for improvement:

- Terminology used
- User Friendliness
- Non standardised information
- Personalised environment
- Credit card back office functionality
- Limited content and number of projects in the database.

Recommended actions that were taken by the technical team:

- Revised terminology approach
- Trace information about the owner of documents in a more effective way
- Improve standardisation of information
- Improve user friendliness, simplify interface
- Improve training
- Improve help file
- Customise first page for the three groups
- Restore demonstration credit card functionality
- Increase the amount of content

The quantitative analysis of the questionnaires showed that for the majority of the results, the service passed successfully the usability criteria. The users in general were pleased with the overall system. The group of questions regarding the general system features gave results significantly over the average while no user failed the system over the general features. However the functionality of the system appeared to be a trade off between the number of features and the usability of the system. A training course could be provided to the novice users since decreasing the features of the system would affect the value of the platform. On the other side, training would increase the cost of the service. Alternatively free on-line training material could be provided to users who may not want to use resources for a formal training course. The addition of more training, enhanced help files, changes in the system terminology, minor modifications of the home page and some changes in the design of user tabs, were sufficient. At the time of the validation the platform contained the details of over 3.600 medical related ongoing projects.

Conclusion and Lessons Learned from Validation and Validation Event

The results have demonstrated that the user requirements were met to a great extent. However the validation also identified some necessary improvements in the functionality aspect of the service. The nature of any user driven service like e-OpenDay makes the optimisation of the service and user satisfaction a continuous process. From a technical focus the platform was adequate to withstand the medium traffic occurred under real conditions.

From a business point of view we can see that the online attendees (n=57) outnumbered attendees present in the conference room (n=35) during the validation event. A number of people attending the event also left the conference room halfway through the event, because of other commitments. These facts confirm the results of the market survey and the business model, which indicated that a large number of healthcare professionals might prefer online attendance rather than commit themselves for a full-day event. Both online and offline attendees also had the chance to view the presentations online at their own convenience.

MARKET SURVEY AND BUSINESS PLAN

The market validation phase of the project had as main goal to prepare the necessary framework for deploying e-OpenDay services. The achievement of this goal required the examination of the market and the identification of the target market segment. An extensive market analysis was performed towards that direction.

Market Analysis

For the purposes of the market analysis a qualitative and quantitative study was performed, identifying and quantifying the potential markets. The attractiveness of the healthcare related market for the introduction of such a service was investigated. The market analysis focused on the e-health technological environment to determine familiarisation of the biomedical and research communities with the concept of using networks and e-tools to assist them in various research tasks, the profile and number of companies involved in research and in academic institutions the profile and number of researchers.

Market Survey Methodology

A questionnaire-based survey was also performed. For the production of the questionnaires the following methodology was followed.

Interviews with key individuals were carried out. Interviews were semi structured and involved discussion/brainstorming sessions. Questions were asked on the viability of the concept in the present e-health scenario in Europe, competition and differentiating strategies, target markets and limitations of the project. Two early versions of the questionnaire were produced and tested with a small number of participants. Two final questionnaires were produced; one intended for researchers and content providers and the other for investors from the industry.

The questionnaires were then distributed to researchers and industry people during conferences, events, using personal and professional contacts and by mail. A number of questionnaires in excess of 250 were collected, and the results were analysed and used in the final version of the Business plan.

Market Survey Results

A thorough market analysis was conducted and conclusions were reached affecting key aspects of the deployment of the service. In particular changes in the e-Health Sector were identified (Markatatos):

- The Internet in the healthcare industry is increasing with a growth of e-commerce in the healthcare sector in Europe. However there is still resistance to adoption of Information Technology (IT) in the health sector but this rapidly decreases.
- e-Health is a fragmented industry with limited effort at integration.
- There is a need to integrate product quality, technology, content and coverage with marketing efforts and customer relationship management.
- e-Health business models tend to be content and transaction-based. Revenue generation is usually from advertising, transaction fees and sponsorships.
- Healthcare is an attractive market for the IT industry. The European Union, individual governments and entrepreneurs are actively promoting e-Health ventures. However, the economic slowdown is adversely impacting on e-Health ventures.
- There are a host of unresolved legal, ethical and political issues related to e-Health.
- Customer preferences and customer acceptance of e-Health by the general public, professionals and the industry are critical factors that will influence the success of Internet ventures in healthcare.
- High quality information and the development of trust are essential for any e-Health venture to gain customer loyalty. Quality

and validity of information and privacy on the Internet is a major concern
- Internet ventures need to build competitive advantage through differentiation of their products and services. The only way to attain defensibility and sustainability in e-commerce ventures is to constantly evolve products, develop customer friendly approaches and content and create sustainable entry barriers early on.

The above conclusions indicated that there is a gap in the ever-changing e-Health sector. Therefore a differentiated service offering quality of content and backed by strong commercial and academic institutions has significant potential in covering the needs of the e-Health actors.

These actors which constitute both the target market of e-OpenDay events and the suppliers of content include the biomedical and pharmaceutical industry, the academic community and individual researchers. The e-OpenDay platform has identified gaps in the market, where it can successfully bridge the needs of researchers to the industry and bring significant advantages to both. More specifically the analysis identified the following gaps in the market:

- Absence of an interactive platform for the research and business communities in health. There are general portals providing scientific information, without the focus on identifying business opportunities and linking the biomedical and healthcare communities with the corresponding industries.
- No visible presentation of ongoing research for funding or commercialization. There is a lack of integration of technology and products along the entire spectrum of e-Health in a single portal.
- Lack of a brokerage and consultation service for these communities backed by the expertise and prestige of an acknowledged

institution such as Royal Brompton and Harefield NHS Trust.

- Absence of a site dedicated to professional education and distant learning offering medical CME accreditation courses. This is of particular interest since there is strong demand for such services across the globe and there is strong potentiality for a service offering them to capitalize a large percentage of this market.

From the market analysis of the European and national markets the size of the market was identified and a competition analysis was also performed. The competition was found to vary, focusing on niche segments of the market, different from those the e-OpenDay aims at, thus providing a valid market opportunity. The size of the market in Europe appeared to be attractive and able to provide the critical user mass for achieving the necessary economies of scale. This would provide the basis for an attractive pricing of the service and move the basis of purchasing decisions away from price focusing them and directing them on the added value features of e-OpenDay.

Finally, as mentioned earlier, two questionnaires were issued in events organized by RBHT and to potential users and content providers from each target market. The results of this analysis were used in the market validation and in the preparation of the second version of the business plan. The questionnaires addressed among others the attractiveness of the service in terms of content volume an issue that might determine the cooperation and partnerships with content providers as well as the range and segments of the market that the e-Openday addresses.

The above conclusions that were derived from the market analysis facilitated the definition of appropriate marketing and promotional strategy, the pricing and sales strategy and will be a valuable contribution in developing the appropriate business models for the successful exploitation of the e-OpenDay concept.

Business Plan

Methodological Approach

A business plan produced at the end of the project, aimed to demonstrate the feasibility of deploying the e-OpenDay and similar services. A group of business experts has evaluated the validity of the plan and their recommendations were taken into consideration. The business plan had input from all the project activities. The user requirements, the results of the market analysis and the system validation were fed in the business plan so that the business model was updated and optimised.

Business Plan Results

The produced plan covered key areas that investors need to consider in evaluating their support for this project. The key sections were:

- The e-OpenDay service – presents an overview of the target market, some key aspects of the e-OpenDay product and services and a description of the e-OpenDay features.
- Results from Market and Business Validation – describes the results of the validation and the conclusions of the market analysis which clearly demonstrates a potential.
- Risk and Contingencies – summarises the key risks associated with the entry of the e-OpenDay product into the market place and strategies for minimising their impact.
- Marketing, Pricing & Sales strategy – set out the various approaches for the successful promotion and marketing of e-OpenDay on a European level.
- Financial Strategy – summarises the results of financial analysis in a financial plan for the next five years, including assumptions and sensitivity analysis, which underpin the business model.

- Business Organization and Management Team - summarises the proposed organizational structure and provides information on service operation, quality and administration.
- Deployment Plan – clearly defines the immediate steps to be followed towards the deployment of the service.

OVERALL IMPACT AND DISSEMINATION OF RESULTS

Evaluation of Impact

The impact assessment of the of the e-OpenDay services was achieved through the design and roll out of the pilot application in the UK, Belgium, Greece, Germany and the Netherlands involving healthcare professionals, industry experts and administrators as users. In all sites, the results emphasised similar impact trends. An information society in constant spreading, busy medical professionals that need to disseminate their research and attend CME courses with limited time and resources, as well as industry people looking for early stage investment opportunities in the healthcare sector.

Conferences

The conferences embraced both national and international events. A wide range of events was identified as suitable for the marketing of e-OpenDay, and a selection of these was targeted during the course of the project. Representatives have attended a number of events in various countries where the project and its objectives were disseminated, either in the form of attending the event and presenting the project, organising a workshop or participating in a more organised and intensive manner, i.e. by reserving a stand and assuring a continuous presence.

- 2nd International Congress "Lung and the Environment 2003": 20-22 June 2003, Nicosia Cyprus
- 3rd International Congress "Lung and the Environment Conference": organised by Royal Brompton & Harefield NHS Trust, the Hellenic thoracic society and the Cypriot Thoracic Society, 4-6 June 2004 in Limassol Cyprus.
- "European Union Funded Projects, Opportunities in the Healthcare Sector", 28 July 2004, Imperial College, London, UK
- Research Open Day, July 8 2003, Hammersmith Hospital, London, UK
- 2nd Joint European/North American Symposium on Congenital Heart Disease in the Adult: 19–20 September 2003 Santorini Island, Greece.
- 'Partnership in Health' event: 17 November 2003, London, UK
- The e-Health Exchange: QEII Conference Centre London, 19 November 2003, UK
- "Intraoperative Echocardiography Course on Coronary Artery Surgery": 4 February 2004, National Heart & Lung Institute, Faculty of Medicine, Imperial College London, UK
- "Advances in Cardiology Course": 5–6 February 2004, National Heart & Lung Institute, Faculty of Medicine, Imperial College London, UK
- Medday event: 28 April 2004, Fraunhofer Institut für Graphische Datenverarbeitung Darmstadt, Germany
- Healthcare Information Exchange: 25 May 2004, London, UK, 10th Annual EARMA Conference, "Research Management and Administration in a Changing World": 24-26 June 2004 in Bucharest, Romania
- Annual meeting of Romanian Society of Cardiology: 15-18 September 2004, Brasov, Romania

EXPLOITATION OF RESULTS

The e-OpenDay business plan is based on the exploitation of services and not on selling an off-the-shelf product. The business runs the service and is responsible to support it regarding information input, technology, infrastructure, labour and operational costs.

e-OpenDay was promoted as a service provider with high reliability and scientific foundation that is always at the front end of medical advances expertise. The pricing policy has been determined on the basis of the market research that was performed during the lifespan of the project. Furthermore it has been validated by all targeted segments at key events as well as by answering questionnaires available on site. e-OpenDay is not a single product or service but offers a bundle of services, which creates more possibilities when deciding a pricing policy. Several factors have determined the pricing of the services including data from market research, the validation of the services as performed both on site and via the questionnaire available on the web, the perceived value of the services on offer, the sales projections and the competitors' products and prices.

A detailed description of the services offered, the percentage of contribution each service makes to the total costs and the contribution of each service to the total sales forecasted is provided below:

High-Value Service

This is the group subscription service, which is associated with a specific number of licences (two hundred licenses and more). It is addressed to all organisations –research or business oriented. It offers members all options for writing to and reading from the e-OpenDay data source. It also enables members to view the events of their choice covered by e-OpenDay.

Virtual Conference Organisation

This service undertakes the overall organisation of a live Open Day event, in conjunction with the organising party, and its coverage over the e-OpenDay network. Services range from setting up guest lists to managing the whole event and web-casting it to all members of the network.

Individual Subscriptions

This service offers read-only privileges to content in the e-OpenDay data store. This excludes content generated by the coverage of live events which can be purchased on a per event basis.

Virtual Conference Attendance

This service is for individuals who are not members of the e-OpenDay network and who wish to view an event web-cast over the network. This is for events which are not subsidized for viewing by their hosts-organisers.

CME Accreditation Courses

e-OpenDay will also offer CME accreditation by enabling participants to view online conferences held at accredited venues in cooperation with accredited institutions in various countries, and will work with them to identify topics and potential participants who could take advantage of these conferences as part of their continuing professional development.

CONCLUSION

It is possible that two distinct worlds, the research and the business communities, can grasp the opportunity to collaborate effectively through a virtual, interactive and real-time platform. The e-OpenDay concept is a way to initiate a new era by building enduring partnerships and motivating both worlds for further interaction and research developments, overcoming the space limitations.

The main objective of the project was to offer a pilot service that can showcase the e-OpenDay market potential and technical feasibility. The service provided two main sets of applications as a virtual platform to the research and business communities in the area of healthcare, aiming to facilitate their interaction and reinforce their co-operation.

The e-OpenDay InfoExchange application consisted of a set of web based interfaces, through which the medical researchers logged-on to submit their work, and at the same time, business partners browsed for interesting project ideas.

The e-OpenDay LiveWeb application facilitated Open Days, Conferences and CME courses, by allowing users to follow these events virtually, web cast live events and record them in order to make them available for playback at a later time for the registered users.

The information content of the pilot platform consisted of:

- Research Open Days (virtual posters)
- Conferences (video presentations and PowerPoint presentations)
- Research Community – Promotion (patents, papers, abstracts, experimental protocols and novel techniques)
- Continuing Medical Education (video presentations, PowerPoint presentations and other learning material)

The e-OpenDay pilot phase concluded to the following statements:

- The pilot services were considered useful and valuable by their end users
- Positive feedback was received by the pre-defined user groups
- Benefits were identified for the medical professionals and the health industry.
- The pilot service set was continuously

evolved and refined throughout the duration of the project, concluding to a configuration with the highest probability of being commercially successful

- Commercial services should contain a critical number of prestigious and active clinical and academic partners, which will become a valuable basis of information providers and scientific experts. The partners should be "networked" and capable in expanding to additional countries
- The key to success will be the effective co-ordination and establishment of a European-wide network of researchers, institutions and funding organisations and the creation of inter-business relationships necessary to generate value from this network

The e-OpenDay project confirmed the fact that health information services are facing rapid development and expansion to wider markets and user groups. A gap in the healthcare information market with exploitation potential has been identified, its requirements and opportunities were defined and a market validation was conducted. The market validation has also revealed some trends which helped to adapt the service according to the needs of the market. By developing a prototype and through user feedback and evaluation processes, a set of services was identified and validated. Based on the project results a model business plan was developed that showcased potential in commercial exploitation. Dissemination to decision makers in health, authorities, enterprise and to the wide public has initiated an interest. Considerable impact was also achieved on the scientific community.

ACKNOWLEDGMENT

e-OpenDay was funded by European Commission, Directorate-General Information Society, under the eTEN scheme and was concluded in January

2005. Project Partners were (i) RBHT (Royal Brompton & Harefield NHS Trust), UK, (ii) ICLC (Imperial College London Consultants), UK, (iii) Hyperwave, Germany, (iv) ATB (Advanced Technologies in Business Ltd), UK, (v) ELYROS (Elyros SA), Belgium, (vi) ING, Greece.

REFERENCES

Antoniadis G., Kofteros S., (2004). *e-OpenDay Project* [Technical Report].

Eng, T. R. (2001). *The eHealth Landscape: A Terrain Map of Emerging Information and Communication Technologies in Health and Health Care*. Princeton, NJ: The Robert Wood Johnson Foundation.

Markatatos G., Atun R., Barthakur N., Jollie C., Kotis T., Stamatopoulos V.G., et al (2004). *e-OpenDay Project Market Analysis*.

Markatatos G., Atun R., Barthakur N., Jollie C., Kotis T., Stamatopoulos V.G., et al (2005). *e-OpenDay Project, Final Business Plan*.

Taylor A.N. (2000). *Royal Brompton Hospital Research Open Day 2000 Proceedings*.

Vasileios G. Stamatopoulos V.G., Karagiannis, G.E. (2005). *e-OpenDay Project Final Report*.

Wutoh, R., Boren, S. A., & Balas, E. A. (2004). eLearning: a review of Internet-based continuing medical education. *The Journal of Continuing Education in the Health Professions*, *24*, 20–30. doi:10.1002/chp.1340240105

KEY TERMS AND DEFINITIONS

Back Office: It a part of most organisations where tasks dedicated to running the company itself take place. The term comes from the building layout of early companies where the front office would contain the sales and other customer-facing staff and the back office would be those manufacturing or developing the products or involved in administration but without being seen by customers.

CME Accreditation: Content for CME programs is developed, reviewed, and delivered by faculty who are experts in their individual clinical areas. Courses have to be accredited in the form of number of credits, by either the European Accreditation Council for Continuing Medical Education (EACCME) or the corresponding American Accreditation Council for Continuing Medical Education (ACCME).

Content Management Server: A content management Server is a computer system used to create, edit, manage, search and publish various kinds of digital media and electronic text. CMSs are frequently used for storing, controlling, versioning, and publishing industry-specific documentation such as image media, audio files, video files, electronic documents, and Web content.

Continuing Medical Education (CME): CME refers to a specific form of continuing education (CE) that helps those in the medical field maintain competence and learn about new and developing areas of their field. These activities may take place as live events, written publications, online programs, audio, video, or other electronic media.

Media Server: A media server is a computer system, providing video on demand, to user computers usually over the Internet, storing various digital media.

Chapter 16
Modelling and Simulation in Biomedical Research

Dolores A. Steinman
University of Toronto, Canada

David A. Steinman
University of Toronto, Canada

ABSTRACT

In the following chapter, the authors will discuss the development of medical imaging and, through specific case studies, its application in elucidating the role of fluid mechanical forces in cardiovascular disease development and therapy (namely the connection between flow patterns and circulatory system disease - atherosclerosis and aneurysms) by means of computational fluid dynamics (CFD). The research carried in the Biomedical Simulation Laboratory can be described as a multi-step process through which, from the reality of the human body through the generation of a mathematical model that is then translated into a visual representation, a refined visual representation easily understandable and used in the clinic is generated. Thus, the authors' daily research generates virtual representations of blood flow that can serve two purposes: a) that of a model for a phenomenon or disease or b) that of a model for an experiment (non-invasive way of determining the best treatment option).

INTRODUCTION

Over the last century society in general and the medical profession in particular has witnessed unimaginable and unprecedented growth, progress, expansion and advances in the technological field, particularly in the area of medical imaging. As we become more intimate with technology we realize how much the understanding of our lives has changed for the better, while at the same time new questions and issues that need addressing where brought to into light. Our particular work relies on the use of medical imaging and computer simulation technologies to visualize a phenomenon otherwise not accessible to the naked eye: blood flow. The following chapter will look at the ways in which these virtual images enable the scientist and the physician to better the understanding of the human body and the ways it functions.

DOI: 10.4018/978-1-60566-768-3.ch016

Mirroring society at large, in the medical world technology is a ubiquitous and versatile presence allowing the real and the virtual merging and blending almost seamlessly. In the following chapter we will also present aspects of this merging encountered in our own research laboratory, as well as briefly reflecting upon the changes in our perception of medical imaging, the developments of the technology over the last half century, but most importantly the major steps forward in terms of its use as a didactic tool as well as aid in both defining a diagnosis and in deciding upon optimal treatment strategies.

BRIEF OVERVIEW OF THE FIELD

The search for the best way of understanding and then explaining motion in general and flow in particular has preoccupied the human mind for centuries. The focus of our research program is blood flow. In itself and in the context of the whole human body function, blood flow is one of the hidden-to-the-naked-eye phenomena, difficult to visualize, expose and explain. The need for a thorough examination lies in the tight connection between arteries, their vessel walls and the blood flowing through. The goal of our research is elucidating the role of fluid mechanic forces in cardiovascular disease and treatment. Arteries adjust their caliber to maintain the level of wall shear stress, the frictional force exerted by flowing blood against the vessel wall. The endothelial cells that line the wall, cells which also respond to and express various molecular factors, mediate this behaviour. Atherosclerosis (fatty deposits in the artery wall) and aneurysms (ballooning of the artery wall) tend to develop at sites of complicated blood flow patterns. Their rupture, the event that precipitates many heart attacks and strokes, is determined in large part by the mechanical forces exerted on them. The success of vascular surgeries or minimally invasive procedures depends on the skill of the surgeon or interventionalist as much

as it depends on the type of intervention or the design of the medical device chosen. The right strategy is crucial in avoiding the induction of further blood-flow-induced complications. With the discovery of the connection between flow patterns and circulatory system disease, the value and significance of computational fluid dynamics (CFD) was acknowledged, and CFD experts have been since involved in the designing and developing of the techniques and devices used in various vascular interventions (such as bypass grafts, stents or coils).

The need for exploring unseen areas of the body is not new: it has been preoccupying anatomists and clinicians since the dawn of medical science, but became the driving force behind perfecting the medical visualization techniques and technology over the last century. It all appears to have begun with Roentgen's discovery of X-rays, which initiated the modern age of medical imaging and set in motion a series of developments and applications that lead to previously unthinkable discoveries in the field. The initial accidental discovery was almost immediately followed by well-designed experiments that helped calibrate and adjust the apparatus and, subsequently, angiography was born from the idea of replacing blood with a radio-opaque dye.[1]

The development of both apparatus and skill enabling direct visual representations of blood opened new paths to a more thorough and detailed study of blood flow in its complexity. Our own work in the Biomedical Simulation Laboratory can be described as a multi-step process whereby from the reality of the human body an easily readable visual model is generated by the computer. To be more precise, data is collected from the patient/subject, and from corroborating information obtained from the medical images (MRI, CT) of the blood vessel, along with the Navier-Stokes equations, we generate a mathematical model for the physiological phenomenon of blood flow; this model is then translated into a visual representation of the phenomenon observed; the last step is the refining of this image to make it understandable by the viewer, in most cases a trained expert (clinician, technologist or medical student). To this translation process, which involves the evolution from a collection of binary data to a sophisticated, clear, accurate visual rendering, the need to merge at both technological and visual vocabulary levels is crucial. By unifying symbols and conventions, medical imaging technology enables the creation of a valid and robust language, which allows not only an accurate representation but also clear and fast communication between the technologist, the clinician and ultimately the patient. This seems a natural path to follow since the ultimate goal is to use all the very complex flow data acquired locally in order to create a visual model as close to the "human subject's body reality" as possible, while presenting it into a form accessible to clinicians and laypeople alike. First asked to the task were the engineers, due to their experience in experimental flow visualization, namely using physical models. As a result, the tendency for these models was to be symmetric and idealized for conceptual simplicity but also for practical reasons of manufacture. However, the engineering way of representing, while perceived as clear and easy to understand within the field, was largely inscrutable to the clinicians, the main users of these models.

As a result of the huge intricacy of the phenomenon itself, blood flow representations have taken on very different forms and shapes over the time. Considering its complexity as well as the multi-faceted aspects of the surrounding tissues, the interactions in the context of the physiology of the human body as a whole, and the challenges of representing it in a clear and accurate way, a transdisciplinary understanding of the issue was critical. To follow and understand the sequence of events, one has to know that the foundation for the mathematical model is Poiseuille's law, which predicts the pressure required to pump a fluid at a desired flow rate through a straight circular pipe of known dimensions. This, relationship, originally determined experimentally by Poiseuille[2], was formalized mathematically by recognizing that the velocity profile under such flow conditions takes the simple shape of a paraboloid. This knowledge of the velocity profile also makes it easy to derive simple formulae for wall shear stress, the frictional force exerted by flowing blood that plays a central role in regulating vascular behaviour. For nearly a century this law, which intentionally ignores the fact that blood actually pulsates with each heart beat, shaped – and, as we will see later, still shapes – the perception of how blood flows.

A century later, the mathematician J. R. Womersley extended Poiseuille's law to pulsating flow in arteries, and his mathematical solution was readily accepted by medical researchers, as a result of it being the fruit of the collaboration with D. A. McDonald, a physiologist colleague. The equations and their solutions were found to be the same as those describing electrical circuits, which, in turn, led to a popularization of electrical analog models of the circulatory system, whereby resistors, capacitors and inductors could be tuned to model the complex relationships between flow (current) and pressure (voltage). As it is well known that arterial diseases affect the nature of the pressure pulse, it was hoped that such models could be used to infer the nature of vascular diseases simply from pressure measurements.

However, by the late 1960's, a new era in blood flow modeling was inaugurated with a number of groups suggesting that the well-known localization of vascular diseases to arterial branches and bends could be explained by the presence of complex blood flow patterns at these sites. This observation, of course, opened up a whole new line of investigation focused on local rather than global blood flow behaviour and brought about the need for a new representation of the phenomenon and the creation of new and structurally different visual models.

USE OF CFD IN THE MEDICAL FELD

As previously indicated, models that engineers perceive as clear and easy to understand rarely suit the clinicians' perspective. Therefore, through collaboration between computer experts, bioengineers and, not in the least, clinicians, computer programs were created that made possible the generation of images that not only accurately reflected the data collected from the patient but also represented it in a clear way familiar to the medical practitioner. Current visual representations are considerably different from their recent predecessors, blood flow imaging having undergone significant, distinct stages brought about by major technological advances over the last couple of decades.

First in line came the conventional X-ray angiogram based on projection of X-rays through body and onto a film. But this particular approach did not allow easy visual access to arteries and blood (due to the presence of "competing" radio-opaque bones). It also limited the projection to a number of accessible planes. The introduction of computers allowed for digital vs. film imaging, and led to digital subtraction angiography, where a "mask" image acquired prior to contrast agent injection is subtracted, digitally, from images acquired during the injection of contrast agent.

The second phase, addressed the planar limitations and, similarly, led to development of Computerized Axial Tomography (CAT) scans, allowing images of non-projectable slices (e.g., axial cuts through the body) to be reconstructed from multiple conventional projections. Nowadays 3D (and emerging 4D) CT scanners provide sufficiently high resolution and image contrast to allow unambiguous detection of different tissues, including blood vessels, without the need for intra-arterial injections. This makes it possible to achieve photorealistic renderings of the digital data.[3] While such medical imaging, along with magnetic resonance imaging (MRI) and ultrasound, are superb

for resolving the vascular anatomy, they remain relatively poor for function, namely, blood flow in our case. So, lastly, computer simulations of blood flow were used to fill the gap.

Beginning with the 1980s and the advent of desktop workstations, computational fluid dynamics (CFD), already an important tool in the design of automobiles and aircraft, was brought in and started to be used extensively in the study of blood flow. As the 1990's approached, medical imaging technologies like Doppler ultrasound and phase contrast MRI began to open up non-invasive windows into real blood flow dynamics, albeit with relatively crude renderings. By the mid-1990s the use of CFD could provide intricate details of blood flow (thus compensating for medical imaging's shortcomings in this area), but typically only for idealized cases. This led to the development of so-called "image-based CFD", whereby anatomical images could be transformed into patient-specific CFD models,[4] thus providing an accessible visual rendering of the "true" blood flow patterns.

VIRTUAL MEDICAL IMAGES: TWO PARADIGMS

An interesting aspect of the computer-generated images is that the simulation allows the move from "direct", yet sometimes limiting, use of patient data to the use of scientific simulation as a mode of investigation, through speculative extrapolation of data. In our daily research life the object of our study lately became, for the most part, the simulated reality of the body. Medical imaging companies are seeking to deploy such seamless computing and imaging technologies, which will further blur the lines between the real and the virtual, allowing previously unimaginable things to be achieved. This will be possible through training the new generation of bioengineers in such manner that their computer skills will be paralleled by their anatomical, physiological and clinical

Figure 1. Two paradigms in virtual imaging. Paradigm 1: the physician is informed by the real and virtual images. Paradigm 2: the physician is informed directly by the virtual patient, now progressively refined until there is no difference between the real and virtual images

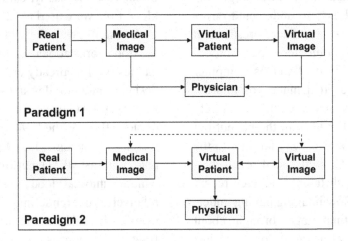

understanding of the phenomena and vice-versa: medical technologists will have better knowledge and understanding of the underlying computer programs. Technological developments are driving this increasingly seamless blurring of the lines between clinical image and simulation and will become second nature to both the computer technologist and the health care provider alike. More broadly, these are examples of what Sarvazyan et al. (1991) call a new philosophy of medical imaging, wherein clinical images are used mainly to refine computer-based models of the patient, from which the diagnoses are ultimately made.

Our daily research requires the translation of patient data into numerical data and then the generation of visual representations of blood flow that can serve two purposes: a) that of a model for a phenomenon or disease, or b) that of a model for an experiment (non-invasive way of determining the best treatment option). Hence, the work our laboratory conducts can be described as falling under the umbrella of the two paradigms summarized in Figure 1. The first paradigm involves the clinician's exposure to conventional visualization drawn from a "direct" medical image of the patient (with an intermediate "virtual" step). The second

paradigm adjust reality through exposing the clinician to a virtual image, based on real-patient data but drawn from a virtual patient, this time.

In the early years, we started by working and developing our models within the first paradigm, using traditional engineering visualizations; however, before we knew it the needs of the clinician, paralleled by the technological developments, caused and supported the shift towards using virtual imaging[5] to represent results. Within this paradigm, virtual imaging is being used as an adjunct to conventional medical imaging. Computer simulations add "knowledge" beyond what can be directly or indirectly visualized. An easy analogy would be polling a few people, and using statistical techniques to infer the opinion of the population. Under correct assumptions and taking all variables into consideration, if the right sample is polled the results of the poll are reasonably accurate. However, the wrong choice of the sample would lead to a totally wrong prediction and, dangerously, the mistake would not become apparent until too late. In our particular case such mistakes or misinterpretations would pose critical problem to the clinician and would have potentially treacherous consequences for the patient.

To summarize, within the first paradigm the clinician would use both real and virtual images to make a diagnosis or plan a treatment strategy, but would have to be aware of the possible traps.

In the case of the second paradigm, in addition to using real images to construct a virtual patient and virtual images as before, the difference between the real and virtual image is assessed and that information is used to intelligently refine the virtual patient. Once the real and virtual images collected from the real patient match the ones form the virtual patient, one can assume the virtual patient is a simulacrum of the real patient, and the clinician examining it may base a correct diagnosis. As just one possible example, we would start with a normal brain virtual image, compared to a real brain image acquired in a patient. Assuming there is a bright spot in the real image, a tumour would be added to the virtual patient until the same bright spot would appear in the virtual image.[6] However, this is extremely challenging in practice because imaging is a "lossy compression" of complex information, and the only way to restore that information is to add more information, say in the form of a model. As long as the model is correct the restoration is reasonable; however, if the model is wrong the reconstructed information is also wrong. In which case the clinician is faced with the same problems as in the first paradigm.

To summarize, within the second paradigm the clinician would rely on virtual images, basing the diagnosis on consulting the 'virtual patient'; in his/her decisions regarding comparative diagnosis or plan a treatment strategy, the clinician would still have to be aware of the possible traps posed by the assumptions and generalizations inherent to the virtual model.

CASE STUDIES

The first case study[7] we are presenting will be discussed in the context of the generalization of individual information, and is that of a patient with a giant aneurysm. The patient was diagnosed by X-ray angiography and treated using a technique in which soft platinum coils are fed through the arteries in the leg, through the aorta, and up into the aneurismal artery. After six months, the coils were found to have compacted – a not uncommon yet potentially dangerous and occasionally fatal complication. Coils work by blocking flow into the aneurysm and thus allowing the blood to slowly clot and seal the aneurysm from the inside. If the coils compact, the blood flow is permitted back in and the aneurysm making it possible for it to regrow and/or potentially rupture. Our CFD-based computer simulation of flow, based on a model of the aneurismal artery reconstructed from the high-resolution 3D X-ray imaging was viewed using a virtual particle visualization technique inspired by engineering visualizations, and it showed that the high velocities during the heart's contraction are in the same direction as the coils were pushed. This piece of information was circumstantial evidence for sure, but an indication of a probable cause for coil compacting. Perhaps given this type of information ahead of time, the clinician might have chosen a different coiling strategy or a more invasive but more robust surgical alternative for the treatment of the aneurysm.

Because of this information's critical content, and its possible use by the clinician in life-or-death decision-making, we had to be absolutely sure of the validity of our simulations and therefore developed "virtual angiography" (Ford et al., 2005), in which a virtual dye is injected into the virtual artery, and then virtually imaged by virtual x-rays. The only reliable way to validate our predicted velocity fields was comparing real and virtual images (the next best thing to the direct comparison), an approach we call "indirect validation". Fortunately, the real angiogram compared well with the virtual one (Figure. 2, panels A vs. B).[8] To reach this result and level of agreement between the real and the virtual angiogram, however, we realized (through trial and error) that the one thing we did have to account for in our virtual patient (namely the CFD simulation) was the fact that the contrast agent is routinely injected at high pressure so as to completely displace the blood, albeit briefly, in order to maximize angiographic image quality. If we did not simulate this in our CFD model, the virtual and real angiograms did not agree as well. This modification of the CFD simulation to account for the injection, recognized only after comparing real and virtual images, reflects the steps in Paradigm 2.

An important point to be made here is that our original experiments with CFD simulations, unperturbed by the contrast agent injection, may in fact be a better reflection of the real patient than the actual medical images themselves (mirroring the actual angiogram situation) (Figure 2, panels C-E).

Another significant case study, and representative of the use of computer-generated images in diagnosis and treatment, is among the first ones in our collaborations with the clinic: that of an atherosclerosis patient who had undergone surgery to remove a plaque at his carotid artery.

One of our neurologist colleagues approached us regarding a routine post-intervention follow-up Doppler (velocity-based) ultrasound exam in which high velocities were reported all along the repaired artery (i.e., internal carotid, IC), suggesting the possibility of a local constriction that was not evident on a B-mode (anatomic) ultrasound examination. Usually, blood flow to the brain – through the internal carotid – is maintained at more or less the same flow rate irrespective of vessel size or physiological state. Since the flow rate is a function of lumen cross-sectional area and velocity, as a result, if the lumen area diminishes velocity increases. Therefore, the recorded values indicated a possible stenosis (lumen area decrease) downstream of which high-speed jets form. These results puzzled the neurologist who had diagnosed the patient and had recommended the original surgery, but who did not participate in the surgery and therefore wasn't aware of the

local conditions and any possible subsequent changes.

As shown in Figure 3, we created a CFD model (simulation) of the patient from a separate MRI scan we carried out. From a sequence of "slices" obtained from the MRI, the artery geometry was reconstructed and then a CFD model generated. The rendition was done using the engineering convention of planes through the model, but colour-coded using the clinical convention of Doppler US: reds and other warm colours for flow to the brain (representing the usual convention of red for arterial flow), blues and cool colours away from the brain (representing the convention of blue venous flow), here highlighting the appearance of any unusually recirculating flow. The CFD model shows high speed but uniformly coloured flow in the internal carotid. The model also clearly shows that the IC was smaller than the external carotid (EC), which is usually not the case. We also noted that there was no major thickening of the artery wall in the IC from the MRI scan shown, and concluded that the high velocities were simply an "artifact" of the smaller artery due either to the way it was surgically put back together, or a possible vasospasm. On the other hand, Doppler US measurements of peak systolic Doppler frequencies from the limited number of locations along the vessel agreed well with CFD-derived values at those same locations, another example of indirect validation against real clinical data.

This case brought to light the use of a patient-tailored simulation and stressed the weight of a 3D/4D visual representation of an anatomical structure in the context of the physiological phenomenon presented, and its interpretation in concert with the clinical data collected from the patient. Also, important to note is the need for the clinician to constantly be vigilant and critical in interpreting patient data.

A further step in the translation back from the generalized model to patient-specific issues (where simulation and reconstruction proved

Figure 2. Case study 1: Real vs. virtual X-ray angiography. Panels A and B show the temporal evolution of x-ray contrast agent injected into a patient with a giant aneurysm (panel A) compared to a virtual angiographic sequence based on virtual contrast agent injected into a CFD model of the same aneurysm (panel B). Panels C and D show the good agreement in colour-coded residence time maps derived from the real and virtual angiographic sequences, respectively. Panel E shows residence times based on a CFD model in which blood flow is undisturbed by the contrast agent injection. Representing the "normal" state of the patient, one could argue that panel E is closer to the actual patient, despite it being virtual

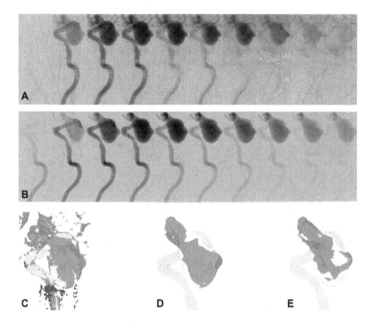

to be instrumental in explaining and exploring misunderstood phenomena and observations) is presented in this last case study. As demonstrated over the years, the "reality" as represented by medical images is not stagnant, but ever changing (or ever evolving) due not only to the evolution of the tools used for investigation of a certain are or phenomenon, but also due to the ability of studying and thus better understanding a multitude of phenomena in their bodily context. Traditionally clinician and medical researcher alike were "blinded" by the widespread clinical assumption of the "fully-developed flow" that must behave according to Poiseuille's law if the artery is relatively long and straight. This assumption, in fact, underlies many well-known basic scientific and clinical imaging studies of vascular disease.

Nevertheless, a number of incidental findings in our research caused us to question it. Most notably, as part of a collaborative study[9] we were provided with MR images of velocity profiles at transverse sections through the common carotid arteries (CCA) of a large number of subjects. Although this artery, which brings blood from the aorta to the brain, is routinely assumed to be long and straight, the velocity profiles we observed were anything like those envisioned by Poiseuille (Figure 4), being decidedly asymmetric. Further investigation of these subjects revealed that in most cases their arteries possessed either subtle curvatures or wiggles, or strong curvatures far upstream from where the MR images were acquired. Using CFD, we were able to show that such anatomic features were more than sufficient

Figure 3. Case study 2: Doppler ultrasound / CFD of a post-surgical patient. Panel A shows a selection of cropped MR images acquired near the level of the carotid artery bifurcation. Panel B shows the lumen boundaries traced from the complete set of images, and reconstructed into a three-dimensional representation of the artery; black boundaries correspond to the images in panel A. Panel C shows the CFD simulation visualized using the engineering convention of colour-coded velocities on selected planes; however, the colour-coding reflects the DUS convention of warm and cool colours for blood moving to and away from the head. Panel D shows peak systolic Doppler frequencies measured during the DUS exam (black numbers) and those "virtually imaged" using the CFD model (red numbers)

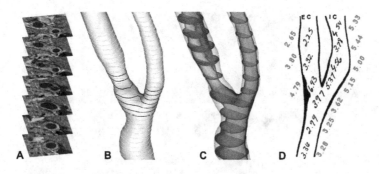

to give rise to the seemingly unusual velocity profiles observed (Lee et al., 2008).

The important implication is that, if faced with an unusual velocity profile at a given point along an artery, the facile assumption is that there is something "wrong" with the subject (immediate suspicion of cardiovascular disease), because the flow profile as shown does not conform to expected flow profiles as described by the conceptually and mathematically simple model presented by Poiseuille. On the other hand, flow-related measurements in seemingly long and straight arteries may harbor large errors, being implicitly based on the assumption of Poiseuille flow within. Either way, it is important to establish a new mental model that supercedes that of Poiseuille in order to adjust to a new reality of the body as better understood through virtual examination.

Moving along the same path, and in continuous effort to replace the prejudice of what blood flow should look like, we have developed a simulacrum of the Doppler Ultrasound (DUS) examination (Figure 5), which engages the user, both visually and aurally in a familiar environment, to help "train" them to break their stereotypes. Despite

being a routine tool, DUS is very hard to interpret; by developing an interactive, real-time DUS simulator we help in training sonographers, but also create a standardized platform to evaluate (accredit) their performance. As a training tool, virtual DUS presents the potential benefit of improving data collection and its interpretation by rendering a representation closer to the actuality of the blood flow in the absence of any interference of tissue between the probe and the blood vessel, the image collected from the virtual patient thus becoming clearer and more accurate.

SUMMARY AND FUTURE DIRECTIONS

The medical images we generate – with the aid of the computer – are truly reflections of a reality that escapes our view and our direct perception. In many ways the role of the image over millennia has evolved without being fundamentally altered. Animal and human representations on cave walls seem to have had an informational purpose as well as a ritual one, while their mere presence was a

Figure 4. Case study 3: Non-Poiseuille flow in the common carotid artery. Panel A shows CCA velocity profiles from 50 ostensibly normal, older individuals, imaged by MRI. The MRI data, normally presented in grayscale, were pseudo-coloured to better highlight the shape of the velocity profiles. For Poiseuille flow, the shapes should be circularly symmetric, which is evident in only a minority of cases. Panel B shows the CCA and carotid bifurcation imaged by MRI, evidencing a mild curvature. In Panel C, an idealized CFD model demonstrates how such mild curvatures can produce velocity profiles like those seen in vivo

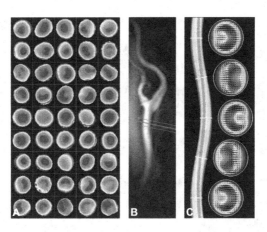

constant reminder of the object represented but not present anymore. Later on, medical images (drawings, woodcuts, engravings) played an explanatory role, but served also as a memorizing tool. Information concerning living bodies through extrapolation of data collected through dissection was not only transmitted from one generation to the next but also fixed into the mind of the medical student throughout the history of his science. Since the Renaissance, students have been confronted with the reality of the body and its interpretations, through visual representations that were images of a "normalized" body. Images became a certain reflection of reality, as perceived by the creator of the image: a corollary of all the information gathered (from direct observation on the human body, but also through extrapolation from studying other living bodies) and a memorizing kit. From Vesalius' Fabrica onwards, anatomical representations become closer to the actual human anatomy owing to confrontation with direct experience (through dissection); however mistakes and misinterpretations were still present and students

continued to propagate the "knowledge" from the printed image and accompanying text and not, as expected, from direct observation of the dissected body (considering that a normalized version of the body is the correct one to perpetuate and study).

As shown in this chapter, the current technologies and medical images mirror the age-old attitude towards anatomical representations. Firstly, the virtual imaging technologies we have discussed are a necessary step towards the stage where the model (virtual patient) is manipulated iteratively until the real and virtual image contents are practically the same. For the patient's sake, the normalized, or generalized image should be just a guiding tool but the patient-tailored image should be the one the clinician relies on when deciding upon the treatment. Secondly (as seen in the example of the clear evidence and proof of the actual blood flow patterns existing within the various arteries depending on their particular geometry) the prejudice of what blood flow supposedly looks like in order to fit the Poiseuille model continues to be prevalent in both clinical

Figure 5. Evolution of virtual Doppler ultrasound. Panel A shows a screen shot from a real Doppler ultrasound examination. Panel B shows a CFD model derived from the same patient shown in Panel A. Here the CFD model is rendered using the CFD convention of planar slices, but is colour-coded using a pseudo-Doppler colour scale to highlight forward (yellow/red) and reverse (blue) flow. In panel C, a sonographer interactively "examines" a virtual patient in the same way that they would examine a real patient. In this early prototype the virtual patient is "placed" in a water-filled peanut-butter jar; ultimately it will "appear" inside a life-like mannequin

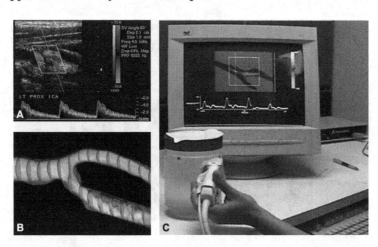

and scientific milieus, and through perseverance and correct application and use of the virtual image a shift should be promoted towards a more realistic interpretation of data.

CONCLUSION

Recognizing the power and authority of the printed medical image, and the tendency of perpetuating the established dogma, the trap we're trying to avoid is creating appealing representations that are misleading or void of meaning; our images aim at being as clear, accurate and precise as possible. Initially computer-generated images were normalized, average, generalized representations of the blood flow; however – as demonstrated in this chapter – the current trend is moving back from the general to the individual! A paradoxically immense step forward in the medical investigation is the simulation that actually brings the clinician closer to the reality of the patient than the direct

investigation would allow. Making the technology easily accessible to physicians brings a lot of good, by enabling them to see the previously unseen; however it carries the hidden danger. Getting closer to the actuality of the body by means of virtual images implies several layers of intervention and requires a very careful and close control. Without appreciating the assumptions and inherent limitations of these technologies, a clinician is exposed to possible misinterpretation due to blind faith in the information received. Virtual medical imaging is no different in that it can give the uninformed user (for example the clinician interacting with the interface) the false sense of reality and truthfulness and potentially lead to great confusion or misunderstandings.

ACKNOWLEDGMENT

We would like to thank the staff, students, collaborators and patients, all too numerous to list here, who participated in the studies described above; and the various agencies that have funded this work over the years.

REFERENCES

Cademartiri, F., Mollet, N., Nieman, K., Krestin, G. P., & de Feyter, P. J. (2003). Images in cardiovascular medicine. Neointimal hyperplasia in carotid stent detected with multislice computed tomography. *Circulation*, *108*(21), e147. doi:10.1161/01.CIR.0000103947.77551.9F

Ford, M. D., Stuhne, G. R., Nikolov, H. N., Habets, D. F., Lownie, S. P., & Holdsworth, D. W. (2005). Virtual angiography for visualization and validation of computational models of aneurysm hemodynamics. *IEEE Transactions on Medical Imaging*, *24*(12), 1586–1592. doi:10.1109/TMI.2005.859204

Ford, M. D., Xie, Y. J., Wasserman, B. A., & Steinman, D. A. (2008). Is flow in the common carotid artery fully developed? *Physiological Measurement*, *29*(11), 1335–1349. doi:10.1088/0967-3334/29/11/008

Glassner, O. (1992). *Wilhelm Conrad Roentgen and the early history of the Roentgen rays.* San Francisco: Norman Publishing.

Lee, S. W., & Steinman, D. A. (2008). Influence of inlet secondary curvature on image-based CFD models of the carotid bifurcation. *Proceedings of the ASME Summer Bioengineering Conference.* New York: ASME Press.

Sarvazyan, A. P., Lizzi, F. L., & Wells, P. N. (1991). A new philosophy of medical imaging. *Medical Hypotheses*, *36*(4), 327–335. doi:10.1016/0306-9877(91)90005-J

Steinman, D. A. (2002). Image-based computational fluid dynamics modeling in realistic arterial geometries. *Annals of Biomedical Engineering*, *30*(4), 483–497. doi:10.1114/1.1467679

Steinman, D. A., Milner, J. S., Norley, C. J., Lownie, S. P., & Holdsworth, D. W. (2003). Image-based computational simulation of flow dynamics in a giant intracranial aneurysm. *AJNR. American Journal of Neuroradiology*, *24*(4), 559–566.

KEY TERMS AND DEFINITIONS

Aneurysm: a focal ballooning of an artery caused by a weakening of the arterial wall.

Atherosclerosis: a disease of the artery characterized by the focal accumulation of fatty material in the blood vessel wall.

Computational Fluid Dynamics (CFD): use of numerical methods and algorithms to solve the Navier Stokes equations that govern fluid flow.

Doppler Ultrasound (DUS): a medical imaging technique that uses reflected sound waves to evaluate blood as it flows through a blood vessel, blood velocities are inferred from measured frequency shifts caused by the Doppler effect as the sound waves reflect off the red blood cells.

Magnetic Resonance Imaging (MRI): a medical imaging technique that uses a powerful magnetic field to align hydrogen atoms in the body, radiofrequency fields and magnetic field gradients are then used to build up images from the signals returned by rotating hydrogen nuclei.

Medical Imaging: The production of visual representations of body parts, tissues, or organs, for use in clinical diagnosis, encompasses x-ray methods, magnetic resonance imaging, single-photon-emission and positron-emission tomography, and ultrasound.

ENDNOTES

[1]   The first angiogram was an X-ray image of an amputated hand into which was injected a radio-opaque mixture; this was apparently done less than a month after the first X-ray image (Glassner, 1992).

2   Poiseuille is a perfect example of trans-disciplinarity, being a physician with a background in engineering. His pioneering experiments of flow in capillary tubes laid the groundwork both for the quantitative study of blood flow and for modern fluid mechanics in general.

3   Excellent examples of this may been seen in Cademartiri et al.'s (2003) images and animations of a stented carotid artery in the neck, acquired via computed tomography.

4   The early development and applications of image-based CFD are reviewed by Steinman (2002).

5   We define virtual imaging as computer-generated images that are traditional medical representation look-alikes, as opposed to the particle path line movies we routinely show and which are virtual versions of traditional experimental engineering visualizations.

6   Of course it could be a different formation (a cyst), in which case other options would be tried and compared for consistency with other medical information (blood work, etc.) As Sarvazyan et al. (1991) point out, the refinement of the virtual patient must be informed by other clinical information beyond the images themselves.

7   This study is detailed in Steinman et al. (2003).

8   Interestingly, these virtual angiography visualizations, originally developed for the purpose of validation, also proved their worth as the most compelling to our clinical collaborators, being "spoken" in their visual language.

9   This study is detailed in Ford et al. (2008).

Chapter 17

Explorative Data Analysis of In-Vitro Neuronal Network Behavior Based on an Unsupervised Learning Approach

A. Maffezzoli
Università degli Studi di Milano-Bicocca, Italy

E. Wanke
Università degli Studi di Milano-Bicocca, Italy

ABSTRACT

In the present chapter authors want to expose new insights in the field of Computational Neuroscience at regard to the study of neuronal networks grown in vitro. Such kind of analyses can exploit the availability of a huge amount of data thanks to the use of Multi Electrode Arrays (MEA), a multi-channel technology which allows capturing the activity of several different neuronal cells for long time recordings. Given the possibility of simultaneous targeting of various sites, neuroscientists are so applying such recent technology for various researches. The chapter begins by giving a brief presentation of MEA technology and of the data produced in output, punctuating some of the pros and cons of MEA recordings. Then we present an overview of the analytical techniques applied in order to extrapolate the hidden information from available data. Then we shall explain the approach we developed and applied on MEAs prepared in our cell culture laboratory, consisting of statistical methods capturing the main features of the spiking, in particular bursting, activity of various neuron, and performing data dimensionality reduction and clustering, in order to classify neurons according to their spiking properties having showed correlated features. Finally the chapter wants to furnish to neuroscientists an overview about the quantitative analysis of in-vitro spiking activity data recorded via MEA technology and to give an example of explorative analysis applied on MEA data. Such study is based on methods from Statistics and Machine Learning or Computer Science but at the same time strictly related to neurophysiological interpretations of the putative pharmacological manipulation of synaptic connections and mode of firing, with the final aim to extract new information and knowledge about neuronal networks behavior and organization.

DOI: 10.4018/978-1-60566-768-3.ch017

INTRODUCTION

The last decade the Neuroscience field of research has taken great advantage by knowledge and methods from electronics, robotics and informatics, especially in order to develop new micro- or nano-technology instrumentations in neurobiology and molecular biology. Such very effective and profitable exchange of know-how is giving birth to a sort of new scientific expertise, also known under the denomination of "Neuroengineering" and "Neuroinformatics". Apart the denomination chosen, the roles of mathematics and engineering science has become very important for neuroscientists, covering from the development of the instrumentation used to a number of various approaches in order to analyze neuronal activity. The last one it is the argument we will discuss in the present Chapter, by introducing the specific case of in-vitro neuronal cultured networks, the data analysis issues of the related recordings and some of the statistical and mathematical methods used to elaborate them, with the aim to extract meaningful and undiscovered models and notions about the neurobiological phenomena involved in such living networks.

At this point it is necessary to introduce the instrumentation used for recording from cultured networks, which is the Multi Electrode Array (MEA) (Maher, 1999; Potter, 2001; Segev, 2004; Baruchi, 2007; Shahaf, 2008; Eckmann, 2008; Shein, 2008), a technology able to complement classical methods in electrophysiology and to solve the considerable limitations given by the single neuron recordings, as from the patch-clamp techniques, given the fact that it is very helpful to retrieve all the information coming from many synaptically connected neurons, with a good resolution both in time and in space, as explained in the following paragraph. The networks of cortical neurons grown on *in-vitro* cultures on MEAs have probably inter-neurons connections much simpler than in vivo population, but despite that they continue to show a behavior showing many of

the original features untouched. Neurons on MEA can be of course stimulated in an electrical way or via the application of drugs, or more generally chemical stimuli, as for our Case Studies.

To this aim, a method and a related software tool, able to evaluate Neuronal Network (NN) features during different environment conditions and stimuli is necessary. Given that MEA data can be studied via signal processing, statistics, pattern recognition or Machine Learning methods, we considered techniques from each one, with special regards to the analysis of the neuronal network activity based on a so-called Unsupervised Learning approach. Related methods and algorithms were implemented in a software framework in Python programming language.

BACKGROUND

Brief Overview of Micro Electrode Array Technology and Data Features

MEA instrumentation comes in various different commercial variants (see http://www.multichannelsystems.com/products-mea/microelectrode-arrays.html), but the standard MEA presents usually 64 electrodes which are spatially displaced in an 8x8 square layout grid, and in particular the four electrodes at the vertices of such square matrix do not record neurons activity. Electrodes have a very small diameter of about 30 μm and each one is 200 μm far away from each other, also in order to prevent that signals recorded by one channel should not be captured, at least with a similar amplitude, by another one. This multiple electrodes instrumentation allows researchers to perform simultaneous mid- or long-term extra-cellular recording from several neuronal sites activities, together to non-invasive stimulation: before an ensemble of neurons cells or tissue slices are placed and cultured on the MEA, then it shall be possible to capture a spontaneous activity of a wide range of excitable cells, e.g. central cortex, retina, peripheral neurons from heart and muscles. A single channel on MEA is able to record the activity of few neurons, placed in the neighborhood, even if it is usual to record up to four cells, up to a relatively high sampling rate of 40 KHz, as in our case.

Spike Sorting Methods

When a time series of analog electrical signals have been recorded by each neuron, role of signal processing and data mining starts to be effective because their techniques are needed in the so-called Spike Sorting tasks. The reason to apply such techniques is easily explained: each channel can record the waveforms of various neurons, as introduced above, but at this time such waveforms are mixed in the single time series, so we want to discriminate waveforms, and so spikes, of the different neurons, or channel units, involved. The task is supported by the fact that each neuron shows a specific shape for its waveform, so in the more general case we have to extract from the entire analog signal and then classify the different spike shapes for each channel. This is the aim of Spike sorting, which helps us with an ensemble of techniques, in order to distinguish or 'sort' the spike waveforms collected by the same electrode but generated by distinct neurons (Lewicki, 1998).

There is a number of useful methods in Spike sorting, e.g. from the simple waveform detection based on a well-fixed threshold value, to the more complex and effective Principal Components Analysis or template matching, the least-squares regression of the estimated spike times against multi-electrode voltages, and clustering algorithms, applied to infer the number and the identity of captured neurons. We will see some of the just reported methods implemented in our work.

Spike Sorting pre-processing can encounter various problems (Meister, 1994), which arise in the following cases:

- when spikes from a single neuron are recorded by different nearby channels: in order to remove subsequent errors, we can consider only the signal with the largest amplitude or the one best resolved in a clustering procedure and discard time-coincident signals in the others channels, or the so called 'cross-talk' spikes;

- if two neurons, both placed near the same channel, show very similar waveforms: consequently their activities shall be indistinguishable and erroneously mixed in a single channel unit. To verify this kind of problem we can calculate the Auto-Correlation Function (ACF) of the times series of the unit, and if the ACF does not show an initial brief silent response, necessarily corresponding to the refractory period, we have to work to disentangle such different sources signals.

- when a single neuron fires spikes of distinct shapes, as example during rapid subsequent bursts trains, and so it will classified in different units: we can check this drawback through the Cross-Correlation Function (CCF) of its signals in the units involved. A similar problem arises when signals come from the axon, even if in this case they have an easily distinguishable shape;

- if two distinct cells fire about coincidently, within a delay of 1-2 ms, so superposing their waveforms with the consequence that clustering processing in Spike Sorting procedures shall fail to identify both neurons. Such problem has been solved by using estimates of the original waveforms calculated a-priori in order to detect the superposition of spikes; then a statistical Maximum-a-Posteriori estimation is performed under the hypothesis that the multi-electrode voltage signal is composed by the linear superposition of Gaussian white noise and spike trains convolved with the spike waveforms above: just cited procedure correctly identifies simultaneous spikes, as confirmed by the correction of Cross-Correlation artifacts (Pillow, 2008).

After Spike Sorting procedures have been performed, we obtain the timestamps of spikes arrival times from each single unit of all channels: in particular only one timestamp corresponds to a spike, so the output is available in the typical form of Point-Process signals. Now we are able to continue further processing, but before it is preferable to spend some words about the neuro-physiological data just obtained. In particular signals we record, as we can see in the rasterplot of Figure 1, are characterized by long silent intervals and bursting events, when many neurons spike about simultaneously at high firing rate. Although such cortical signals shows a great variability, specially during rest or spontaneous activity condition (Softky, 1993; Shadlen, 1998), it is quite evident that such long silent intervals are not strictly interesting, because they seem not transmitting relevant information, e.g. in term of firing rate. On the contrary a lot more significant experimental intervals seem to be represented by the bursts (Berry II, 1997), which convey information through high rate spiking activity, also known as Intra-Burst Spike Rate as explained below. Of course it is not only the burst firing rate the most prominent feature of the information transmitted, and there is a long debate about the question, which we can reassume in the distinguished Rate or Time Coding hypotheses on neurons behavior (Rieke, 1997); at the moment we want to give more details about the burst events just introduced, and which have been the material of our work, given the fact that we discarded a priori the signals from inter-burst experimental time intervals.

Brief Description of Neuronal Burst Spiking Activity

The main information needed to perform a quantitative study of neuronal data is given by the timing if spike events, and in particular we are interested in the burst events, which usually capture the almost total amount of spikes counts during both spontaneous and stimulated activity in our *in-vitro* cortical cultivations. Here we want to give a brief overview about the burst spiking activity phenomenon and some evidence of its importance in neurophysiological experiments, with the aim to furnish a support to better understand the bursting concerted activity of a neurons population, exactly the one from which we extracted the timestamps processed in our works.

It is well know that neurons can tend to fire action potentials in brief events of high-frequency discharge, or the so-called bursts, and it is very probable that bursts have an important role in synaptic plasticity, selective communication between neurons, dysfunctional states and in sensory information transmission.

Bursts can reflect in animals transient and high-frequency activation of neurons generated simply by particular sensory inputs or, in a more complex manner, by specific intrinsic cellular mechanisms which have been well characterized in vitro. For the latter intrinsic mechanism and in the in vivo events, inputs are localized onto the dendrites of bursting neurons, they originate from feedback loops, and the generation of bursts is related to interactions between sensory variables, intrinsic cellular properties and network characteristics (Krahe, 2004). In support to the important role of bursts to the sensory information transmission, there are various studies confirming that presynaptic bursts are able to improve the reliability of information transmission in spite of unreliable synapses, and also to enhance the efficiency of such transmission, as tested during in vivo sensory experiments (Krahe, 2004). In-vitro experiments trying to define the role of bursts as information-

Figure 1. Rasterplot representing the patterns of spiking activity of different neurons under spontaneous conditions. From left to right, and with even finer time scale as pointed by upper arrows, the four panels show the burst structure of a population of neurons. Each row in the rasterplot corresponds to timestamps from a single neuron and each column of almost synchronous timestamps to a burst. Blue lines in the third and fourth panel give the start and end timestamps within a burst, and arrows show spike pattern features of a burst, as the duration (BD), the intra-burst spike rate (IBSR), the number of spikes (SN), and spike-frequency adaptation (SFA) (adapted from Maffezzoli, 2008).

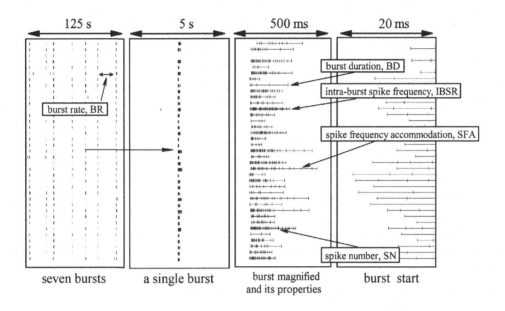

carriers in presence of synaptic transmission unreliability showed the following: if a synapse is unable to transmit a spike even if depolarized by a pre-synaptic single spike, it has been found that many spikes fired within a brief time support a spike postsynaptic response (Thomson, 1997), so burst spikes allow information transmission across synapses more reliably than isolated spikes, when isolated spikes can be considered as noise suppressed by such synapses (Lisman, 1997); so bursts are also considered as performing a sort of noise filtering, because they are able to transmit as much information as tonic spikes, but at a higher signal-to-noise ratio. Evidence for this hypothesis comes from primary visual and auditory cortices, as tested in the shaper, or less wide, tuning curve of bursting cells during sensory or stimuli coding (Eggermont, 1996).

Many efforts in the study of sensory coding are involved to investigate about how to consider a single burst as specific event: in particular researchers are asking if a burst has to be considered only as a unitary event or if we should also know its internal temporal structure, by which extract more information to carry to postsynaptic neurons. There are actually several hypotheses about which features of burst's structure are important to understand information transmission (at regard see also the graphic representation of some of these in Figure 1):

- the temporal duration of a burst (BD);
- the number of all spikes in a burst (SN), or the number of spikes inside a burst;
- the mean spike firing frequency in a burst (IBSR), charged to the process of

postsynaptic target selection and related of course to the two above. A compartmental neuronal model should implement this 'selective communication' between neurons by a filtering mechanism tuned to specific IBSRs (Krahe, 2004);

- Statistical moments of various order of the Inter Spike Intervals (ISI) distribution;
- The Inter Burst Interval (IBI), or the time intervals lasting between two subsequent burst events.
- other features of bursts have been identified in the latency of the first spike fired, where the corresponding neuron is also called the burst leader (Middlebrooks, 1994; Richmond, 2004; Eckmann, 2008; Shahaf, 2008). These works, opening the way to other studies investigating the existence of more complex coding schemes based on the timings of entire intra-burst spike pattern (VanRullen, 2001; Thorpe, 2001; Wohlgemuth, 2007), indicate that the relative timing of spiking across neurons can better characterize stimulus parameter and identify it, even if conclusions at regard depend to stimuli and cells involved.

DATA ANALYSIS OF IN-VITRO NEURONAL NETWORK ACTIVITY

Aims

Analysis of the concerted activity of a population of neurons represents the new field of study in Neuroscience, and this direction is simply explained by the fact that only the investigation of neurons collective behavior can help to replicate, even if in a small scale of course, and hopefully to better understand the complex organization of that more complex and bigger neuronal network present in living systems and above all constituting the organic bricks of the brain. Here the aims of a computational neuroscientists, and of

neurophysiologists, biologists, physicists, mathematicians and engineers are to work together in order to extract new knowledge about the behavior, and stimuli responses and dynamic adaptive capabilities of Neuronal Networks (NN). MEA instrumentation gives the chance to overcome the limitations of single neuron registration technique, as patch-clamp, to control the experimental environment or the kind of stimulation to apply on the NN. At the same time we have to say that *in-vitro* cultured NN on MEA represent a good but not a perfect approximation to the original tissue in the brain: first of all it is impossible to record from the thousands of *in-vitro* cultivated neurons, but generally up to one hundred neurons, and the *in-vitro* cultivation is unable to replicate the complex and well structured organization of living networks, e.g. in our case the synaptical links between cortical neurons are mechanically dissociated by means of trituration. Notwithstanding such limitations do not represent an obstacle to identify in multi-channels recording a source if rich and fascinating information to analyze NN behavior, also taking in account that the continual progresses in nano-tecnology will help to build soon larger and more precise MEA substrates.

From the point of view of data processing in Neuroscience, neuroscientists and neurophysiologists require to perform analyses by applying well-established statistical and signal processing paradigms as well as data mining and pattern recognition methods wherever possible. Several standard statistical procedures, widely used in other fields of science have found their way into various applications in Neuroscience together to mathematical models of neurons behavior, in such a way that we can distinguish two main approaches in Computational Neuroscience: a 'computational oriented' approach, concerning mainly with neural information processing and how to implement various forms of adaptive behavior in neural networks with dynamic synapses responses and which is the one we applied; and a 'biophysical oriented' approach, concerned with the intrinsic dynamics of single cells or small ensemble of these, with effects on the entire network behavior and described by biophysical Compartmental model of neurons, e.g. the Firing-Rate models, Integrate-and-Fire networks or Synfire Chains (Vogels, 2005; Goodman, 2008). We developed our study based on the first approach, as presented below.

Methods Outline

Starting from the instrumentation just introduced above, we remember that data from MEA are sampled at up to 40 KHz, but it is usual to work only on spike events data, or the timestamps of neurons spiking during the experimental time interval in question. The last data representation is also called in statistics as Point Processes data, and there are many methods, involving primarily the analysis of more or less complex probability distribution modeling such distributed spiking activity in a reliable manner: the most applied are the Poisson discrete probability function for the counts of timestamps in limited time intervals, or the Exponential continuous probability density function together to the Normal density and mixtures of them applied on the ISIs distribution (Mood, 1974). Unfortunately it is difficult to know or infer from small samples data the type and the parameters of such probability functions in a significant manner, and so it is practice to build a non-parametric and empirical probability distribution by performing a so-called binning procedure on the timestamps data, by counting how many spikes there are in subsequent and non-overlapping very small time intervals, the so called bins (Abbott, 2001).

So starting by the 'binned' spiking activity representation, researchers can continue their investigation with the aim to obtain new information about neuronal networks cultivated in vitro. Now it is possible to apply many different statistical models and indexes, and techniques from Information Theory, and Linear and Non-linear

signal processing and methods from Machine Learning. In our work we studied MEA data based on the last family of techniques, in order to extract from such multivariate sets of long recording data new insights by considering an Unsupervised Learning approach. In particular we performed classification, representation and pattern recognition techniques, e.g. Clustering and Principal Component Analysis, and pruning and Outliers detection operations to explore the rich amount of data by taking no a priori knowledge about the same, but considering only the knowledge about the neurophysiological basis of our neuronal recordings, as demonstrated by the application of algorithms for the burst detection and burst features extraction.

Case Study of Neurons Classification Based on Their Burst Features

In the following we will present the methods we developed for the analysis of the multivariate neuronal data described above.

At the begin timestamps data for all MEA units, stored originally in a file with the .plx extension by the software Offline Sorter, Plexon Inc., are read thanks to a Matlab® script we wrote, which exploits the Plexon Inc. Software Development Kit (SDK) expressly furnished by the producer for Matlab® users in order to store all the timestamps in a ASCII comma-delimited text file with a .csv extension: timestamps are written as textual float values, by preserving their original sampling precision, and, even if a binary storage would be more efficient the textual data have the advantage that they can be inspected through any software editor. In the following these data are read and processed by a software tool we wrote ad-hoc in Python language, as underlined in better detail at the end of the Chapter. Among various functions, this tool is able to perform burst detection for each unit, by performing the MaxInterval Burst Detection algorithm of Plexon® Neuroexplorer,

Figure 2. The timestamps of an exemplary burst extracted from the rasterplot for a single neuron, and during distinct experimental conditions characterized by drug stimuli at different concentration: there is the 'con' or control stage, when recording the spontaneous spiking activity, and the application of 200 nM (middle) and 3 mM gabazine. For each experimental step above, quantitative values of the burst features investigated are listed (adapted from Maffezzoli, 2008).

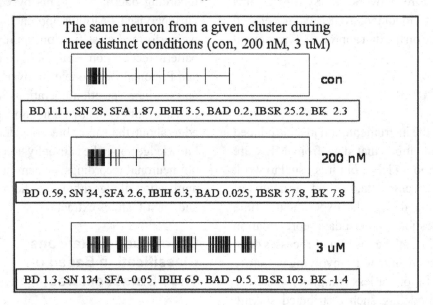

the must-have on the field of neurophysiological data analysis packages with its rich set of analysis options.

Then we extract for each burst specific features, among those we listed in the previous paragraph; the features we used are the following, getting above explained acronyms: BD, SN, IBSR, intra-burst ISIs, IBI. Then we calculated the mean value of BD, SN, IBSR from all bursts, and we used IBIs to calculate a quantitative term, in the Figure called short IBI/tot IBI, giving the fraction of bursts having an IBI under a given threshold, and we used the ISIs from all burst to calculated the skewness and kurtosis, terms derived from third and fourth order statistical moments (Mood, 1974; Papoulis, 2002) and here called Spike Frequency Adaption (SFA) and Burst Kurtosis (BK) respectively. Also we calculated from ISIs the Burst Adaptation Rate, a term estimating if the spike rate increases or decreases during a burst.

Such results are collected for all experimental stages, both spontaneous or stimulated by drug application activities, as showed in Figure 2.

The analysis then proceeds by selecting, between the many available, only one experimental stage and then:

- we applied the Principal Component Analysis (PCA) method (Johnson, 2002) on burst features data from the just selected experimental interval, usually taking in account a number of components conveying at least the 80% of the total variance of data in order to work on data in the most significant and informative PCA coordinates;

- and finally we proceeded in our Unsupervised Learning approach, by performing on data in output from PCA the widely used k-means clustering algorithm (Duda, 2000), with the aim to inspect the

Figure 3. An excerpt from rasterplot of a single exemplary bursts during control condition (first column), 0.2 (middle) and 10 uM gabazine (right) stimulation. When searching for three clusters and obtained the related classification of various neurons, their spike patterns can be plotted as showed here for a single burst: the upper horizontal panel gives traces for neurons of class 1, whether the middle one for neurons from class 2, and finally the lower one for neurons from the third class. We outline as activities of all neurons in each class are repeatedly similar along the three different pharmacological conditions (adapted from Maffezzoli, 2008).

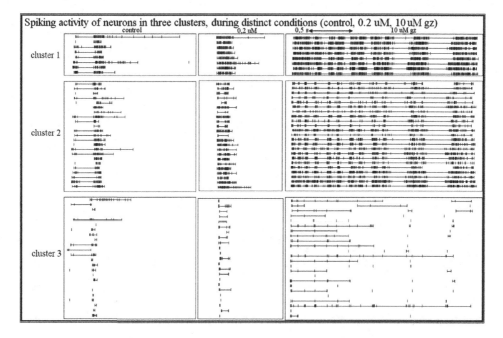

presence of groups, or better families, of neurons bursting in a similar manner during the given experimental stage, as quantitatively classified thanks to the burst features we extracted at begin, and as showed in Figure 3 and Figure 4.

Finally we want to underline that methods here described have been written in Python programming language within a software framework developed ex-novo for the purposes of the present study and implementing many unsupervised learning algorithms. Python is a powerful, dynamically typed, interpreted language with a clear and simple syntax, giving to programmers the chance to rapidly write easily understandable and reusable

code. In particular our software tool performs all the processing functions and algorithms above cited, e.g. PCA and k-means clustering algorithms, criteria evaluating the goodness of clustering and an Outliers detection technique, as well as read input data of MEA timestamps from ASCII files, perform burst detection algorithm and burst features extraction, produce reports for all results obtained in formatted output. Almost all methods have been written ex-novo and sometimes we exploited the complete, very efficient and fast Python scientific libraries of the Scipy/Numpy package (Jones, 2001), and graphical results can be plotted by using the Matplotlib libraries in Python (Barrett, 2005).

Figure 4. An example of classification of neurons in five class based on the height burst features explained in the text. In each panel features for each cluster found (cluster 1 corresponding to black square, cluster 2 to red circle, cluster 3 to green triangle, cluster 4 to blue triangle and cluster 5 to magenta diamond) have been plotted as a function of the concentration of gabazine. We want to point out that each feature, except SFA, behaves differently along the distinct experimental conditions, so confirming the goodness of the classification of neurons performed based on burst features quantitative values calculated at the begin and the clusters number choosen before to start the clustering processing. (adapted from Maffezzoli, 2008).

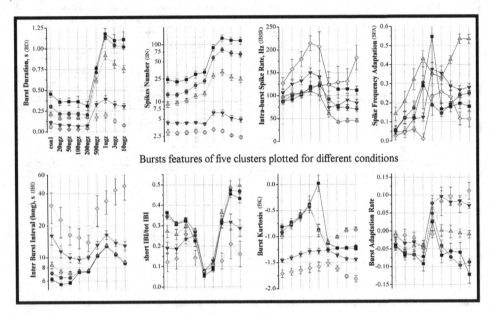

REFERENCES

Abbott, L. F., & Dayan, P. (2001). *Theoretical neuroscience: computational and mathematical modeling of neural systems.* Cambridge, MA: MIT Press.

Barrett, P., Hunter, J., Miller, J. T., Hsu, J.-C., & Greenfield, P. (2005). Matplotlib - A portable Python plotting package. *Astronomical Data Analysis Software and Systems XIV, 347,* 91–95.

Baruchi, I., & Ben-Jacob, E. (2007). Towards neuro-memory-chip: Imprinting multiple memories in cultured neural networks. *Physical Review E: Statistical, Nonlinear, and Soft Matter Physics, 75,* 050901. doi:10.1103/PhysRevE.75.050901

Berry, M. J. II, Warland, D. K., & Meister, M. (1997). The structure and precision of. retinal spike trains. *Proceedings of the National Academy of Sciences of the United States of America, 94,* 5411–5416. doi:10.1073/pnas.94.10.5411

Duda, O. R., Hart, P. E., & Stork, D. G. (2000). *Pattern Classification.* New York: Wiley-Interscience.

Eckmann, J.P., Shimshon, Jacobi, S., Marom, S., Moses, E., Zbinden, C. (2008). Leader neurons in population bursts of 2D living neural networks. *New Journal of Physics, 10,* 19. doi:10.1088/1367-2630/10/1/015011

Eggermont, J. J., & Smith, G. M. (1996). Burst-firing sharpens frequency-tuning in primary auditory cortex. *Neuroreport*, *7*, 753–757. doi:10.1097/00001756-199602290-00018

Goodman, D., & Brette, R. (2008). Brian: a simulator for spiking neural networks in Python. *Front Neuroinform*, *2*, 5. doi:10.3389/neuro.11.005.2008

Johnson, R. A., & Wichern, D. W. (2002). *Applied Multivariate Statistical Analysis* (5th Ed.). Upper Saddle River, NJ: Prentice-Hall.

Jones, E., Oliphant, T., Peterson, P., et al. (2001). *SciPy: Open source scientific tools for Python.* Retrieved from http://www.scipy.org/

Krahe, R., & Gabbiani, F. (2004). Burst firing in sensory systems. *Nature Reviews. Neuroscience*, *5*, 13–23. doi:10.1038/nrn1296

Lewicki, M. S. (1998). A review of methods for spike sorting: the detection and classification of neural action potentials. *Network (Bristol, England)*, *9*, R53–R78. doi:10.1088/0954-898X/9/4/001

Lisman, J. E. (1997). Bursts as a unit of neural information: making unreliable synapses reliable. *Trends in Neurosciences*, *20*, 38–43. doi:10.1016/S0166-2236(96)10070-9

Maffezzoli, A., Gullo, F., & Wanke, E. (2008). *Detection of intrinsically different clusters of firing neurons in long-term mice cortical networks.* 6th Int. Meeting on Substrate-Integrated Microelectrodes (MEA Meeting 2008), July 8–11, Reutlingen, Germany.

Maher, M. P., Pine, J., Wright, J., & Tai, Y.-C. (1999). The Neurochip: a new multielectrode device for stimulating and recording from cultured neurons. *Journal of Neuroscience Methods*, *87*(1), 45–56. doi:10.1016/S0165-0270(98)00156-3

MEA types of Multi Channel Systems (n.d.). Retrieved from http://www.multichannelsystems.com/products-mea/microelectrode-arrays.html

Meister, M., Pine, J., & Baylor, D. A. (1994). Multineuronal signals from the retina: acquisition and analysis. *Journal of Neuroscience Methods*, *51*, 95–106. doi:10.1016/0165-0270(94)90030-2

Middlebrooks, J. C., Clock, A. E., Xu, L., & Green, D. M. (1994). A panoramic code for sound location by cortical neurons. *Science*, *264*, 842–844. doi:10.1126/science.8171339

Mood, A. M., Graybill, F. A., & Boes, D. C. (1974). *Introduction to the Theory of Statistics.* Columbus, OH: McGraw-Hill, Inc.

Papoulis, A., & Pillai, S. U. (2002). *Probability, Random Variables and Stochastic Processes.* Columbus, OH: McGraw-Hill, Inc.

Pillow, J., Shlens, J., Paninski, L., Sher, A., Litke, A., Chichilnisky, E. J., & Simoncelli, E. (2008). Spatio-temporal correlations and visual signalling in a complete neuronal population. *Nature*, *454*, 995–999. doi:10.1038/nature07140

Potter, S. M. (2001). Distributed processing in cultured neuronal networks. *Progress in Brain Research*, *130*, 49–62. doi:10.1016/S0079-6123(01)30005-5

Richmond, B., & Wiener, M. (2004). Recruitment order: a powerful neural ensemble code. *Nature Neuroscience*, *7*, 97–98. doi:10.1038/nn0204-97

Rieke, F., Warland, D., Van Steveninck, R. de R., & Bialek, W. (1997). *Spikes: Exploring the Neural Code.* Cambridge, MA: MIT Press.

Segev R., Baruchi, I., Hulata, E., Ben-Jacob, E. (2004). Hidden Neuronal Correlations in Cultured Networks. *Phys. Rev. Lett.* *92*(11): 118102(1)-118102(4).

Shadlen, M. N., & Newsome, W. T. (1998). The variable discharge of cortical neurons: implications for connectivity, computation and information coding. *The Journal of Neuroscience, 18,* 3870–3896.

Shahaf, G., Eytan, D., Gal, A., Kermany, E., & Lyakhov, V. (2008). Order-Based Representation in Random Networks of Cortical Neurons. *PLoS Computational Biology, 4*(11), e1000228. doi:10.1371/journal.pcbi.1000228

Shein, M., Volman, V., Raichman, N., Hanein, Y., & Ben-Jacob, E. (2008). Management of synchronized network activity by highly active neurons. *Physical Biology, 5*(3), 36008. doi:10.1088/1478-3975/5/3/036008

Softky, W. R., & Koch, C. (1993). The highly irregular firing of cortical cells is inconsistent with temporal integration of random EPSPs. *The Journal of Neuroscience, 13,* 334–350.

Thomson, A. M. (1997). Activity-dependent properties of synaptic transmission at two classes of connections made by rat neocortical pyramidal axons in vitro. *The Journal of Physiology, 502,* 131–147. doi:10.1111/j.1469-7793.1997.131bl.x

Thorpe, S., Delorme, A., & VanRullen, R. (2001). Spike based strategies for rapid processing. *Neural Networks, 14*(6-7), 715–726. doi:10.1016/S0893-6080(01)00083-1

Van Rullen, R., & Thorpe, S. J. (2001). Rate coding versus temporal order coding: what the retinal ganglion cells tell the visual cortex. *Neural Computation, 13,* 1255–1283. doi:10.1162/08997660152002852

Vogels, T. P., Rajan, K., & Abbott, L. F. (2005). Neural network dynamics. *Annual Review of Neuroscience, 28,* 357–376. doi:10.1146/annurev.neuro.28.061604.135637

Wohlgemuth, S., & Ronacher, B. (2007). Auditory discrimination of amplitude modulations based on metric distances of spike trains. *Journal of Neurophysiology, 97,* 3082–3092. doi:10.1152/jn.01235.2006

KEY TERMS AND THEIR DEFINITION

Burst: the intermittent ignition of one or many neurons firing action-potential sequences at a high spiking rate. Although the biophysical mechanisms of burst firing are nowadays object of study, it is well understood that burst have a prominent role in higher brain centers activities, as much as in sensory information transmission. Sometimes it is also called as Spike volley.

Compartmental Model: a computer implemented model, which replicates the behavior of a neuron by dividing it into small electrical compartments and then simulates the transmission of electrical signals inside the neuron and across its membrane surface outside the same.

MEA: A micro-electrode (or multi-electrode) array (MEA) is a technology composed by several, typically at least 64, electrodes or channels, which are able to simultaneously record at high temporal resolution or to stimulate the electrical activity from several sites.

Network Model: A quantitative model replicating the behavior of a number of neurons connected via synapses. Such model is applied, specially in compartmental modeling, to study the neural activity of a population of neurons when spiking spontaneously or under a stimulated condition with modified neurons and synapses parameters.

Spike Sorting: an ensemble of mathematical, statistical and Machine Learning methods applied on raw waveforms of neuronal electrical activity captured by a single recording channel, in order to classify and then distinguish the waveforms of spikes fired by distinct neurons or units of the channel above.

Spike Time: The time instant when an action potential occurs, calculated from a stimulus onset, experiment start time or another event, and strictly depending to the temporal resolution used to capture the action potential.

Spikes: An action potential generated by a neuronal cell and consisting in a very brief, generally 1 millisecond long, electrical signal travelling along its axon and transmitting very quickly its signal to other neurons via a synapse. An action potential starts from a rapid change in the voltage sign from negative to positive, and then rapidly coming back to the original polarity: such specific cycle of the potential in time shows a voltage membrane depolarization, and then a repolarization followed by an undershoot. After

a spike has been fired, a neuron is unable to fire immediately after, but after a physiological time delay known as refractory period, generally with a duration of 2-5 ms.

Spike-Time Histogram: A mathematical and graphical technique used to estimate the behavior, precisely the spike number or the firing rate, of one or more neurons as a function of time, by counting, or averaging across trails or neurons number, the spikes counts in subsequent time-bins. If normalizing by the bin time length, the instantaneous firing rate is obtained. Often it is calculated after a precise stimuli and for many trials, in this case getting the denomination of Post-Stimulus Time histogram (PSTH).

Timestamp: A unitary temporal event corresponding to a spike fired by a recorded neuron.

Chapter 18
Ubiquitous Healthcare

Theodor Panagiotakopoulos
University of Patras, Greece

Maria-Anna Fengou
University of Patras, Greece

Dimitrios Lymberopoulos
University of Patras, Greece

Eduard Babulak
University of the South Pacific, Fiji

ABSTRACT

We have already moved away from traditional desktop-based computer technologies towards ubiquitous computing environments that progressively exist in our daily activity. This chapter introduces the concept of ubiquitous computing in the domain of healthcare as well as the prevalent technology its implementation depends on. This technology, named context-awareness, and a generic system for its realization are comprehensively described. Furthermore, the authors outline the main services that a context-aware system can provide and concluding they discuss the impact of ubiquitous computing in the healthcare domain. The authors aim at providing an overview of the technological proceedings in this area and through this understanding assist researchers to their brainstorming.

UBIQUITOUS COMPUTING

The future of computing has been painted by many visionaries. It was coined as ubiquitous computing by Mark Weiser; D.A. Norman introduced invisible computing, IBM promotes pervasive computing; Sybase calls it mobile embedded computing; and Sun uses the term Post-PC era (Milojicic et al., 2001). Each one perceives differently this new technologi-cal region and as a result focuses on a different aspect of it. The prevailing name is ubiquitous computing even if pervasive computing is also used alternately. A common feature of all these diverse approaches though is the center of gravity is moving now from technology to users and services (Milojicic et al., 2001). All promote the current trend through which communication systems and their components are built aiming to offer services based on the analysis of the context of their users and adapt to the specific preferences of each user. Adaptation is further done

DOI: 10.4018/978-1-60566-768-3.ch018

regarding the potential of the environment through which a user communicates and the situation he finds himself in.

According to M. Weiser (1991), one of the first that introduced the term ubiquitous computing, the most profound technologies are those that disappear. Such a disappearance is a fundamental consequence not only of technology but of human psychology. The main concept of ubiquitous computing is that users will interface with services and computing hardware and software will be transparent (Kumar et al., 2003). The realization of this concept will be supported with multiple devices, either mobile or embedded in almost every type of physical space and device people interact with such as homes, offices, cars, tools, appliances, clothing and various consumer goods - all communicating through increasingly interconnected networks. The functionality is not based on each device separately but on the interaction of all of them. The objective is that devices fit the human environment instead of demanding from humans to enter theirs (Weiser, 1991). This is succeeded by providing relevant information in the right form, at the time and place it is needed. This is the key to customized and personalized information systems that remain invisible until needed (Ley, 2007). Ubiquitous computing has many potential applications, from health and home care to the domain of advertisements (Kohda & Endo, 1996), tourist guides (Cheverst, Davies, & Michell, 2002), real estate brokers (Boddupalli et al., 2003) and intelligent transport systems (Parliamentary Office of Science and Technology, UK, 2006).

Ubiquitous computing is often used interchangeably with two other terms: pervasive computing and ambient intelligence, even if they have a different notion. There is a confusion concerning the way these terms are used and the concept they reflect. In reality, they differ in a sufficient level to be distinct from each other.

Ubiquitous computing is considered to be the integration of mobility and pervasive computing functionality (Lyytinen & Yoo, 2002). Mobile computing expands our capability to be productive and to communicate independently of the device's location. The flaw of mobile computing is that computing device cannot seamlessly and flexibly obtain information about the context in which computing takes place and adjust it accordingly. This deficiency is covered by pervasive computing which is an area with computing devices which are embedded in the natural environment and interact with it. Figure 1, which depicts the difference between ubiquitous and pervasive computing, follows (Lyytinen & Yoo, 2002):

Correspondingly, ubiquitous computing and ambient intelligence (AmI) imply a slightly different focus: ubiquitous highlights the "everywhere" technological features of this new vision whereas ambient focus on populating the environment with smart devices and making information services accessible everywhere. AmI is a term used more in Europe and ubiquitous computing more in USA. It should be noticed that this vision consists not only of the computing part that the term ubiquitous mainly reflects but also the ability for communication, the facility for use and the unobtrusive interfaces. Therefore, the term AmI is preferred in Europe and can be defined as the convergence of three fundamental technologies: ubiquitous computing (including miniaturization of computing capabilities, new materials, sensing technologies), ubiquitous communication and intelligent user-friendly interfaces (Punie, 2003).

Several umbrella projects in leading technological organizations have already worked on and are still exploring ubiquitous computing. Xerox's Palo Alto Research Center (PARC), for example, has been working on pervasive computing applications since the 1980s. Carnegie Mellon University's Human Computer Interaction Institute (HCII) is working on similar research in their *Project Aura*, whose stated goal is "to provide each user with an invisible halo of computing and information services that persists regardless of location." The

Figure 1. Difference between ubiquitous and pervasive computing

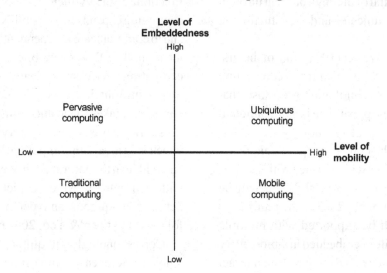

Massachusetts Institute of Technology (MIT) has a project called *Oxygen*. MIT named their project after that substance because they envision a future of ubiquitous computing devices as freely available and easily accessible as oxygen is today (Beth Archibald Tang, 2004).

Observing all these projects that have been developed, an inference could be expressed with certainty: the realization of ubiquitous computing requires the collaboration of many ICT areas. Its implementation depends on several factors such as (Weiser, 1991; Kjeldskov & Skov, 2007; Kunze, Grobmann, Stork, & Muller-Glaser, 2002; Ley, 2007):

- Specific application devices (smaller, lower power processors)
- Physical and cognitive embedded systems
- Ubiquitous connectivity
- Interoperability (seamless networks, self-configuring)
- Intelligent systems including sensor networks, context awareness, data handling
- Security and reliability

Several key research problems arise that designers and engineers have to deal with such

as the adoption of appropriate communication protocols. The traditional study of designers on the efficiency of software may not be sufficient anymore. Designers have to study in detail work activities, the way people interact with the environment as well as their use of technology (Kjeldskov & Skov, 2007). Thus, they have to design systems that does not disrupt but support the natural workflow of the user. Designers can be informed for the existence of such problems by the use of continuous usability evaluations (Satyanarayanan, 2001).

However, the thorniest problems in a ubiquitous computing environment are privacy, security and safety. This issue stems from its ability to gather sensitive data (user's everyday interactions, movements, preferences and attitudes), retrieve and use information from large databases of already stored data, and finally alter the environment via actuating devices. There are three measures that are cited as effective to give protection to a ubiquitous system: the volume of transmitted data should be limited to the minimum, transmitted data should first be encrypted and security should be considered an integral element of computing devices (Wright, Gutwirth, Friedewald, Vildjiounaite, & Punie, 2008).

UBIQUITOUS COMPUTING
IN HEALTHCARE

Nowadays, healthcare industry is confronting a number of challenges. The most significant is the high cost which is proportional to the demographics of the population. The wide growth of population ageing in many nations and the diffusion of several chronic diseases result in a high amount of people that will need care and assistance services while being away form healthcare provisioning centers. Other challenges are the growing incidence of medical errors and the lack of medical staff especially in rural and underserved urban areas. Healthcare workers are supposed under pressure to provide better services to more people using limited financial and human resources. Another important issue is the fact that in many cases people suffering from various disorders such as. epilepsy, need immediate treatment which is not facilitated by the current healthcare infrastructure. In addition, there is increased need by a considerable amount of people, who do not necessarily suffer from some kind of disorder to monitor their health condition in order to remain healthy and ensure a better quality of life.

A way to significantly eliminate all these thorny issues is to increasingly use the information technology and more specifically the ubiquitous computing technology (Kunze, Grobmann, Stork, & Muller-Glaser, 2002). The wide scale deployment of wireless networks will improve the communication among patients, physicians and other healthcare workers and healthcare service providers (Yuan, Guan, Lee, & Lee, 2007). A ubiquitous environment enables the delivery of accurate medical information anytime, anywhere, thereby reducing medical errors and improving access to medical information. Its implementation eventually reduces the long-term costs and improves quality of service of already existing medical services while offering a set of new services tailored to the preferences of each patient (Varshney, 2003).

Users of ubiquitous computing systems in the healthcare domain are frequently on the move while at the same time relying increasingly on centralized computerized information. They are most likely to use different points of access such as PDAs, laptops or PCs to solve a given task (Kjeldskov & Skov, 2007). These points of access collect information provided by diverse sensing devices and transmit it to the central provider's premises for further evaluation, processing and storage. Depending on the capabilities of the access device, an initial processing and interpretation of the collected information might take place as well.

The function of this system requires the seamless integration between the different devices, both mobile and stationary terminals, comprising the system as a whole. The enabling technology for such communication systems is the so called context-aware technology which emerged as key technology to realize ubiquitous computing. Exploiting information provided through sensors, either internal or external to the communication system, it has the potential to facilitate applications that are aware of their environment and enables them to adapt to the current context according to the preferences of the user.

Depending on application there is a wide range of services that can be offered to patients and physicians. The most important are: mobile telemedicine, patient monitoring, location-based medical services, emergency response and management, pervasive access to medical data, personalized monitoring and lifestyle incentive management (Varshney, 2003).

Several frameworks have been developed aiming to realize ubiquitous computing and addressing different application domains of medicine such as pathology, cardiology and psychology. Such a framework is the Arrhythmia Monitoring System (AMS) which is a working wireless telemetry system test bed developed at NASA and Case Western Reserve University's Heart & Vascular Center. AMS collects real-time ECG signals from mobile or homebound patients, combines GPS location data, and transmits both to a remote station for display and monitoring. The end-to-end system architecture is described below. The *wearable server* is a small data collection and communication device the patient wears that transmits signals to a *central server* sitting in close proximity to the patient. The central server performs several functions, including data compression, location awareness via GPS signals, and rudimentary arrhythmia detection. It also serves as a wireless gateway to a long-distance cellular communication network. Data is routed over the Internet to the *call center*, where medical professionals monitor the ECG signal and respond to alerts (Liszka et al., 2004).

According to many researches existing in the literature, there are many issues that must be undertaken so that the applications of ubiquitous computing will be efficient, effective and commercially accepted. For example, applications or services of ubiquitous computing must be minimally intrusive. In other words, they have to provide the appropriate information, in the right form, at the time and place it is needed. Furthermore, ubiquitous networking provides ubiquitous access to central information services but this has to be achieved with low power consumption and ensuring of a high level security (Kunze, Grobmann, Stork, & Muller-Glaser, 2002). Privacy and security are two important issues in the healthcare domain. Healthcare information should be available anytime, anywhere, but only to authorized persons. This information could be abused for instance by insurance companies in refusing coverage for people with poor health or be other corporations with similar profits. (Varshney, 2003). Thus, there should be clear and strict guidelines and control on who can access such data.

In the following sections context-aware technology will be introduced. Furthermore, we will analyze the domains this technology is implemented and the benefits this implementation offers to all involved players. A context-aware system

architecture will be illustrated and a generic context-aware framework will be described. Finally, we will outline various types of context-aware and discuss the impact ubiquitous computing has in the domain of healthcare.

CONTEXT-AWARENESS IN HEALTHCARE

Context-awareness is a concept that had been described much time ago, but there was not the adequate technological infrastructure (such as wireless technology, mobile tools, sensors, wearable devices, handheld computers) to support its implementation in applications (Bricon-Souf & Newman, 2007).The architecture of context-aware computing applications is still an active research area with no standard available for the development of the applications. There are also several issues to be encountered that concern the captors, the mobile communication and the security of such systems. Besides the problems that need solution and the fact that there are not any standards supporting the implementation of context-awareness, this technology has already been incorporated in many applications. Some of these are met in shopping (as assistance) (Kjeldskov & Graham, 2003), in mobile tourist guides (Cheverst, Davies, & Michell, 2002), mobile communications (Pandev, Ghosa, & Mukheriee, 2002) and in healthcare.

According to Rakotonirainy et al. (2000), context-awareness allows an entity to adapt its behavior to the circumstances in which it finds itself. In other words, context-awareness is a rather general concept that refers to the capability of a system to be aware of its physical and logical environment and to intelligently react according to this awareness. There can be one type of classification according to the level of interactivity in context-aware mobile systems. These interactivity levels (Barkhuus & Dey, 2003) consider different degree of user involvement in discovering and reacting. The first level which is called personalization lets the user specify his personal settings. The second one, that is a circumstance in which the application presents context to the user with no automatic updates, is defined as passive context-awareness. Finally, in the third level which is named active context-awareness the application autonomously changes the application according to sensed information.

In the healthcare domain, context-awareness implies the capability of the system to acquire, interpret the information and then perform the suitable actions with objective the provision of the best possible treatment. The incorporation of context–aware technology in ubiquitous environments leads to a health system which is mostly based on early diagnosis and proactive treatment. The treatment is now shifted to self-care, mobile care, home care and preventive care. With the wide-area connectivity and information handling, already existing and emerging ICT technologies offer, services supporting these objectives can be delivered to the patients wherever they are; in their house, in a room, in a vehicle or even in the road in an emergency situation. Personalization is the final step in order to provide context-aware services and applications customized to every individual's needs either this is a patient or a medical expert.

The content of the context-aware medical services and the kind of personalization highly depend on the concrete attributes such as location, physical conditions, time and activity of a patient or in other words on the context of a patient. Such attributes form the patient's context and characterize the complete state of a patient. A patient context describes aspects of the situation a patient has been, is or will be in.

The healthcare domain can be divided into three major application areas where ubiquitous computing is employed. The first is the professional environments such as hospitals, regional health centers and private nursing homes where ubiquitous computing is mostly used to serve

medical experts. The second is a home-based health provision where healthcare services are patient-oriented while the third application area consists of those patients that want or need to be constantly monitored while keeping their mobility. Before examining context-awareness and its implementation in these application areas, it is necessary to define the meaning of the term context.

An Overview of Context

The concept of context intersects with a diverse range of research areas such as artificial intelligence, (ontology) and knowledge representation. Its definition is more likely to change according to the application domain and related purposes. According to Dey and Abowd (1999), context is any information that can be used to characterize the situation of an entity (in our case the entity is a patient). This definition indicates a very general attitude towards context that can be applied to a wide field of intended application environments.

Several classifications of the notion context have been proposed from different research groups. According to Kunze et al. (2002), the context can be classified into low-level context and high-level context. The term low-level context represents the information that can be directly sensed by sensors while high-level context refers to the context information derived from the combination of different information sources. Panagiotakopoulos et al. (2008) indicated active context as a third context type for this classification. Active context refers to an instance of a context when it is adapted to a certain environment at a certain moment in time. Another classification can be into primary and secondary context types. The primary context types are the types of contextual information that are identified as more important than others due to practical reasons. For example, these primary context types can be location, identity, activity and time (Rakotonirainy, Loke, & Fitzpatrick, 2000). The secondary context is that information that can be derived from the primary one but itself is not part of it.

According to the domain context-awareness is applied, different context parameters are considered to be of great importance. For example, in the advertisement paradigm, mood, location, time and lifestyle information are considered to be some of the main context elements to deliver the advertisement respective services while in the mobile communications paradigm, location, speed, velocity, direction and network resource requirements are fundamental for service provisioning. In healthcare, what matters is the patient context. It is significant to categorize it in order to provide structured and organized processing methods. Following the context categorization introduced by the Wireless World Research Forum, Working Group 2, (2003) we consider a generic classification of a patient context depicted in Figure 2:

The patient context provides the overall context information and is consisted of the following context notions:

- Environment context captures these entities that surround the user. These entities can be objects interacting with the patient and physical conditions such as temperature, humidity light and noise.
- Identity context describes the physiological and psychological state of the patient. The first part contains information such as oxygen level, blood pressure, temperature and pulse while the second one contains information related to the patient's present mood and stress level. It also refers to demographic and personal information of a patient such as name, age, gender and address.
- Activity context depicts the task the patient performs and his role in it. Hobbies, scheduled activities and related actions are also part of the activity context.
- Temporal context captures attributes like time, day, month etc.
- Spatial context provides information such as location, direction and speed. It also defines the patient's social arenas described with a name such as home, work and entertainment.
- Social context can contain information about individuals interacting with the patient and the role the user plays in this context. A role can additionally be played in a social arena.
- Device context refers to the display, processing and memory capabilities of the device a patient uses to communicate with a health care service provider. It also contains information about the surrounding networks, their availability and their resources

In every context-aware healthcare service/application the first thing to be done is to define the aspects of the context knowledge which they are interested in. For example, in the cardiological paradigm, social context is of minor importance while in the psychology paradigm is fundamental. It must also be mentioned that some of the context parameters are quite static such as demographic data while the most of them are rather dynamic such as location and time.

The patient context is useful in three cases: presentation of information and medical services to a patient, execution of the medical service and

Figure 2. The patient context classification

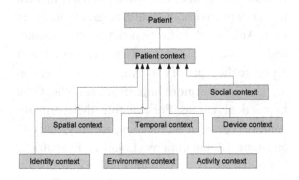

tagging of context to information for later retrieval. Hence, acquisition of such context information can be profitably exploited by a context-aware system basically for the three objectives. In a proactive manner, the system could provide prediction of potential critical episodes along with feedback to the patients regarding the actions they should perform in order to avoid them. In a reactive manner, it could alert authorized persons of evolving critical patients. Finally, context data could be stored in a database, so that context history is maintained for future processing.

Professional Healthcare Environments

A fundamental spatial application area of ubiquitous context-aware technology is professional healthcare environments. Even if it is difficult to predict the final form that future hospitals will have, this new technology supports a new way of envisaging its function, which will be deeply different from the current one. Healthcare work, for instance, in hospitals, is typically characterized by very complex and specialized procedures in which technology may contribute to achieve better performance, reduction of errors made in treatment of patients and reduction of economical costs (Kjeldskov & Skov, 2007). Another significant issue constitutes the cooperation between health care professionals which represents a large part of their activity. This cooperation, direct or indirect, is indispensable but a problem with transmission of information between them still induces breakdowns in communication (Grosjean & Lacoste, 1999). This operational void provokes rare but serious errors which even become fatal and obviously exist in all hospitals (Kohn, Corrigan, & Donaldson, 2000). However, the advent of context-awareness in current technologies in combination with new mobile tools and communication abilities will help health care professionals to manage their tasks while increasing the quality of patient care (Bricon-Souf & Newman, 2007).

There are several concepts that have been developed in order to support the activity of healthcare workers and to improve patient's care. In some medical systems, doctors can review and update the patient's medical record from any location using a handheld device (Varshney, 2003). This facility eliminates the need for time-consuming and errors associated with handwritten instructions. Depending on the infrastructure, there is the ability to transmit critical information about victims in a hospital before they arrive. This allows the specialists to prepare the patient's treatment or even provide immediate instructions for treatment if it is necessary. In addition, the system could obtain real-time traffic information from wireless networks to efficiently route the emergency vehicles to victims and then to the closest available hospital (Varshney, 2003). Moreover, with remote monitoring patients that are no longer in danger can be transferred earlier to other units within a hospital, allowing the medical staff to handle other more urgent cases. This new concept of function limits the healthcare cost and improves the overall quality of service.

Some hospital-based prototypes have already been developed for this particular domain. Major context-aware systems applied to professional healthcare environments include the following:

- In Denmark (Bardram, 2004; Bardram, 2003), in the center of Pervasive Health care the proposed prototype includes a context-aware hospital bed with an incorporated display ready for use both from the patient and the clinicians according to their needs. This bed is "intelligent" enough to know who is using it and what surrounds it (person or object). This system also includes a context-aware pill container that is aware of the patient and a context-aware Electronic Patient Record. The bed displays information according to the context it conceives.

- The University of Cambridge, in UK (Mitchell, Spiteri, Bates, & Coulouris, 2000), has developed a prototype which allows the localization of a team member and initiates a audio-video conference from the nearest point with clinicians who are notified of the call through an active badge that they carry.
- In CICESE, in Mexico (Munoz et al., 2003), the idea of the project is the extension of instant messaging to add context-awareness as part of the message. This function is incorporated in handheld devices which provide information on clinical patient record.
- MobileWARD (Kjeldskov & Skov, 2004) developed in Aalborg, in Denmark is a prototype designed to support the morning procedure tasks while roaming in hospital ward, and is able to display patient lists and patient's information according to the context. This handheld device presents information depending on the location of the nurse and the time of the day.
- CAMMD (O' Sullivan et al., 2006) is an agent-based architectural solution which stands for Context-Aware Mobile Medical Devices. This framework facilitates the proactive communication of patient records to a portable device based upon the active context of its medical practitioner.

Most projects do not use a structured context model and this is probably observed due to the fact that the services to be delivered require rather few context parameters to be realized. For example, MobileWARD exploits information regarding location, time and identity of surrounding people while "The Hospital of the future" prototype exploits information of location and surrounding objects and people.

All these prototypes deliver medical expert related services and thus are not providing any type of direct services to the patients. The main objective is to relieve medical practitioners from the burden of their work by replacing traditional paper based information retrieval with the computer aid information access. Furthermore, they address one crucial problem that medical practitioners face while conducting their everyday activities. That is the limited information access while being away from their offices and more especially the insufficient information access at the points of care. Finally, they provide location and time dependent information decreasing the amount of information that the medical staff has to process. For instance, MobileWARD provides medical staff the capability of accessing information while on the move regarding only the very specific patients lying in a specific ward at a certain point in time.

However, even if much investigation has already been carried out in this domain, there are specific features "unique" to the rhythm of health care work (particularly in hospitals) that will have yet to be explored to a sufficient level of granularity (Bricon-Souf & Newman, 2007; Reddy & Dourish, 2002). It is significant for designers to focus on the usability of the developed prototype and not only on its innovation. The technology should be adapted to the medical staff's way of working instead forcing the medical staff to understand and adapt to the new technology. For that reason, before the release and during the design of such systems usability and interaction evaluations should be executed.

Home-Based Health Provision

The increased life expectancy and the decreasing birth rates in most of the developed countries result in an increasingly aging population and have created amplified needs for health treatment. Simultaneously, chronic diseases have become often phenomenon in the current societies. Until recently, the ability to live independently in home required either good physical and mental health or such an economical comfort that would allow the patient to be cared at home (Kentta et al., 2007). However, with the advent of ubiquitous computing and the use of context-aware technology at home, health treatment will be shifted to a more proactive function. This addresses a general concept of maintaining the wellness of society or improving the general health of society. Healthcare services are provided to people that are not necessarily ill but are already well and want to stay that way by following specific diet, nutrition, lifestyle and exercise. Such homes are called Smart Homes; the underlying technology within the Smart Home environment comprises of three main components (Augusto, Nugent, Martin, & Olphert, 2005): communications, computing and user interfaces.

In a typical scenario of a Smart Home which supports health monitoring systems the house is equipped with a number of different sensors placed in everyday appliances, like the fridge, cooker or the door and can send information to healthcare providers. There are sensors for monitoring risk situations (e.g. fire alarms, smoke alarms, anti-burglar alarm), for monitoring activities (e.g. movement detectors, specialized tags, cooker sensors, water sensors) and non-intrusive biosensors for monitoring some health parameters which will facilitate the medical management (Augusto & Nugent, 2004). It has both internal and external networks which allow the interaction and remote control of systems as well as the access to services and information databases from within and beyond the home. This technology is discreetly positioned in the home environment and may unobtrusively gather information relating to the general activity and either alter the environmental context or inform the care providers. The implementation of such technologies will support long-term care service provision while enabling cost savings and effective resource management (i.e. doctors, biomedical instruments).

The services (Paganelli, Spinicci, & Giuli, 2008) that can be provided are classified into three categories: services for emergency detection

and management, user assistance services such as taking medication, monitoring vital signs and services which enable better life quality. These services are supposed to improve the quality life of both patients and their relatives who possibly would be obliged to limit their daily activities and terminate their social lives for a period (O'Sullivan et al., 2006). The delivery of an extensible and effective set of care services which allow patients to be assisted at home in a familiar environment depends strongly on the collaboration of all the involved people (patient, family and especially health care providers).

Several researches and projects have been developed in this field from many institutions. One of the first relevant trials is the Aware Home Initiative developed in the Georgia Institute of Technology where the main objectives have been to gain awareness of inhabitant activities and to enable services for maintaining independence and quality of life for an ageing population (Abowd, Bobick, Essa, Mynatt & Rogers, 2002). Another contribution is the INHOME project (Korhonen, Paavilainen, & Särela, 2003) which aims at providing means for improving the quality of life of elderly people at home by developing technologies for managing their domestic environment and enhancing their autonomy and safety at home. Various examples exist in the field of assistive technologies and home-based health monitoring systems: Vivago® is an alarm system which provides long-term user activity monitoring and alarm notification (Särela et al., 2003); the CareMedia system (Bharucha, 2006) uses multimedia information to track user activities.

All these approaches aiming at a home-based healthcare provision, illustrate a group of objectives. Some of the most important are: the improvement of health and well-being, social support, provision of a better quality of life, reinforcement of independent living and establishment of a well defined and efficient care model. While some of these objectives are addressed to a satisfactory level, many problems have to be overcome so that a home care framework can be effectively and efficiently implemented in real world situations.

A first issue that needs to be addressed is the lack of cooperation between the different stakeholders towards establishing a widely accepted care model. Furthermore, there is a gap between emerging home-based care solutions and needs and requirements of patients and healthcare providers. The designers should deeply study and understand the human behavior that they intend to complement in order to be productive. Another significant issue is interoperability between hardware and software vendors as well as between them and the service providers. Interoperability problems are often met even among diverse health care providers. In addition, the sensing devices should be designed as simple as the appliances people use in their daily lives and be seamlessly integrated to the existing home environment so as to be usable. Finally, the way the content of services is delivered to the home sustained patients must be non invasive and at any occasion adapt to the patient's context not interrupting his everyday life flow.

Wearable Systems

There is a considerable amount of people, suffering of some kind of disorder or not, that need to be more interactive and more conscious of their own health condition (Paradiso, Belloc, Loriga, & Taccini, 2005) in a daily basis without limiting their mobility and independency. In some occasions it is required a continuous monitoring of a patient's health condition so as to provide immediate care in case of a critical episode. Some indicating examples are cardiological disorders, epileptic patients and pregnants. Consequently, there is a need for a new technology based on context and mobility support, a technology different for the one used in desktop environments. In order to address this need, wearable systems have been developed. These systems combine the advances of telecommunication, microelectronics and information technology (Park & Jayaraman, 2003). Such innovative systems are able to provide improved health care to individuals while their continuous communication to a monitoring clinical center inspires safety to them.

A wearable context-aware system for ubiquitous healthcare consists of three parts: wearable sensor systems, wearable computers and communication modules. In general, several sensors transmit the measured physiological and context data over a body area network (BAN) to a computing unit (e.g. a PDA), which fuses the sensor data out of them, estimates the health status and communicates with the surrounding networks.

Kang et al. (2006) proposed a wearable context-aware system, where the wearable systems are connected to wearable computers via Bluetooth or Zigbee communication. The measured bio-signals are transmitted to wearable computers and processed into preliminary contexts. The wearable computer processes the gathered data into a context. The preliminary context can be converted to higher level context by reasoning engine or other methods that will be studied in detail in section "Context Reasoning". The wearable context-aware system is connected with service providers via a network, and ubiquitous healthcare services are provided via some service provision mechanisms.

The services that may be delivered are classified into two categories (Kang, Lee, Ko, Kang, & Lee, 2006): local services and remote services. The local services are provided to users by the wearable computer which is located in the personal area network according to the identified contexts whereas remote services are provided by remote servers operated by service providers. In this case, the contexts that are needed for the remote services should be transferred via the wide area network from the wearable computer to the remote servers.

Obviously, to ensure patient mobility, attention has to be paid in the design of the vital sensors and the wearable computer which functions as a base station. Designers should consider body placement, human movement, weight, size and other similar constraints. However, a significant problem designers have to deal with is the power supply. With the rapidly increasing performance of processors, memories and other components their power consumption is also rising. Therefore, low power consumption is the major design constraint for wearable system environment (Kunze, Grobmann, Stork, & Muller-Glaser, 2002).

Several research teams are working in this field. Basically, experimental solutions have been proposed and a lot of effort is yet needed in order to realize wearable context aware systems in real world situations.

CONTEXT-AWARE SYSTEM OVERVIEW

Health care systems can integrate context-awareness computing, not only to explore new tools but also to propose useful and acceptable improvement of the existing ones. A generic architecture of a context-aware system is illustrated in Figure 3.

The context aware system depicted in the above figure spans a network comprised of sensor systems, local servers (LS), communication networks and central medical servers (MS). The sensor networks consist of several sensors placed

on the patient's body and the LS. Sensors are connected with each other and LS via Zigbee or Bluetooth communication protocols. They have the responsibility to collect context information and transmit them to the LS.

The LS that can be implemented on a personal digital assistant (PDA), cell phone, desktop PC or home gateway, sets up and controls the sensor network communicating with the MS through various networks (e.g. Internet, GPRS, 2G, 3G, WLAN). It handles the sensor registration (type and number of sensors), initialization (e.g. specify sampling frequency and mode of operation), customization (e.g. run user specific calibration or user specific signal processing procedure upload) as well as the dynamical configuration of the sensor network according to the services needs. This server provides visual and/or audio interface to the user and transfers the context information to the healthcare center. The LS gathers all the data from the sensors, processing them and reasons about the patient's context. Then, it transmits the reason context to the MS for further processing and medical history purposes. Furthermore, it has the responsibility to communicate the actions that must be fulfilled by the system as that the delivered service adapts to the requirements of the detected patient's context. Physician's instructions and patient's preferences follow these requirements.

Figure 3. Context-aware system architecture

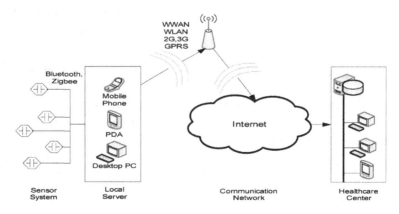

The MS keeps the electronic medical records of registered patients as well as their respected profiles which contain LS related data, context related data, application specific data, patient and physician preferences etc. The patient has not any access to the MS while the authorized physician has full access having the opportunity to review the context history of the patient forward new instructions to them such as medicine and exercises and alter the thresholds of the measured values that indicate some kind of abnormality. Patients interact with MS in an indirect manner discussing with their attendant physician and reaching an agreement for the changes that should be performed to their profiles. Still patients have no access to the MS and the physician is responsible to update their profiles. Medical experts can access medical servers through desktop PCs, laptops or other mobile devices.

Some main objectives that such a system has to address are to:

- support data collection from internal and external -to the system- sensors
- data archiving aiming to a complete context history
- be universal enough to cover the common needs of the patients
- be sufficient flexible to adjust to the specific needs and preferences of every patient
- be easily reconfigurable to better address the different needs when the health condition of a patient changes
- facilitate natural interaction between patient an d system
- be unobtrusive
- provide adjustable trade-off between privacy and health monitoring, which can be file-tuned in timely manner when a need arises
- offer personalized services which content always apply to the present needs of the patient

Sensor Networks

A series of technological advances especially in electronics has led to the miniaturization of electrical devices. The creation of small and intelligent sensors in combination with the increasing use of networks and the growing interest for remote health monitoring have developed the Body Area Networks or BANs. BAN is a system that interconnects several devices (especially sensors), which usually communicate wirelessly, worn either on or implanted in the human body in order to share information and resources. When one refers to a type of BAN which applies in telemedicine and mobile health (m-health), it is preferred the use of the term Body Area Sensor Network or BASN. In medical applications, each node comprises a biosensor or a medical device with a sensing unit. In short, BASN are also called Body Sensor Networks (BSN). (Poon & Zhang, 2006). This type of networks can be utilized to support a provision of medical services, which content is determined by a set of physiological values, to patients suffering from chronic diseases such as diabetes, asthma and heart attack. In addition, they can provide interfaces for the disabled, for diagnostics, for drug administration in hospitals, for tele-monitoring of human physiological data and as aid for rehabilitation (Latre, Poorter, Moerman, & Demeester, 2007a).

Various types of sensors can be integrated in a BSN or be external to it yet contribute to acquire the context knowledge required by a context aware service. The following list summarizes some of the potential sensor types:

- body sensors measuring physiological values such as ECG, EMG, EDA, glucose, blood pressure, heart rate
- position sensors providing information regarding location and direction
- inertial sensors sensing accelerations and rotations
- visual sensors such as cameras

- sound sensors such as microphones and hydrophones
- sensors measuring environmental parameters such as noise level, humidity, temperature and level of lightning
- motions sensors such as speedometers

All the potential context-aware services that collect data through sensor networks can be classified into four categories (Latre, Poorter, Moerman, & Demeester, 2007a) according their requirements (data rate and QoS):

1. Low data rate and low reliability
2. Low data rate and high reliability
3. High data rate and low reliability
4. High data rate and high reliability

This classification indicates that the traffic in these systems is heterogeneous. For this reason, appropriate protocols should be implemented. The first communication protocols have already been proposed; they are divided in two categories: those related with the intra-body communication (control the information handling between the sensors and the personal devices (Latre et al., 2007b) and those related to the extra-body communication (ensures communication between the personal device and an external network (Milenkovic, Otto, & Jovanov, 2006)). New protocols will ease the use of BSNs and the development of new applications.

Depending on the context-aware service/application, the data information can be delivered in a different way (Latre, Poorter, Moerman, & Demeester, 2007a):

- Continuously or periodically with small intervals
- Demand driven; data is sent only when it is needed
- Event driven; data is sent whenever an event occurs (when a threshold is crossed)
- Hybrid; a combination of the above types

There is a set of very important issues that must be considered during designing and development of context-aware systems that naturally incorporate one or more sensor systems. The first issue is the energy consumption both of sensors and the unit that collects thw raw data they transmit. The last three ways of delivering data, mentioned above, contribute to the efforts of reducing of energy consumption as well as to network resources savings which are significant problems in these systems. Moreover, the use of appropriate protocols may also reduce the power supply while maintaining reliability and delay in a satisfied level (Latre, Poorter, Moerman, & Demeester, 2007a). Most of the sensor networks also use multi-hop routing which increases the reliability and connectivity of the network and lowers the energy consumption. Because the information is confidential and private, the incorporation of encryption, key establishment and authentication is indispensable. Less threatening but still considered as a problem is the possibility of interference between BSNs of different individuals when in the future many people have their own BSNs (Poon & Zhang, 2006). Usability is not less significant. It should be noticed that this system will be used either by patients or medical staff and not by technology experts. Designers should enable the network configure itself automatically (Latre, Poorter, Moerman, & Demeester, 2007a): i.e. whenever a node is put on the body and turned on, it joins the network.

Another important issue is the size of sensors; they should be as tiny as possible because they are mainly intended to be wearable so as to facilitate mobility. A new approach is the integration of textile sensors into functional clothes. Such sensor systems, which are designed to be minimally invasive, are based on smart textile technologies, where conductive and piezoresistive materials in the form of fiber are used to realize clothes, in which knitted fabric sensors and electrodes are distributed and connected to an electronic portable unit (Paradiso, Belloc, Loriga, & Taccini, 2005).

They are able to detect, acquire and transmit physiological signs. Nanotechnology is also an emerging technology which promises to offer new sensor devices applied in BSN.

Context-Aware Framework

In this section, we will describe in detail a context-aware framework which aims to offer the appropriate functionality a context-aware system requires. Many scenarios exist in the literature regarding the technology and the software components needed for such a framework to be realized. Nevertheless, regardless of the implemented software technology, some core functions, a context-aware framework should perform, are identified by the majority of the research activities in this domain. These functions are knowledge representation, context reasoning and service personalization.

Paganelli et al (Paganelli, Spinicci, & Giuli, 2008) proposed a web-based platform while trying to support general purpose and healthcare-related context-aware services for home-based assistance. They used semantic web technologies for data representation, context processing and service personalization. Couderc and Kermarrec (1999) proposed an infrastructure, to improve the level of service taking into account context information,

introducing the concept of contextual objects in distributed context-aware information systems. Agent technology, as an alternative to the client-server approaches, has been also used (O' Sullivan et al., 2006) as the enabling middleware utilized by the distributed components within context-aware systems. This technology is seen as a highly suitable paradigm and inter-connection infrastructure for the development of mobile context-aware systems (O' Sullivan et al., 2006) (Figure 4).

In 2004 a new architecture emerged, giving the opportunity to deploy a wide range of context-aware services both in local and mobile networks. Since then, it has been widely used by many research efforts. This architecture is provided by the OSGi service platform (Gu, Pung, & Zhang, 2004) which offers a standardized, component-oriented, computing environment for networked services. OSGi delivers a common platform for service providers, content providers, software and hardware developers to deploy, integrate and manage services to a wide range of environments in a coordinated way. Furthermore, OSGi framework offers many benefits such as standardized hardware abstraction, flexible integration platform based on open standards, hosting of diverse services from diverse providers on a single gateway platform and various levels of system security. In

Figure 4. Context-aware framework

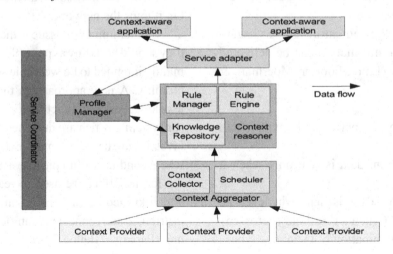

fig. 4 a context-aware framework is presented, where we looked forward to involve several components proposed in a wide range of frameworks regarding context-awareness in healthcare.

The framework illustrated in the above figure consists of the following components:

- **Context providers:** they obtain the context data from sensing devices converting them to OWL objects. These service components provide the low level contexts.
- **Context aggregator:** this component comprises of two software entities, the context collector and the scheduler. Scheduler allocates corresponding timers and pointers triggering data acquisition. Context collector gathers all the context data from the context providers. It processes the collected data and reasons, depending on the application, about the redundant and the useful data with which the context reasoner will be fed.
- **Context reasoner:** it is the central component of this framework and consists of three sub-components:
  - *Knowledge repository:* this is a knowledge base containing context models and instances stored in a local database. It also provides interfaces to other components to access context knowledge.
  - *Rule manager:* it has the responsibility of maintaining and updating (implicitly or explicitly) subscribed rules and transforming them into rules that can be handled by the rule-engine.
  - *Rule-engine:* based on the rules, the identified low contexts and the profile information it provides the high level contexts interpreting the low level ones. A generic rule-engine like JESS (Friedman-Hill, 2005) is commonly used to perform the reasoning task.

When the high-level context is obtained, the context reasoner transmits it to the service adapter.

- **Profile manager:** this component has the responsibility to download, store and keep the patient's user profile updated, in order to deliver it to the other components of the active context-aware framework whenever they request it. Finally, it transmits the reasoned active context back to the MS, so that a context history is maintained.
- **Service adapter:** based on the reasoned context and the profile information it performs a reasoning process similar to the one performed in the context reasoner determining the actions that should be done. It compares the context data of the reasoned context with the user profile data, applying semantic matching. The outcomes are communicated to the appropriate service components so that the service will change its behavior according to the requirements of the reasoned context and the patient's and physician's preferences.
- **Service coordinator:** the context-aware framework consists of several components that need to interact with each other. The service coordinator provides the functionality for this interaction by processing queries for contextual data between the framework's components. It supports discovery and selection of context providers, context reasoners, context aggregators, profile managers and service adapters, so that they can be located by services and applications.

The OSGi platform is a Java based component-oriented framework. Every component of the context-aware framework can be developed in a bundle format according to the OSGi specifications. OSGi framework enables the downloading of a bundle to a remote gateway (the LS in our case) and handles the life-cycle management as well as the management of the installation and update of each bundle within the context-aware framework.

It can lie concluded that there are numerous challenges, design and development issues in implementing such a framework. These mainly depend on the available network resources, storing and processing capabilities of the MBU, context and profile data processing policies and security issues placed by patients and attendant physicians.

Context Modeling

In order to describe context in a semantic level and get to understand how the structuring of context should be, we need to model it. Context modeling is a key feature in context-aware systems; it provides context in a structured form for the development of intelligent services, where information is situation dependent. The role of context modeling is to describe the relationship between the vocabulary and the concept of knowledge in a domain. It should provide application adaptability, resource awareness, mobile service, semantic service adaptability, code generation and context aware interfaces (Preuveneers, 2004).

The current approaches to modeling context for ubiquitous computing are classified by the scheme of data structures which are used to exchange contextual information in the respective system (Strang & Linnoff-Popien, 2004). The context modeling techniques that exist are: key-value modeling, mark-up scheme modeling, graphical modeling, object-oriented modeling, logic-based modeling and ontology-based modeling.

The model of key-value pairs is considered to be the most simple data structure for modeling contextual information. It is usually used in distributed service frameworks. Schilt et al (Schilit, Adams, & Want, 1994) used it to model context by providing the value of context information to an application as an environmental variable.

Key-value pairs may be easy to manage but do not possess the capability for sophisticated structuring for enabling efficient context retrieval algorithms.

All the markup scheme models are based on a hierarchical data structure which consists of markup tags with attribute and content (Strang & Linnoff-Popien, 2004). An example of this approach, are the Comprehensive Structured Context Profiles (CSCP) by Held et al. (2002). There are several context modeling approaches in the markup scheme category. Some of them are capable of enabling context-awareness to applications and other parts of ubiquitous computing infrastructure (Indulska et al., 2003). A known example of graphical model is the Unified Modeling Language (UML). Even if it has a generic structure, it is not really applicable in context modeling in the domain of healthcare.

Another model is the object-oriented. Its intention is to exploit the main benefits of every object oriented approach (encapsulation, reusability) to cover parts of the problem arising from the dynamic nature of the context in a ubiquitous environment. Access to contextual information is provided only through specified interfaces. An approach within the object-oriented model is the Active Object Model of the GUIDE project (Cheverst, Mitchell, & Davies, 1999). In a logic based model, the context is defined as facts, expressions and rules. Common to all logic based models is the high degree of formality. Several approaches have been proposed in this type of model.

Ontology models are enough applicable in ubiquitous environments and especially in the domain of medicine. They are considered to be the best candidate for expressing context and domain knowledge. Ontologies depict the shared understanding of a domain and count classes, inheritance, relationships between classes and instances as some of their major components. The ability to reason over relationships defined in an ontology and, therefore, relate instances to their abstracted types is the primary benefit of using ontologies. Some other benefits when using ontology are the following (Wang, Zhang, Gu, & Pung, 2004):

- It enables computational entities to have a common set of concepts about context.
- It can exploit various existing logic reasoning mechanisms.
- It allows the knowledge reuse.

However, there are some weaknesses such as consensus must be reached about their contents and the fact that maintenance may become arbitrarily hard.

Several ontology languages exist to design the ontology based context model, such as RDFS, DAML+OIL and OWL which is derived from the previous. In the healthcare domain OWL has a significant advantage from the other ontology languages because of its increased expressivity. Domains like healthcare that lend naturally to formalization benefit greatly from such languages. An example of a context modeling approach using OWL is CONON by Wang et al. (2004). Considering the context classification, presented in section "An overview of context", a modeling approach using OWL is illustrated in Figure 5

(Panagiotakopoulos, Lymberopoulos, & Manwlessos, 2008):

The context model is divided into two layers. The upper layer, where the set of ontologies is stable and reusable and common for all the application domains within healthcare and the lower layer where the ontologies depend on the application domain. The overall context information is provided by the Patient ontology. Each patient has its own instance regarding the class Patient. In this case, the ontologies of the upper class define the concepts of activity, location, time, identity, environment, social and device while the lower layer hosts the properties of these classes. Which ontologies will finally be included in the ontology-based context model, depends on the domain and mainly on the application of concern.

Context Reasoning

Context reasoning has two aspects: checking the consistency of context and converting low-level explicit context to high-level implicit context (Wang, Zhang, Gu, & Pung, 2004). The low-level or preliminary context is the context obtained by directly measured values. In order to array loose pieces of information provided by the low level

Figure 5. Ontology-based context model

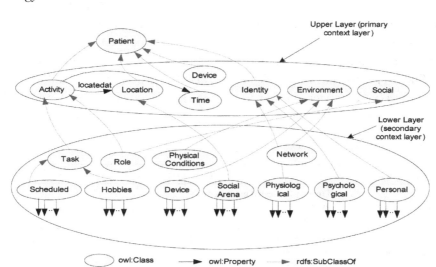

contexts so that they can be used in healthcare services a conversion, in a semantic level, from low level contexts to high level contexts must be performed. This is achieved through reasoning processes which are based on the implemented context model. To be used in healthcare services, this low-level context should be converted to high-level context according to the context model and the inference mechanism. Context reasoning is the part of system that enables service personalization (i.e. specific actions performed over identified contexts) which will be the subject of the next section.

Depending on the context model used to a context-aware system, there are certain reasoning mechanisms that can be applied or a combination of them. The reasoning mechanisms can be classified into three categories:

*   User-defined rule-based reasoning; rules are written in different types of logic like description logic, first-order logic, temporal logic, fuzzy logic, etc.
*   Learning mechanisms (Bayesian networks, neural networks, reinforcement learning)
*   Ontology reasoning

What rule-based reasoning mechanisms will be used depends on the logic requirements. In detail, when there is concern about the temporal sequence in which various events occur, it would need to use some form of temporal logic to express the rules. When generic conditions need to be expressed, it would need to use some form of first order logic (Table 1). When there is a need for more expressive power (i.e. characterizing transi-

tive relations), it would need higher order logic. When there is a need to handle uncertainty, it may require some form of fuzzy logic. (Ranganathan & Campbell, 2003)

Diverse machine learning techniques can also be used to deal with context, depending on the kind of concept to be learned. When one wants to learn the appropriate action to perform in different states in an online, interactive manner, it could use reinforcement learning or neural networks. When one needs to learn the conditional probabilities of different events, Bayesian learning is appropriate. (Ranganathan & Campbell, 2003)

Ontology reasoning mechanisms support constructs for describing relationships between classes such as inheritance relationship, equality relation and inverse relation (Gu, Pung, & Zhang, 2004; Ko, 2007). Such systems are based on a specific language like OWL. In Figure 6 (Wang, Zhang, Gu, & Pung, 2004), some ontology reasoning rules, based on OWL, are presented.

Even if an ontology reasoning mechanism is complete, experiments have shown that executing ontological reasoning at the time of the service request is unfeasible due to the response query time. This time is correlated to the number of instances of the examined ontology class as well as to the depth of the class within the ontology hierarchy (Wang, Zhang, Gu, & Pung, 2004; Agostini, Bettini, & Riboni, 2006).

The kind of reasoning mechanism that eventually will be used is determined not only by its appropriateness according to the expressivity but also by its performance and the requirements of the application in which it is applied. Rule-based

Table 1. Rule-based context reasoninga

Type of application	High-level context generating rules
Smart home-related application	If (patient is located in kitchen) ∧ (oven is turned on) Then HighLevelContext = Cooking
Healthcare-related application	If (patient's heart rate > 140) ∧ (patient's blood pressure > 17) Then HighLevelContext = CriticalSituation

Figure 6. Ontology-based context reasoning

Transitive-Property	$(?P \text{ rdf:type owl:TransitiveProperty}) \wedge (?A\ ?P\ ?B) \wedge (?B\ ?P\ ?C) \Rightarrow (?A\ ?P\ ?C)$
subClassOf	$(?a \text{ rdfs:subClassOf } ?b) \wedge (?b \text{ rdfs:subClassOf } ?c) \Rightarrow (?a \text{ rdfs:subClassOf } ?c)$
subProperty-Of	$(?a \text{ rdfs:subPropertyOf } ?b) \wedge (?b \text{ rdfs:subPropertyOf } ?c) \Rightarrow (?a \text{ rdfs:subPropertyOf } ?c)$
disjointWith	$(?C \text{ owl:disjointWith } ?D) \wedge (?X \text{ rdf:type } ?C) \wedge (?Y \text{ rdf:type } ?D) \Rightarrow (?X \text{ owl:differentFrom } ?Y)$
inverseOf	$(?P \text{ owl:inverseOf } ?Q) \wedge (?X\ ?P\ ?Y) \Rightarrow (?Y\ ?Q\ ?X)$

reasoning mechanism is commonly used in context aware health care solutions over context ontology instances.

Service Personalization

Personalization is about building customer loyalty by building a meaningful one-to-one relationship; by understanding the needs of each individual and helping satisfy a goal that efficiently and knowledgeably addresses each individual's need in a given context (Riecken, 2004). This process is divided into two discrete stages. The first contains the actions that must be taken in order to provide a service tailored to the preferences and needs of each user. Each user, either he is a medical expert or a patient, has different habits and necessities of life. Respectively, each disorder has to be treated in a different way depending on the case history and the idiosyncrasy of each patient. So the type of services to be delivered as well as the content of them should be adapted to each patient. The second stage deals with the way the service will be presented to the user depending on time, his location, his surroundings and the device he possesses which will host this information. In both stages, context reasoning is mainly applied but with different objectives from the ones described in the previous section.

- Explicitly; user-defined rule based approach is followed where the rules are determined by the user. The negative aspect of this approach is that in a temporal extent the user's current needs would differ and the system will not cover the user's requirements anymore.
- Implicitly; self learning mechanisms in combination with explicit feedback is used. In this case, the adaptation which emerged may not meet the profile of the user and his requirements from the system. The system functions autonomously without user's interference; a feature which can be turned into a negative aspect.
- A combination of the previous; the hybrid approach is considered to be the rather well adapted method to the user's needs. It conflates the positive aspects of the two first methods. It optimizes the operation of the system while the possibility to gratify its potential users augments.

Figure 7 shows the personalization process for context-aware healthcare services/applications. The component that performs personalization is the service adapter which was initially presented in "Context-aware framework" and is further examined here. Service adapter consists of three

Figure 7. Context-aware healthcare personalization

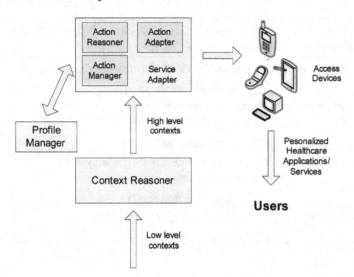

elements: the action manager, the action reasoner and the action adapter.

The action manager has the responsibility to maintain and update rules regarding the actions that have to be performed in a specific context. These rules reflect the actual preferences and needs of patients and medical experts which are provided by the profile manager. It also translates these rules into rules that can be handled by the action reasoner. The action reasoner performs the reasoning process taking into account the identified high level context and a set of rules provided by the action manager. Finally, the action adapter determines the presentation of the service to the user. It performs service adaptation based on the identified high level context and a set of rules provided by the action manager.

In the presented personalization process, rule based approaches basically used a set of condition-actions rules. These rules can be obtained explicitly by patients and physicians or imlicitly using learning techniques such as association rules algorithms. Concluding, there can be used several other techniques apart from the rule based method similar to those described in the previous section.

CONTEXT-AWARE SERVICES

The emergence of context awareness and its implementation in the domain of healthcare improved the level of service of already existed services and supported the creation of several new healthcare services. A generic classification of context aware services consists of two parts:

- Proactive services which are services aiming to an early diagnosis and prediction of potential critical episodes providing motivational feedback to the patients
- Reactive services through which authorized persons are alerted in case of involving emergency situations.

In a more detailed view, context-awareness in conjunction with advanced ICT technologies offer a wide range of services and applications. The most important of them are the following:

- Tele-monitoring services self or persistent assessment of patient condition and provision of feedback to authorized persons.
- Self health services assisting patients to

identify and change the harmful and un-helpful habits/attitudes that strengthen their disorders. These services provide real time and situation dependent guidance and support.

- Notification services providing healthcare suggestions such as a healthy lifestyle, diet and exercise. These services are commonly called reminders as they remind patients of specific actions that must be taken in a periodical basis.
- Emergency treatment services through which authorized persons are alerted and either real time communication between them and patients is established or emergency actions are performed.
- Archiving services offering context information archiving for medical history or further processing purposes

The aforementioned services offered by the implementation of context-aware technology in combination with many other services offered by the ubiquitous infrastructure (such as pervasive access to medical records) create a powerful set of services both for patients and medical experts. All these new services have a great impact both in healthcare and social level which will be in the concluded section.

DISCUSSION AND CONCLUDED REMARKS

In this chapter we discussed about ubiquitous healthcare and context-aware computing which is the enabling technology for ubiquitous environments. Ubiquitous computing depicts a vision of the future information society where humans will be surrounded by intelligent interfaces supported by computing and networking technology that is everywhere, embedded in everyday objects and smart materials. From the healthcare perspective, this vision enables preventive care decreasing the

possibility of critical episodes and serious harmful situation. Furthermore, medical staff has the opportunity to access medical information contradicted to the traditional health infrastructure. Emergency situations receive immediate treatment and medical errors are reduced by providing a shared communication infrastructure.

From a socio-economic perspective, the emphasis will be on user-friendliness, user-empowerment and support for human interactions. According to ITU, ubiquitous computing will contribute to the dissemination of advanced technological knowledge as well as improve consultations and second opinions. The usage of user-friendly, natural interfaces will make it easier and faster to everyone to learn to use services offered by ubiquitous computing. Moreover, universal care provision and a much broader reach in rural and remote areas is achieved. Regular or on demand health care becomes available in remote areas which help to slow population migration or attract people back to previously abandoned areas. Cost savings of regarding hospitalization and travel to far-away doctors and hospitals, reduced waiting lists, stress reduction, training and education are also some important socio-economic benefits.

Apart from the individual, healthcare and socio-economic impact of ubiquitous computing in healthcare, there are some problems that need solution so that ubiquitous healthcare applications/services will be used in real world situations and become commercial products. A first issue is the fact that many sensing devices, wearable systems and feedback (suggested actions, reminders, etc.) become intrusive and impede the flow of the patients' daily living. According to European Union, the developing information and knowledge society should be for everybody. Some criteria such as age, gender, education, income and family composition usually influence negatively not only user acceptance but also user's attitude towards the knowledge of new technologies and the availability of resources (Punie, 2003). Another feature of a ubiquitous environment is its ability to detect and monitor constantly what people are doing in their daily lives, both offline and online in order to provide the adequate services. This capability could be considered intrusion into the individual's privacy. Personal information will be digitally gathered, stored and may disclosed to other services. For that reason, it is indispensable the input of borders in monitoring and surveillance as well as the protection of personal information (Bohn, Coroama, Langheinrich, Mattern, & Rohs, 2005).

The whole research in this field proposes human-centered design, development guidelines and attention to social matters in order to ameliorate the effort and the realization of this vision.

REFERENCES

Abowd, G. D., Bobick, I., Essa, I., Mynatt, E., & Rogers, W. (2002). *The Aware Home: Developing Technologies for Successful Aging.* Paper presented at the AAAI-02 Workshop on Automation as Caregiver: The Role of Intelligent Technology in Elder Care, Edmonton, Alberta, Canada.

Agostini, A., Bettini, C., & Riboni, D. (2006). Experience report: Ontological Reasoning for Context-Aware Internet Services. In *the 4th Annual IEEE International Conference on Pervasive Computing and Communications* (pp. 5-12). Pisa, Italy: IEEE.

Archibald Tang, B. (2004). *What is pervasive computing?* Retrieved October 24, 2008, from http://searchnetworking.techtarget.com/sDefinition/0,sid7_gci759337,00.html

Augusto, J. C., Nugent, C., Martin, S., & Olphert, C. (2005). Software and Knowledge Engineering Aspects of Smart Homes Applied to Health. In C. D. Nugent, P. J. McCullagh, E. T. McAdams, & A. Lymberis (eds.), *Personalised Health Management Systems - The Integration of Innovative Sensing, Textile, Information and Communication Technologies, Studies in Health Technology and Informatics* (Vol. 117, pp. 164-171). Singapore: IOS Press.

Augusto, J. C., & Nugent, C. D. (2004). A New Architecture for Smart Homes Based on ADB and Temporal Reasoning. In D. Zhang, M. Mokhtari (Eds.), *Toward a Human Friendly Assistive Environment (Proceedings of 2nd International Conference on Smart homes and health Telematic, ICOST2004), Assistive Technology Research Series* (Vol. 14, pp. 106-113). Singapore: IOS Press.

Bardram, J. (2003). *Hospitals of the future— ubiquitous computing support for medical work.* In the *2nd international workshop on ubiquitous computing for pervasive healthcare applications* (Ubihealth 2003 proceedings), Seattle, WA.

Bardram, J. (2004). Applications of context-aware computing in hospital work—examples and design principles. In the *2004 ACM symposium on Applied computing.* Nicosia, Cyprus: ACM.

Barkhuus, L., & Dey, A. (2003). *Is context-aware computing taking control away from the user? Three levels of interactivity examined.* Paper presented at the UbiComp2003 conference (pp. 149-156), Seattle, WA, USA.

Bharucha, A. J. (2006). *CareMedia: Automated Video and Sensor Analysis for Geriatric Care.* Presented at the Geriatric Care Annual Meeting of the American Association for Geriatric Psychiatry, San Juan, Puerto Rico.

Boddupalli, P., Al-Bin-Ali, F., Davies, N., Friday, A., Storz, O., & Wu, M. (2003). Payment support in ubiquitous computing environments. In the *5ᵗʰ Workshop on Mobile Computing Systems and Applications (WMCSA)*. Monterey, CA: IEEE.

Bohn, J., Coroama, V., Langheinrich, M., Mattern, F., & Rohs, M. (2005). Social, Economic, and Ethical Implications of Ambient Intelligence and Ubiquitous Computing. In W. Weber, J. M. Rabaey, E. Aarts (eds.) *Ambient Intelligence* (pp 5-29). Berlin: Springer.

Bricon-Souf, N., & Newman, C. R. (2007). Context-awareness in health care: A review. *International Journal of Medical Informatics, 76*(1), 2–12. doi:10.1016/j.ijmedinf.2006.01.003

Cheverst, K., Davies, N., & Michell, K. (2002). *A reflective study of the GUIDE system.* Paper presented at the 1st workshop on mobile tourism support (pp.17-23).

Cheverst, K., Mitchell, K., & Davies, N. (1999). Design of an object model for a context sensitive tourist GUIDE. *Computers & Graphics, 23*(6), 883–891. doi:10.1016/S0097-8493(99)00119-3

Couderc, P., & Kermarrec, A.-M. (1999). Improving Level of Service for Mobile Users Using Context-Awareness. In *The 18th IEEE Symposium on Reliable Distributed Systems* (pp. 24-33). Lausanne, Switzerland: IEEE.

Dey, G. O., & Abowd, G. D. (1999). *Towards a Better Understanding of Context and Context-Awareness.* Paper presented at the 1ˢᵗ International Symposium of Handheld and Ubiquitous Computing (pp. 304-307), Karlsruhe, Germany.

Friedman-Hill, E. (2005). *Jess, the rule engine for the java platform.* Livermore, CA: Sandia National Laboratories.

Grosjean, M., & Lacoste, M. (1999*). Communication et intelligence collective: le travail a l'hopital.* Paris: Presses universitaires de France.

Gu, T., Pung, H. K., & Zhang, D. Q. (2004). Toward an OSGi-Based Infrastructure for Context-Aware Applications. *IEEE Pervasive Computing / IEEE Computer Society [and] IEEE Communications Society, 3*(4), 66–74. doi:10.1109/MPRV.2004.19

Held, A., Buchholz, S., & Schill, A. (2002). Modeling of context information for pervasive computing applications. In the *6ᵗʰ World Multiconference on Systemics, Cybernetics and Informatics* (SCI), Orland, FL.

Indulska, J., Robinsona, R., Rakotonirainy, A., & Henricksen, K. (2003). Experiences in using cc/pp in context-aware systems. In M.-S. Chen, P. K. Chrysanthis, M. Sloman, & A. Zaslavsky (Eds.), *MDM2003,* (LNCS Vol. 2574, pp. 247–261). Melbourne, Australia: Springer.

Kang, D., Lee, H., Ko, E., Kang, K., & LEE, J. (2006). A Wearable Context Aware System for Ubiquitous Healthcare. In *the 28ᵗʰ IEEE EMBS Annual International Conference.* New York: IEEE.

Kentta, O., Merilahti, J., Petakoski-Kult, T., Ikonen, V., & Korhonen, I. (2007). Evaluation of Technology-Based Service Scenarios for Supporting Independent Living. In *The 29ᵗʰ Annual International Conference of the IEEE EMBS.* San Diego, CA: IEEE.

Kjeldskov, J., & Graham, C. (2003). A review of mobile HCI research methods. In L. Chittaro (Ed.) *5ᵗʰ International Symposium on Human-Computer Interaction with Mobile Devices and Services, LNCS* (2795: 317-335). Udine, Italy: Springer.

Kjeldskov, J., & Skov, M. (2004). Supporting work activities in healthcare by mobile electronic patient records. In M. Masoodian, S. Jones, & B. Rogers (Eds.) *6th Asia–Pacific Conference on Human–Computer Interaction,* (LNCS Vol. 3101, pp. 191-200). Rotorva, New Zealand: Elsevier Science.

Kjeldskov, J., & Skov, M. B. (2007). Exploring context-awareness for ubiquitous computing in the healthcare domain. *Personal and Ubiquitous Computing, 11*(7), 549–562. doi:10.1007/s00779-006-0112-5

Ko, E. J., Lee, H. J., & Lee, J. W. (2007). Ontology-Based Context Modeling and Reasoning for U-HealthCare. *IEICE Transactions on Information and Systems . E (Norwalk, Conn.), 90-D*(8), 1262–1270.

Kohda, Y., & Endo, S. (1996). Ubiquitous advertising on the WWW: Merging advertisement on the browser. *Computer Networks and ISDN Systems, 28*(7-11), 1493-1499.

Kohn, L. T., Corrigan, J. M., & Donaldson, M. S. (Eds.). (2000). *To Err is Human: Building a Safer Health System.* Washington, DC: National Academy Press.

Korhonen, I., Paavilainen, P., & Särela, A. (2003). Application of ubiquitous computing technologies for support of independent living of the elderly in real life settings. In *The 2nd International Workshop on Ubiquitous Computing for Pervasive Healthcare Applications* (Ubihealth 2003 Proceedings), Seattle, WA.

Kumar, M., Shirazi, B. A., Das, S. K., Sung, B. Y., Levine, D., & Singhal, M. (2003). PICO: A Middleware Framework for Pervasive Computing. *Pervasive computing, 2*(3), 72-79.

Kunze, C., Grobmann, U., Stork, W., & Muller-Glaser, K. D. (2002). Application of ubiquitous computing in personal health monitoring systems. *Biomedizinische Technik. Biomedical Engineering, 47*(1), 360–362. doi:10.1515/bmte.2002.47.s1a.360

Latre, B., Braem, B., Moerman, I., Blondia, C., Reusens, E., Joseph, W., & Demeester, P. (2007b). A low-delay protocol for multihop wireless body area networks. In *The 4th Annual International Conference on Mobile and Ubiquitous Systems: Networking & Services* (pp. 1-8). Philadelphia: IEEE.

Latre, B., Poorter, E., Moerman, I., & Demeester, P. (2007a). MOFBAN: A Lightweight Modular Framework for Body Area Networks. In T.-W. Kuo et al. (Eds.) 2007 IFIP Int. Conf. on *Embedded and Ubiquitous Computing - EUC 2007* (4808, pp. 610-622). Taipei, Taiwan: Springer.

Ley, D. (2007). Ubiquitous computing. []. Coventry, UK: Becta.]. *Emerging Technologies for Learning, 2,* 64–79.

Liszka, K. J., Mackin, M. A., Lichter, M. J., York, D. W., Pillai, D., & Rosenbaum, D. S. (2004). Keeping a Beat on the Heart. *Pervasive Computing, 3*(12), 43–49.

Lyytinen, K., & Yoo, Y. (2002). Issues and Challenges in Ubiquitous Computing. *Communications of the ACM, 45*(12), 62–96. doi:10.1145/585597.585616

Milenkovic, A., Otto, C., & Jovanov, E., (2006). Wireless sensor networks for personal health monitoring: Issues and an implementation. *Computer Communications, Wireless Sensor Networks and Wired/Wireless Internet Communications, 29*(13-14), 2521–2533.

Milojicic, D., Messer, A., Bernadat, P., Greenberg, I., Fu, G., Spinczyk, O., et al. (2001). *Pervasive Services Infrastructure* (Tech. Report). HP Labs.

Mitchell, S., Spiteri, M., Bates, J., & Coulouris, G. (2000). Context aware multimedia computing in the intelligent hospital. In the *9th ACM SIGOPS European Workshop*. Denmark: ACM.

Munoz, M., Rodriguez, M., Favela, J., Martinez-Garcia, A., & Gonzalez, V. (2003). Context-aware mobile communication in Hospitals. *IEEE Computer*, *36*(9), 38–46.

O' Sullivan, T., O'Donoghue, J., Herbert, J., & Studdert, R. (2006). CAMMD: Context-Aware Mobile Medical Devices. *Journal of Universal Computer Science*, *12*(1), 45–58.

Paganelli, F., Spinicci, E., & Giuli, D. (2008). ERMHAN: A Context-Aware Service Platform to Support Continuous Care Networks for Home-Based Assistance. *International Journal of Telemedicine and Applications*, (Vol. 2008).

Panagiotakopoulos, T. H., Lymberopoulos, D. K., & Manwlessos, G. M. (2008). Monitoring of patients suffering from special phobias exploiting context and profile information. In the *8th IEEE International Conference on Bioinformatics and BioEngineering (BIBE)*. Athens, Greece: IEEE.

Pandev, V., Ghosa, D., & Mukheriee, B. (2002). Exploring user profiles to support differentiated services in next-generation wireless networks. *IEEE Network*, *18*(5), 40–48. doi:10.1109/MNET.2004.1337734

Paradiso, R., Belloc, C., Loriga, G., & Taccini, N. (2005). Wearable Healthcare Systems, New Frontiers of e-Textile. *Studies in Health Technology and Informatics*, *117*, 9–16.

Park, S., & Jayaraman, S. (2003). Enhancing the quality of life through wearable technology. *IEEE Engineering in Medicine and Biology Magazine*, *22*(3), 41–48. doi:10.1109/MEMB.2003.1213625

Parliamentary Office of Science and Technology. UK (2006). *Pervasive Computing, 263*.

Poon, C. C. Y., & Zhang, Y. (2006). A Novel Biometrics Method to Secure Wireless Body Area Sensor Networks for Telemedicine and M-Health. *IEEE Communications Magazine*, *44*(4), 73–81. doi:10.1109/MCOM.2006.1632652

Preuveneers, D., Van Den Bergh, J., Wagelaar, D., Georges, A., Rigole, P., Clerckx, T., & Berbers, Y. Coninx K., Jonckers V., & De Bosschere K. (2004). Towards an extensible context ontology for ambient intelligence. In P. Markopoulos, B. Eggen, E. Aarts, & J. L. Crowley (Eds.), *2nd European Symposium on ambient Intelligence, Ambient Intelligence* (3295, 138-159). Eindhoven: Springer.

Punie, Y. (2003). *A social and technological view of Ambient Intelligence in Everyday Life: What bends the trend?* European Media, Technology and Everyday Life Research Network, (EMTEL2) Key Deliverable Work Package 2, EC DG-JRC IPTS, Sevilla.

Rakotonirainy, A., Loke, S. W., & Fitzpatrick, G. (2000). Context-Awareness for the Mobile Environment. *Position Paper for CHI2000 Workshop on "The What, Who, Where, When, Why, and How of Context Awareness"*. The Hague, The Netherlands.

Ranganathan, A., & Campbell, R. H. (2003). A Middleware for Context-Aware Agents in Ubiquitous Computing Environments. In M. Endler, & D. Schmidt (Eds.), *ACM/IFIP/USENIX International Middleware Conference, LNCS* (2672: 143-161). Rio de Janeiro, Brazil: Springer.

Reddy, M., & Dourish, P. (2002). A finger on the pulse: temporal rhythms and information seeking in medical work. In the *ACM Conference on Computer Supported Cooperative Work*, CSCW. New Orleans, Louisiana: ACM.

Riecken, D. (2004). Personalized Views of Personalization. *Communications of the ACM*, *43*(3), 74–82.

Sarela, A., Korhonen, I., Lotjonen, J., Sola, M., & Myllymaki, M. (2003). IST Vivago - an intelligent social and remote wellness monitoring system for the elderly. In the conference on *Information Technology Applications in Biomedicine* (4th International IEEE EMBS proceedings). Birmingham, UK: IEEE.

Satyanarayanan, M. (2001). Pervasive Computing: Vision and Challenges. *IEEE Personal Communications*, *8*(4), 10–17. doi:10.1109/98.943998

Schilit, B. N., Adams, N. L., & Want, R. (1994). Context-aware computing applications. Paper presented at the *IEEE Workshop on Mobile Computing Systems and* Applications, Santa Cruz, CA, USA.

Strang, T., & Linnoff-Popien, C. (2004). *A Context Modeling Survey*. Presented at the Workshop on Advanced Context Modelling, Reasoning and Management (proceedings of the 6th International Conference on Ubiquitous Computing – UbiComp), Nottingham, England.

Varshney, U. (2003). Pervasive Healthcare. *IEEE Computer*, *36*(12), 138–140.

Wang, X. H., Zhang, D. Q., Gu, T., & Pung, H. K. (2004). Ontology Based Context Modeling and Reasoning using OWL. In the *2nd IEEE Annual Conference on Pervasive Computing and Communications* (pp. 18-22). Orlando, FL: IEEE.

Weiser, M. (1991). The Computer for the 21st Century. *Scientific American*, *265*(3), 66–75.

Wireless World Research Forum. Working Group 2, (2003). Service Architectures for the Wireless World. Whitepaper on Service Adaptability, version 1.0. Retrieved June 23, 2007, from http://www.wireless-world-research.org.

Wright, D., Gutwirth, S., Friedewald, M., Vildjiounaite, E., & Punie, Y. (Eds.). (2008). *Safeguards in a World of Ambient Intelligence*. Springer.

Yuan, W., Guan, D., Lee, S., & Lee, Y. (2007). The Role of Trust in Ubiquitous Healthcare. In the *9th International Conference on e-Health Networking, Application and Services*. Taipei, Taiwan: IEEE.

Chapter 19
E–Health Project Implementation:
Privacy and Security Measures and Policies

Konstantinos Siassiakos
University of Piraeus, Greece

Athina Lazakidou
University of Peloponnese, Greece

ABSTRACT

Privacy includes the right of individuals and organizations to determine for themselves when, how and to what extent information about them is communicated to others. The growing need of managing large amounts of medical data raises important legal and ethical challenges. E-Health systems must be capable of adhering to clearly defined security policies based upon legal requirements, regulations and standards while catering for dynamic healthcare and professional needs. Such security policies, incorporating enterprise level principles of privacy, integrity and availability, coupled with appropriate audit and control processes, must be able to be clearly defined by enterprise management with the understanding that such policy will be reliably and continuously enforced. This chapter addresses the issue of identifying and fulfilling security requirements for critical applications in the e-health domain. In this chapter the authors describe the main privacy and security measures that may be taken by the implementation of e-health projects.

INTRODUCTION

The introduction of technology changed how physicians and other health organizations keep personal health information. In now day's medical data are being kept in a computer, so we talk for an Electronic Patient Record (EPR) and not for a printed medical record. Health information systems rely upon a computerised infrastructure. The development of internet technology and web-based applications made health information more accessible than ever before - from many locations by multiple Health providers and health plans. In the near future, the Internet will probably be the platform of choice for processing health transactions and communicating information and data. But along with this accessibility come increased threats to the security of health information. And those who would steal,

DOI: 10.4018/978-1-60566-768-3.ch019

divert, alter, or misuse personal information are becoming even more skilled at finding what they want and covering their tracks.

PRIVACY AND SECURE INFORMATION EXCHANGE

Health information exchange refers to the sharing of clinical and administrative data across the boundaries of health care institutions and other health data repositories. Electronic information sharing is called electronic health information exchange. Many stakeholder groups (payers, patients, providers, and others) realize that if data could be more readily shared, the safety, quality, and cost of health care processes would improve. From a cultural and technical standpoint, sharing health data is not easy. Stakeholders have competing priorities. Financial concerns, unresolved issues related to rights to access data, and privacy and security issues are among some of the hardest challenges to overcome.

Privacy and security measures are of great concern in all technology sectors, thus leading to ever-evolving, ever-improving protections becoming available. Certainly, public entities must make the most of these developments. In fact, while it is challenging to protect the security of electronic records, it is practically impossible to protect the security of paper records. Electronic records, which can be encrypted and password-protected, are more secure than paper records, less likely to be lost, misfiled, or damaged, and are capable of being backed up. Families must be assured that information provided to the government will only be exchanged with their consent and that, when shared, will be protected from misuse during the transfer.

Medical Data Privacy

Today, individual health and medical data can be collected, collated, stored, analyzed and distributed in unprecedented quantities over the Internet and put to diverse uses for the ease of medical practice. Confidentiality in recording patient information and transferring this information is of utmost importance in protecting patient privacy. These should comply with the Health Insurance Portability and Accountability Act of 1996 protocols protecting patient records. E-health involves new forms of patient-provider interaction, which pose new challenges and threats to privacy issues.

Healthcare is experiencing unprecedented growth in the number and variety of e-health practices being adopted as computer technology and internet network connectivity become increasingly affordable. Data holders operating autonomously, and with limited knowledge, are left with the difficulty of releasing information that does not compromise privacy and confidentiality.

Methodology

There is a set of security services needed for realizing trustworthy e-health solutions. Those security services must be comprehensively integrated in the e-health application. Furthermore, a set of infrastructure services has to be specified and implemented. For keeping the solutions future-proof, they have to comply with architectural principles and paradigms.

The methodology developed is based on 3 key assumptions. The first assumption is that, in order for stakeholders to trust electronic health information exchange, decisions about how to protect the privacy and security of health information should be made at the local community level. Second, to accomplish this goal, discussions must take place to develop an understanding of the current landscape and the variation that exists between organizations within each state and, ultimately, across states. Finally, stakeholders at the state and community levels, including patients and consumers, must be involved in identifying the current variation, understanding the rationale that underlies the current business practices, deciding what the privacy and security requirements are, and developing solutions to achieve broad-based acceptance.

Most stakeholders agree that electronic health information exchange is beneficial. Widespread electronic health information exchange is expected to improve continuity of care across health care providers; reduce medical errors; avoid costly duplicate testing; eliminate unnecessary hospitalizations; increase consumer convenience, eliminate repetitive registration and permission forms; provide life-saving early detection of an infectious disease outbreak as anonymous data from emergency rooms is sent to public health systems instantly; and ensure that patients' health information is available when needed.

Three basic assumptions underlie this effort:

- It is valuable to identify good practices and solutions that have the potential to accelerate nationwide electronic health information exchange, particularly with respect to privacy and security questions for consideration and adoption by communities and states.
- Health care is local and the solutions to improving health care should accommodate community variation.
- Stakeholders at the state and community levels, including patients and consumers, must be involved in developing solutions to achieve acceptance.

The main Domains of Privacy and Security are the following:

- User and Entity Authentication
- Authorization and Access Control
- Patient and Provider Identification
- Transmission Security
- Information Protection
- Information Audits
- Administrative and Physical Safeguards

- Country Law
- Use and Disclosure Policy
- Privacy-enhanced Techniques

There are many situations in which privacy can be an issue, accordingly Privacy-enhanced Techniques research covers many different areas, including:

- Anonymous communication (anonymous retailers, anonymous surfing, etc.),
- Anonymous transactions,
- Anonymous publication and storage,
- Anonymous credentials,
- Anonymity in files and databases.

This paper focuses at medical applications, in which privacy issues are raised by the information content of the stored data, hence only the latter techniques are discussed. Privacy-enhancing techniques for privacy protection within databases help to protect the privacy of a subject of a database record (i.e. a person or organization listed in the database). Simply put, these Privacy-enhanced Techniques allow to store relevant and useful information in a way that noone can ever find out, who the information is actually about. Examples of these techniques are (Claerhout et al, 2005):

- "Hard" de-identification by the owner of the data;
- Various types of anonymisation and/or pseudonymisation;
- Privacy risk assessment techniques;
- Controlled database alteration (modification, swapping or dilution of data);
- Data flow segmentation;
- Privacy-enhancing intelligent software agents for databases.

Today, Privacy-enhanced Techniques technology has already proven its usefulness for privacy protection in health-related marketing and research data collection (De Meyer et al, 2002). Focus lays on pseudonymisation techniques, and complementary Privacy-enhanced Techniques enhancing their effectiveness.

PSEUDONYMISATION TECHNIQUES

Pseudonymisation refers to privacy-enhancing techniques and methods used to replace the true (nominative) identities of individuals or organizations in databases by pseudo-identities (pseudo-IDs) that cannot be linked directly to their corresponding nominative identities (DeMoor et al, 2003). When data is being pseudonymised, identifiers and "payload data" (non-identifying data) are separated. The pseudonymisation process translates the given identifiers into a pseudo-ID by using secure, dynamic and preferably irreversible cryptographic techniques (the identifier transformation process should not be performed with translation tables). For an observer, the resulting pseudo-IDs are thus represented by complete random selections of characters. This transformation can be implemented differently according to the project requirements (Claerhout et al, 2005). Pseudonymisation can:

- always map a given identifier with the same pseudo-ID;
- map a given identifier with a different pseudo-ID;
- time-dependant (e.g. always varying or changing over specified time intervals);
- location-dependant (e.g. changing when the
- data comes from different places);
- content-dependant (e.g. changing according to
- the content);

Pseudonymisation is used in data collection scenarios where large amounts of data from different sources are gathered for statistical processing and data mining (e.g. research studies). In contrast with horizontal types of data exchange (e.g. for direct care), vertical communication scenarios (e.g. in the context of disease management studies and other research) do not require identities as such: here pseudonymisation can help find solutions. It is a powerful and flexible tool for privacy protection in databases, which is able to reconcile the two following conflicting requirements: the adequate protection of individuals and organizations with respect to their identity and privacy, and the possibility of linking data associated with the same data subject (through the pseudo-IDs) irrespective of the collection time and place. Because of this flexibility, however, correct use of pseudonymisation technology is not as straightforward as often suggested. Careless use of pseudonymisation technology could lead to a false feeling of privacy protection. Many more aspects of the pseudonymisation process are closely linked and key to ensuring optimum privacy protection, as for example, the location of the identifier and payload processing, the number of steps in which the pseudonymisation is performed, the use of trusted third parties (TTPs). The latter is an important aspect, because use of a trusted third party for performing the pseudonymisation process offers some clear advantages:

1. As the communicating parties (data sources and collectors) not always trust each other, trust can be established indirectly by use of a third, independent party. The parties are then bound by a code of conduct, as specified in a privacy and security policy agreement they agree on with the TTP.
2. Use of a TTP offers the only reliable protection against several types of attacks on the pseudonymisation process.
3. Complementary privacy measures (PETs) and data-processing features can easily be implemented, e.g. controlled reversibility without endangering privacy.

Organizations willing to act as pseudonymisation TTP need to satisfy some important requirements like, e.g. they should be strictly independent;be able to guarantee security and trustworthiness of their methods (openness), software modules, platforms and infrastructure;be able to provide professional expertise related to the domain where the pseudonymisation is being performed;implement monitoring and quality assurance services and programs; etc (Claerhout et al, 2005).

CONCLUSION

Advances in technology, especially in the provision of broadband communication networks allows for the seamless transmission of medical data, which together with the widespread provision of public education on healthcare matters, make it certain that the accessibility and quality of healthcare services of the future should be both uniform and ubiquitous. Privacy includes the right of individuals and organizations to determine for themselves when, how and to what extent information about them is communicated to others. Various privacy-enhancing technologies exist that can be used for the correct treatment of sensitive data in medicine. Advanced pseudonymisation techniques can provide optimal privacy protection of individuals while still allowing the grouping of data collected over different time periods (cf. longitudinal studies) and from different sites (cf. multi-centre studies).

REFERENCES

Blobel, B. (2007). Comparing approaches for advanced e-health security infrastructures. *International Journal of Medical Informatics, 76* (5 - 6), 454 - 459.

Claerhout, B., & DeMoor, G. J. E. (2005). Privacy protection for clinical and genomic data: The use of privacy-enhancing techniques in medicine. *International Journal of Medical Informatics, 74,* 257–265. doi:10.1016/j.ijmedinf.2004.03.008

De Meyer, F., Claerhout, B., & De Moor, G. J. E. (2002). The PRIDEH project: taking up privacy protection services in e-health. In *Proceedings MIC 2002 "Health Continuum and Data Exchange"* (pp. 171-177). Amsterdam: IOS Press.

De Moor, G. J. E., Claerhout, B., & De Meyer, F. (2003). Privacy enhancing techniques: the key to secure communication and management of clinical and genomic data. *Methods of Information in Medicine, 42,* 148–153.

Dritsas, S., Gymnopoulos, L., Karyda, M., Balopoulos, T., Kokolakis, S., Lambrinoudakis, C. & Katsikas, S. (2006). A knowledge-based approach to security requirements for e-health applications, ejeta. *The electronic Journal of E-commerce Tools and Applications, 2*(1).

Gritzalis, S., Iliadis, J., Gritzalis, D., Spinellis, D., & Katsikas, S. (1999). Developing Secure Web-based Medical Applications. *Medical Informatics journal, 24*(1), 75-90.

Narasimhan, V. L., & Croll, P. R. (2008). Communicating Security Policies to Trusted e-Health Information Systems: A Specification Process based Approach. In R. Merrell, R. A. Cooper (Ed.) *Proceedings of Telehealth and Assistive Technologies (TeleHealth/AT 2008).*

Podichetty, V.K, Biscup, R.S. (2003). E-Health: A New Approach in Healthcare Practice. *The Internet Journal of Allied Health Sciences and Practice, 1*(2).

Schumacher, M., Fernandez, B. E., Hybertson, D., Buschmann, F., & Peter Sommerlad, P. (2006). *Security Patterns: Integrating Security and Systems Engineering.* Hoboken, NJ: Wiley.

Shepperd, S., Charnock, D., & Gann, B. (1999). Helping patients access high-quality health information. *BMJ (Clinical Research Ed.), 319,* 764–766.

Siponen, M. (2005). Analysis of modern IS security development approaches: towards the next generation of social and adaptable ISS methods. *Information and Organization, 15*(4), 339–375. doi:10.1016/j.infoandorg.2004.11.001

KEY TERMS AND DEFINITIONS

Availability: The prevention of unauthorised withholding of information.

Confidentiality: The prevention of unauthorised disclosure of information.

ePHI (electronic Protected Health Information): Information specifically identifying a person that is, stored electronically or sent or shared electronically.

Health Care: Care, services, or supplies related to the health of an individual. Health care includes, but is not limited to, the following: (a) preventive, diagnostic, therapeutic, rehabilitative, maintenance, or palliative care, and counselling, service, assessment, or procedure with respect to the physical or mental condition, or functional status, of an individual or that affects the structure or function of the body, and (b) sale or dispensing of a drug, device equipment, or other item in accordance with a prescription.

Health Information: The term 'health information' means any information, whether oral or recorded in any form or medium, that is created or received by a health care provider, health plan, public health authority, employer, life insurer, school or university, or health care clearinghouse, and relates to the past, present, or future physical or mental health or condition of an individual, the provision of health care to an individual, or the past, present, or future payment for the provision of health care to an individual.

HIPAA: The Health Insurance Portability and Accountability Act of 1996.

Information Security: A collection of policies, procedures and safeguards that help maintain the integrity and availability of information systems and controls access to their contents.

Integrity: The prevention of unauthorised modification of information. It is important to maintain information integrity as any changes in data can have an impact on healthcare.

Security Policy: The framework within which an organization establishes needed levels of information security to achieve the desired confidentiality goals. A policy is a statement of information values, protection responsibilities, and organization commitment for a system. The American Health Information Management Association recommends that security policies apply to all employees, medical staff members, volunteers, students, faculty, independent contractors, and agents.

Compilation of References

A. Tirado-Ramos, P.M.A. Sloot, A.G. Hoekstra, M. Bubak, (2004), "An Integrative Approach to High-Performance Biomedical Problem Solving Environments on the Grid", Parallel Computing, Special issue on High-Performance Parallel Bio-computing.

Abbott, L. F., & Dayan, P. (2001). *Theoretical neuroscience: computational and mathematical modeling of neural systems*. Cambridge, MA: MIT Press.

Abowd, G. D., Bobick, I., Essa, I., Mynatt, E., & Rogers, W. (2002). *The Aware Home: Developing Technologies for Successful Aging*. Paper presented at the AAAI-02 Workshop on Automation as Caregiver: The Role of Intelligent Technology in Elder Care, Edmonton, Alberta, Canada.

Abrahams, J. P. (1993). *Compression of X-ray images*. Jt CCP4 ESF–EACBM Newsl. *Protein Crystallogr., 28*, 3–4.

Abrahams, J. P., & Leslie, A. G. W. (1996). Methods used in the structure determination of bovine mitochondrial F_1 ATPase. *Acta Crystallographica. Section D, Biological Crystallography, 52*, 30–42. doi:10.1107/S0907444995008754

Abramson, D., Giddy, J., & Kotler, L. (2000), "High Performance Parametric Modeling with Nimrod/G: Killer Application for the Global Grid?" IPDPS'2000, Mexico, USA.

Adams, P. D., Grosse-Kunstleve, R. W., Hung, L. W., Ioerger, T. R., McCoy, A. J., & Moriarty, N. W. (2002). PHENIX: building new software for automated crystallographic structure determination. *Acta Crystallographica. Section D, Biological Crystallography, 58*, 1948–1954. doi:10.1107/S0907444902016657

Agostini, A., Bettini, C., & Riboni, D. (2006). Experience report: Ontological Reasoning for Context-Aware Internet Services. In *the 4th Annual IEEE International Conference on Pervasive Computing and Communications* (pp. 5-12). Pisa, Italy: IEEE.

Aiello, L., Callerano, J., Gardner, T., King, D., Blankenship, G., Ferris, F., & Klein, R. (1998). Diabetic retinopathy. *Diabetes Care, 21*, 143–156.

Akerstrom, B., Flower, D. R., & Salier, J. P. (2000). Lipocalins: unity in diversity. *Biochimica et Biophysica Acta, 1482*, 1–8.

Ala-Korpela, M., Changani, K.K., Hiltunen, Y., Bell, J.D., & Fuller, B.J., Bryant, et al. (1997). Assessment of quantitative artificial neural network analysis in a metabolically dynamic ex vivo 31P NMR pig liver study. *Magnetic Resonance in Medicine, 38*, 840–844. doi:10.1002/mrm.1910380522

Allen, J., Oates, C., Chishti, A., Ahmed, I., Talbot, D., & Murray, A. (2006). Thermography and colour duplex ultrasound assessments of arterio-venous fistula function in renal patients. *Physiological Measurement, 27*, 51–60. doi:10.1088/0967-3334/27/1/005

Allen, J., Oates, C., Chishti, A., Schaefer, G., Zhu, S., Ahmed, I., et al. (2005). Renal fistula assessment using automated thermal imaging. *3rd European Medical and Biological Engineering Conference*.

Aloisio, G., Cafaro, M., Fiore, S., & Mirto, M. (2006), "A Split & Merge Data Management Architecture for a Grid Environment", 19th IEEE Symposium on Computer-Based Medical Systems.

Alzheimer's Association. (2009). *What is Alzheimer's*. Retrieved from http://www.alz.org/alzheimers_disease_what_is_alzheimers.asp

Alzheimer's disease. (2009). Retrieved from http://en.wikipedia.org/wiki/Alzheimer%27s_disease

Amouh, T., Gemo, M., Macq, B., Vanderdonckt, J., Wahed, A., & Reynaert, M. S. (2005). Versatile Clinical Information System Design for Emergency Departments. *IEEE Transactions on Information Technology in Biomedicine*, *9*(2), 174–183. doi:10.1109/TITB.2005.847159

Anbar, N., Milescu, L., Naumov, A., Brown, C., Button, T., Carly, C., & AlDulaimi, K. (2001). Detection of cancerous breasts by dynamic area telethermometry. *IEEE Engineering in Medicine and Biology Magazine*, *20*(5), 80–91. doi:10.1109/51.956823

Anderson, R. M. (1988). The epidemiology of HIV infection: Variable incubation plus infectious periods and heterogeneity in sexual activity. *Journal of the Royal Statistical Society. Series A, (Statistics in Society)*, *151*(1), 66–98. doi:10.2307/2982185

Antoniadis G., Kofteros S., (2004). *e-OpenDay Project* [Technical Report].

Archibald Tang, B. (2004). *What is pervasive computing?* Retrieved October 24, 2008, from http://searchnetworking.techtarget.com/sDefinition/0,sid7_gci759337,00.html

Ardaiz, L., Diaz de Cerio, L., Gallardo, A., Messeguer, R., & Sanjeevan, K. (2004), "ULabGrid Framework for Computationally Intensive Remote and Collaborative Learning Laboratories", IEEE International Symposium on Cluster Computing and the Grid.

Atala, A. (2004). Tissue engineering and regenerative medicine: concepts for clinical application. *Rejuvenation Research*, *7*, 15–31. doi:10.1089/154916804323105053

Audet, J. (2004). Stem cell bioengineering for regenerative medicine. *Expert Opinion on Biological Therapy*, *4*, 631–644. doi:10.1517/14712598.4.5.631

Augusto, J. C., & Nugent, C. D. (2004). A New Architecture for Smart Homes Based on ADB and Temporal Reasoning. In D. Zhang, M. Mokhtari (Eds.), *Toward a Human Friendly Assistive Environment (Proceedings of 2nd International Conference on Smart homes and health Telematic, ICOST2004), Assistive Technology Research Series* (Vol. 14, pp. 106-113). Singapore: IOS Press.

Augusto, J. C., Nugent, C., Martin, S., & Olphert, C. (2005). Software and Knowledge Engineering Aspects of Smart Homes Applied to Health. In C. D. Nugent, P. J. McCullagh, E. T. McAdams, & A. Lymberis (eds.), *Personalised Health Management Systems - The Integration of Innovative Sensing, Textile, Information and Communication Technologies, Studies in Health Technology and Informatics* (Vol. 117, pp. 164-171). Singapore: IOS Press.

Bachmair, A., Finley, D., & Varshavsky, A. (1986). In vivo half-life of a protein is a function of its amino-terminal residue. *Science*, *234*(4773), 179–186. doi:10.1126/science.3018930

Bacon, D. J., & Anderson, W. F. (1988). A fast algorithm for rendering space-filling molecule pictures. *Journal of Molecular Graphics*, *6*, 219–220. doi:10.1016/S0263-7855(98)80030-1

Baldazzi, V., Castiglione, F., & Bernaschi, M. (2006). An enhanced agent based model of the immune system response. *Cellular Immunology*, *244*, 77–79. doi:10.1016/j.cellimm.2006.12.006

Bardram, J. (2003). *Hospitals of the future—ubiquitous computing support for medical work*. In the *2nd international workshop on ubiquitous computing for pervasive healthcare applications* (Ubihealth 2003 proceedings), Seattle, WA.

Bardram, J. (2004). Applications of context-aware computing in hospital work—examples and design principles. In the *2004 ACM symposium on Applied computing*. Nicosia, Cyprus: ACM.

Barkhuus, L., & Dey, A. (2003). *Is context-aware computing taking control away from the user? Three levels of interactivity examined*. Paper presented at the UbiComp2003 conference (pp. 149-156), Seattle, WA, USA.

Barrett, P., Hunter, J., Miller, J. T., Hsu, J.-C., & Greenfield, P. (2005). Matplotlib - A portable Python plotting package. *Astronomical Data Analysis Software and Systems XIV*, *347*, 91–95.

Baruchi, I., & Ben-Jacob, E. (2007). Towards neuro-memory-chip: Imprinting multiple memories in cultured neural networks. *Physical Review E: Statistical, Nonlin-*

ear, and Soft Matter Physics, 75, 050901. doi:10.1103/PhysRevE.75.050901

Bassett, L. W. (1992). Mammographic analysis of calcifications. *Radiologic Clinics of North America, 30*(1), 93–105.

Baxevanis, A. D., & Francis Ouellete, B. F. (2001). *Bioinformatics: Practical Guide to the Analysis of Genes and Proteins*. Hoboken, NJ: Wiley Interscience Press.

Belliveau, J. W., Kennedy, D. N. Jr, McKinstry, R. C., Buchbinder, B. R., Weisskoff, R. M., & Cohen, M. S. (1991). Functional mapping of the human visual cortex by magnetic resonance imaging. *Science, 254*(5032), 716–719. doi:10.1126/science.1948051

Benvenuto, N., & Piazza, F. (1992). On the complex backpropagation algorithm. *IEEE Transactions on Signal Processing, 40*(4), 967–969. doi:10.1109/78.127967

Benyoussef, A., HafidAllah, N. E., ElKenz, A., Ez-Zahraouy, H., Loulidi, M. (2003). Dynamics of HIV infection on 2D cellular automata. *Physica A, 322*, 506–520. doi:10.1016/S0378-4371(02)01915-5

Berg, W. A., D'Orsi, C. J., Jackson, V. P., Bassett, L. W., Beam, C. A., Lewis, R. S., & Crewson, P. E. (2002). Does training in the Breast Imaging Reporting and Data System (BI-RADS) improve biopsy recommendations or feature analysis agreement with experienced breast imagers at mammography? *Radiology, 224*(3), 871–880. doi:10.1148/radiol.2243011626

Berlekamp, E. R., Conway, J. H., & Guy, R. K. (2004). *Winning Ways for your Mathematical Plays* (2nd edition). Wellesley, MA: A. K. Peters Ltd.

Berman, H. M. (2008). The Protein Data Bank: a historical perspective. *Acta Crystallographica. Section A, Foundations of Crystallography, 64*, 88–95. doi:10.1107/S0108767307035623

Berman, H. M., Westbrook, J., Feng, Z., Gilliland, G., Bhat, T. N., & Weissig, H. (2000). The Protein Data Bank. *Nucleic Acids Research, 28*, 235–242. doi:10.1093/nar/28.1.235

Bernaschi, M., & Castiglione, F. (2001). Design and implementation of an immune system simulator. *Computers in Biology and Medicine, 31*, 303–331. doi:10.1016/S0010-4825(01)00011-7

Berry, M. J. II, Warland, D. K., & Meister, M. (1997). The structure and precision of. retinal spike trains. *Proceedings of the National Academy of Sciences of the United States of America, 94*, 5411–5416. doi:10.1073/pnas.94.10.5411

Beste, G., Schmidt, F. S., Stibora, T., & Skerra, A. (1999). Small antibody-like proteins with prescribed ligand specificities derived from the lipocalin fold. *Proceedings of the National Academy of Sciences of the United States of America, 96*, 1898–1903. doi:10.1073/pnas.96.5.1898

Bethanis, K., Tzamalis, P., Hountas, A., & Tsoucaris, G. (2002). Ab initio determination of a crystal structure by means of the Schrödinger equation. *Acta Crystallographica. Section A, Foundations of Crystallography, 58*, 265–269. doi:10.1107/S0108767302003781

Bethanis, K., Tzamalis, P., Hountas, A., & Tsoucaris, G. (2008). Convergence study of a Schrödinger-equation algorithm and structure-factor determination from the wavefunction, *Acta Crystallographica. Section A, Foundations of Crystallography, 64*, 450–458. doi:10.1107/S0108767308010416

Bethanis, K., Tzamalis, P., Hountas, A., Mishnev, A. F., & Tsoucaris, G. (2000). Upgrading the twin variables algorithm for large structures. *Acta Crystallographica. Section A, Foundations of Crystallography, 56*, 105–111. doi:10.1107/S0108767399013355

Bharucha, A. J. (2006). *CareMedia: Automated Video and Sensor Analysis for Geriatric Care*. Presented at the Geriatric Care Annual Meeting of the American Association for Geriatric Psychiatry, San Juan, Puerto Rico.

Bhatia, S. N., & Chen, C. S. (1999). Tissue engineering at the micro scale. *Biomedical Microdevices, 2*, 131–144. doi:10.1023/A:1009949704750

Birney, E. (2001). Hidden Markov Models in Biological Sequence Analysis. *IBM Journal of Research and Development, 45*(3/4).

Bishop, C. M. (1994). Novelty Detection and Neural Network Validation. *IEE Proceedings. Vision Image and Signal Processing, 141*, 217–222.

Bishop, C. M. (1995). *Neural Networks for Pattern Recognition*. Oxford, UK: Clarendon Press.

Bishop, C. M. (2002). *Neural networks for pattern recognition*. Oxford, UK: Oxford University Press.

Bishop, R. E. (2000). The bacterial lipocalins. *Biochimica et Biophysica Acta, 1482*, 73–83.

Blobel, B. (2007). Comparing approaches for advanced e-health security infrastructures. *International Journal of Medical Informatics, 76* (5 - 6), 454 - 459.

Blow, D. (2002), *Outline of Crystallography for Biologists*, Oxford University Press

Bocchi, L., Coppini, G., Nori, J., & Valli, G. (2004). Detection of single and clustered microcalcifications in mammograms using fractals models and neural networks. *Medical Engineering & Physics, 26*(4), 303–312. doi:10.1016/j.medengphy.2003.11.009

Boddupalli, P., Al-Bin-Ali, F., Davies, N., Friday, A., Storz, O., & Wu, M. (2003). Payment support in ubiquitous computing environments. In the *5th Workshop on Mobile Computing Systems and Applications (WMCSA)*. Monterey, CA: IEEE.

Bohn, J., Coroama, V., Langheinrich, M., Mattern, F., & Rohs, M. (2005). Social, Economic, and Ethical Implications of Ambient Intelligence and Ubiquitous Computing. In W. Weber, J. M. Rabaey, E. Aarts (eds.) *Ambient Intelligence* (pp 5-29). Berlin: Springer.

Bohringer, C., & Rutherford, T. F. (2008). Combining bottom-up and top-down. *Energy Economics, 30*(2), 574–596. doi:10.1016/j.eneco.2007.03.004

Bork, P., Holm, L., & Sander, C. (1994). The immunoglobulin fold structural classification, sequence patterns and common core. *Journal of Molecular Biology, 242*, 309–320.

Boss, G., Malladi, P., Quan, D., Legregni, L., & Hall, H. (2007), "Cloud Computing", http://www.ibm.com/developerworks/websphere/zones/hipods/

Bott, O. J. (2004). Electronic Health Record: Standardization and Implementation. In *2nd OpenECG Workshop*, Berlin, Germany, (pp. 57-60).

Bourenkov, G. P., & Popov, A. N. (2006). A quantitative approach to data-collection strategies. *Acta Crystallographica, D62*, 58–64.

Brehm, M., Pinto, A., Daniels, K., Schneck, J., Welsh, R., & Selin, L. (2002). T cell immunodominance and maintenance of memory regulated by unexpectedly cross-reactive pathogens. *Nature Immunology, 3*, 627–634.

Bricogne, G. (1988). *A Baysian statistical theory of the phase problem. i. a multichannel maximum entropy* formalism for constructing generalised joint probability distributions of structure factors. *Acta Crystallographica. Section A, Foundations of Crystallography, 44*, 517–545. doi:10.1107/S010876738800354X

Bricogne, G. (1997a). Ab initio macromolecular phasing: a blueprint for an expert system based on structure factor statistics with built in stereochemistry. *Methods in Enzymology, 277*, 14–19. doi:10.1016/S0076-6879(97)77004-6

Bricogne, G. (1997b). Efficient sampling methods for combinations of signs, phases, hyperphases, and molecular orientations. *Methods in Enzymology, 276*, 424–448. doi:10.1016/S0076-6879(97)76070-1

Bricon-Souf, N., & Newman, C. R. (2007). Context-awareness in health care: A review. *International Journal of Medical Informatics, 76*(1), 2–12. doi:10.1016/j.ijmedinf.2006.01.003

Brooke, J. (1996). *SUS: a "quick and dirty" usability scale.*

Brown, A. S. (2002). Consolidation theory and retrograde amnesia in humans. *Psychonomic Bulletin & Review, 9*, 403–425.

Brünger, A. T., Adams, P. D., Clore, G. M., DeLano, W. L., Gros, P., & Grosse-Kunstleve, R. W. (1998). Crystallography & NMR System (CNS): a new software suite for macromolecular structure determination. *Acta Crystallographica. Section D, Biological Crystallography, 54*, 905–921. doi:10.1107/S0907444998003254

Buchbinder, S. S., Leichter, I. S., Lederman, R. B., Novak, B., Bamberger, P. N., Coopersmith, H., & Fields, S. I. (2002). Can the size of microcalcifications predict malignancy of clusters at mammography? *Academic Radiology, 9*(1), 18–25. doi:10.1016/S1076-6332(03)80293-3

Burla, M. C., Caliandro, R., Camalli, M., Carrozzini, B., Cascarano, G. L., & De Caro, C. (2007). IL MILIONE: a suite of computer programs for crystal structure solution of proteins. *Journal of Applied Crystallography, 40,* 609–613. doi:10.1107/S0021889807010941

Burla, M. C., Camalli, M., Carrozzini, B., Cascarano, G. L., De Caro, C., & Giacovazzo, C. (2005). SIR2004: an improved tool for crystal structure determination and refinement. *Journal of Applied Crystallography, 38,* 381–388. doi:10.1107/S002188980403225X

Burnett, M. N., & Johnson, C. K. (1996), *ORTEPIII: Oak Ridge thermal ellipsoid plot program for crystal structure illustrations.* Report ORNL-6895. Oak Ridge National Laboratory, Tennessee, USA.

Burns, J. (2005). *Emergent networks in immune system shape space.* PhD thesis, Dublin City University, School of Computing.

Burns, J., & Ruskin, H. J. (2004). *Network topology in immune system shape space (. LNCS, 3038,* 1094–1101.

Cademartiri, F., Mollet, N., Nieman, K., Krestin, G. P., & de Feyter, P. J. (2003). Images in cardiovascular medicine. Neointimal hyperplasia in carotid stent detected with multislice computed tomography. *Circulation, 108*(21), e147. doi:10.1161/01.CIR.0000103947.77551.9F

Cahill, R. N. P., Frost, H., & Trnka, Z. (1976). The effects of antigen on the migration of recirculating lymphocytes through single lymph node. *The Journal of Experimental Medicine, 143,* 870–888. doi:10.1084/jem.143.4.870

Cammarata, S., McArthur, D., & Steeb, R. (1983). Strategies of cooperation in distributed problem solving. In *Proceedings of the Eighth International Joint Conference on Artificial Intelligence (IJCAI-83),* Karlsruhe, Germany.

Cancer Facts and Figures. (2004). American Cancer Society. Retrieved from http://www.cancer.org/downloads/STT/CAFF_finalPWSecured.pdf

Carter, D. R., & Beaupré, G. S. (2001). *Skeletal Function and Form: Mechanobiology of Skeletal Development, Aging, and Regeneration.* Cambridge, UK: Cambridge University Press.

Castiglione, F., Poccia, F., D'Offizi, G., & Bernaschi, M. (2004). Mutation, fitness, viral diversity, and predictive markers of disease progression in a computational model of HIV type 1 infection. *AIDS Research and Human Retroviruses, 20*(12), 1314–1323. doi:10.1089/aid.2004.20.1314

Castrillo, J. I., Hayes, A., Mohammed, S., Gaskell, S. J., & Oliver, S. G. (2003). An optimized protocol for metabolome analysis in yeast using direct infusion electrospray mass spectrometry. *Phytochemistry, 62,* 929–937.

Celada, F., & Seiden, P. E. (1992). A computer model of cellular interactions in the immune system. *Immunology Today, 13*(2), 56–62. doi:10.1016/0167-5699(92)90135-T

Chan, H. P., Lo, S.-C., Sahiner, B., Lam, K. L., & Helvie, M. A. (1995). Computer-aided detection of mammographic microcalcifications: pattern recognition with an artificial neural network. *Medical Physics, 22,* 1555–1567. doi:10.1118/1.597428

Chang, Y. H., Zheng, B., Good, W. F., & Gur, D. (1998). Identification of clustered microcalcifications on digitized mammograms using morphology and topography-based computer-aided detection schemes. A preliminary experiment. *Investigative Radiology, 33*(10), 746–751. doi:10.1097/00004424-199810000-00006

Chapekar, M. S. (1996). Regulatory concerns in the development of biologic-biomaterial combinations. *Journal of Biomedical Materials Research, 33,* 199–203. doi:10.1002/(SICI)1097-4636(199623)33:3<199::AID-JBM10>3.0.CO;2-C

Chapekar, M. S. (2000). Tissue engineering: challenges and opportunities. *Journal of Biomedical Materials Research, 53,* 615–620. doi:10.1002/1097-4636(2000)53:6<617::AID-JBM1>3.0.CO;2-C

Cheah, C. M., Chua, C. K., Leong, K. F., Cheong, C. H., & Naing, M. W. (2004). Automatic algorithm for generating complex polyhedral scaffold structures for tissue engineering. *Tissue Engineering, 10,* 595–610. doi:10.1089/107632704323061951

Chen, R., Enberg, G., & Klein, G. (2007). Julius--a template based supplementary electronic health record system. *BMC Medical Informatics and Decision Making, 7*–10.

Cheverst, K., Davies, N., & Michell, K. (2002). *A reflective study of the GUIDE system.* Paper presented at the 1st workshop on mobile tourism support (pp.17-23).

Cheverst, K., Mitchell, K., & Davies, N. (1999). Design of an object model for a context sensitive tourist GUIDE. *Computers & Graphics, 23*(6), 883–891. doi:10.1016/S0097-8493(99)00119-3

Chew, S. J., Cheng, H. M., Lam, D. S. C., Cheng, A. C. K., Leung, A. T. S., & Chua, J. K. H. (1998). OphthWeb-cost-effective telemedicine for ophthalmology. *HKMJ*, *4*(3), 300–304.

Cheynier, R., Henrichwark, S., Hadida, F., Pelletier, E., Oksenhendler, E., Autran, B., & Wain-Hobson, S. (1994). HIV and T cell expansion in splenic white pulps is accompanied by infiltration of HIV-specific cytotoxic T lymphocytes. *Cell*, *78*(3), 373–387. doi:10.1016/0092-8674(94)90417-0

Choi, J., Yoo, S., Park, H., & Chun, J. (2006). MobileMed: A PDA-Based Mobile Clinical Information System. *IEEE Transactions on Information Technology in Biomedicine*, *10*(3), 627–635. doi:10.1109/TITB.2006.874201

Chronaki, E., Lelis, P., Demou, C., Tsiknakis, M., & Orphanoudakis, S. C. (2001). An HL7/CDA framework for the design and deployment of Telemedicine services. *Proceedings of the 23ʳᵈ Annual EMBS International Conference*, October 25-28, Istanbul, Turkey, (pp. 3504-3507).

Chu, T. M. G., Halloran, J. W., Hollister, S. J., & Feinberg, S. E. (2001). Hydroxyapatite implants with designed internal architecture. *Journal of Materials Science. Materials in Medicine*, *12*, 471–478. doi:10.1023/A:1011203226053

Churchill, E. F., Nelson, L., & Denoue, L. (2003). Multimedia fliers: Information sharing with digital community bulletin boards. In *Proceedings of the Communities and Technologies* (pp. 19-21). Amsterdam: Kluwer.

Ciardelli, G., Chiono, V., Cristallini, C., Barbani, N., Ahluwalia, A., & Vozzi, G. (2004). Innovative tissue engineering structures through advanced manufacturing technologies. *Journal of Materials Science. Materials in Medicine*, *15*, 305–310. doi:10.1023/B:JMSM.0000021092.03087.d4

Ciechanover, A.L., Schwartz. (1989). How are substrates recognized by the ubiquitin-mediated proteolytic system? *Trends in Biochemical Sciences*, *14*(12), 483–488. doi:10.1016/0968-0004(89)90180-1

Claerhout, B., & DeMoor, G. J. E. (2005). Privacy protection for clinical and genomic data: The use of privacy-enhancing techniques in medicine. *International Journal of Medical Informatics*, *74*, 257–265. doi:10.1016/j.ijmedinf.2004.03.008

Clos, A., Schaefer, G., & Nolle, L. (2007). Exudate detection in eye digital fundus images using neural networks. In *13ᵗʰ Int. MENDEL Conference on Soft Computing*, (pp. 121-127).

Clote, P. (2000). *Computational Molecular Biology*. Hoboken, NJ: Wiley.

Collaborative Computational Project Number 4 *Acta Cryst.* (1994), *D50*, pp. 760-763.

Constable, I., Yogesan, K., Eikelboom, R., Barry, C., & Cuypers, M. (2000). Fred Hollows lecture: digital screening for eye disease. *Clin. Exp. Ophthalmol.*, *28*, 129–132. doi:10.1046/j.1442-9071.2000.00309.x

Cooley, J. W., & Tukey, J. W. (1965). An algorithm for machine calculation of complex Fourier series. *Mathematics of Computation*, *19*(90), 297–301. doi:10.2307/2003354

Cooley, T., & Micheli-Tzanakou, E. (1998). Classification of Mammograms Using an Intelligent Computer System. *Journal of Intelligent Systems*, *8*(1/2), 1–54.

Cornette, J. L., Cease, K. B., Margalit, H., Spouge, J. L., Berzofsky, J. A., & DeLisi, C. (1987). Hydrophobicity scales and computational techniques for detecting amphipathic structures in proteins. *Journal of Molecular Biology*, *195*(3), 659–685. doi:10.1016/0022-2836(87)90189-6

Corning, P. A. (2002). The re-emergence of "emergence": a venerable concept in search of a theory. *Complexity*, *7*(6), 18–30. doi:10.1002/cplx.10043

Couderc, P., & Kermarrec, A.-M. (1999). Improving Level of Service for Mobile Users Using Context-Awareness. In *The 18th IEEE Symposium on Reliable Distributed Systems* (pp. 24-33). Lausanne, Switzerland: IEEE.

Cowtan, K. (1994). DM: an automated procedure for phase improvement by density modification. CCP4 ESF-EACBM Newsl. *Protein Crystallogr.*, *31*, 34–38.

Cowtan, K. (2003). An Overview of some developments in Crystallographic Computing Methods worldwide. *Crystallography Reviews, Vol.*, *9*(1), 73–80. doi:10.1080/0889311031000069326

Crowther, J. R. (Ed.). (1995). *ELISA: Theory and Practice*. Totowa, NJ: Humana Press.

Crutchfield, J. P. (1994). The calculi of emergence: computation, dynamics and induction. *Physica D. Nonlinear Phenomena, 75*(1-3), 11–54. doi:10.1016/0167-2789(94)90273-9

Dauter, Z., Wilson, K. S., Sieker, L. C., Meyer, J., & Moulis, J. M. (1997). Atomic resolution (0.94Å) structure of Clostridium acidurici ferredoxin. Detailed geometry of [4Fe-4S] clusters in a protein. *Biochemistry, 36,* 16065–16073. doi:10.1021/bi972155y

David Sprott and Lawrence Wilkes. (2004), "Understanding Service-Oriented Architecture", CBDI Forum

David, H.A., Hinton, G.E. & Sejnowski, T.J. (1985). A Learning Algorithm for Boltzmann Machines. *Cognitive Science: A Multidisciplinary Journal, 9*(1), 149-169.

Dayhoff, J. E. (1989). *Neural Network Architectures: An Introduction.* New York: Van Nostrand Reinhold.

de La Fortelle, E., & Bricogne, G. (1997). Maximum-likelihood heavy-atom parameter refinement in the MIR and MAD methods. *Methods in Enzymology, 276,* 472–494. doi:10.1016/S0076-6879(97)76073-7

De Meyer, F., Claerhout, B., & De Moor, G. J. E. (2002). The PRIDEH project: taking up privacy protection services in e-health. In *Proceedings MIC 2002 "Health Continuum and Data Exchange"* (pp. 171-177). Amsterdam: IOS Press.

De Moor, G. J. E., Claerhout, B., & De Meyer, F. (2003). Privacy enhancing techniques: the key to secure communication and management of clinical and genomic data. *Methods of Information in Medicine, 42,* 148–153.

De Wolf, T., & Holvoet, T. (2005). *Emergence versus Self-Organisation: different concepts but promising when combined.* (. LNCS, 3464, 1–15.

DeGrado, W. F. (1997). PROTEIN DESIGN: Enhanced: Proteins from Scratch. *Science, 278,* 80–81. doi:10.1126/science.278.5335.80

Dengler, J., Behrens, J., & Desaga, J. F. (1993). Segmentation of microcalcifications in mammograms. *IEEE Transactions on Medical Imaging, 12,* 634–642. doi:10.1109/42.251111

Derome, A. E. (1987). *Modern NMR Techniques for Chemistry Research (Organic Chemistry Series, Vol 6).* Oxford, UK: Pergamon Press.

Des Higgins & Taylor. W. (2000). *Bioinformatics: Sequence, structure, and databanks.* Oxford, UK: Oxford University Press.

Desjarlais, J. R., & Mayo, S. L. (2002). Computational protein design. *Current Opinion in Structural Biology, 12,* 429–430. doi:10.1016/S0959-440X(02)00343-3

Dey, G. O., & Abowd, G. D. (1999). *Towards a Better Understanding of Context and Context-Awareness.* Paper presented at the 1st International Symposium of Handheld and Ubiquitous Computing (pp. 304-307), Karlsruhe, Germany.

Diestel, R. (2005). *Graph Theory* (Graduate Texts in Mathematics, Volume 173). New York: Springer.

Dionysiou, D., Stamatakos, G., & Uzunoglu, N. (2004). A four-dimensional simulation model of tumour response to radiotherapy in vivo: parametric validation considering radiosensitivity, genetic profile and fractionation. *Journal of Theoretical Biology, 230*(Issue 1), 1–20. doi:10.1016/j.jtbi.2004.03.024

Dodson, E. (2003). Is it Jolly SAD? *Acta Crystallographica. Section D, Biological Crystallography, 59,* 1958–1965. doi:10.1107/S0907444903020936

Doi, K., Giger, M. L., Nishikawa, R. M., & Schmidt, R. A. (1997). Computer-Aided Diagnosis of Breast Cancer on Mammograms. *Breast Cancer (Tokyo, Japan), 4*(4), 228–233. doi:10.1007/BF02966511

Doig, A. J., (n.d.). *Protein Engineering - Introduction & Lecture Summaries.* 2PAB: Physical and Analytical Biochemistry Lectures, Department of Biomolecular Sciences, UMIST.

Dolin, R. H., Alschuler, L., Boyer, S., Behlen, F. M., Biron, P. V., & Shabo, A. (2006). HL7 Clinical Document Architecture, Release 2. *Journal of the American Medical Informatics Association, 13,* 30–39. doi:10.1197/jamia.M1888

Drenth, J. (1994), *Principles of Protein X-ray Crystallography,* Springer-Verlag

Dritsas, S., Gymnopoulos, L., Karyda, M., Balopoulos, T., Kokolakis, S., Lambrinoudakis, C. & Katsikas, S. (2006). A knowledge-based approach to security requirements for e-health applications, ejeta. *The electronic Journal of E-commerce Tools and Applications, 2*(1).

Ducruix, A., & Giece, R. (1999), *Crystallization of Nucleic Acids and Proteins*, Oxford University Press

Duda, O. R., Hart, P. E., & Stork, D. G. (2000). *Pattern Classification*. New York: Wiley-Interscience.

Durfee, E. H. (1998). *Coordination of distributed problem solvers*. Boston: Kluwer Academic Publishers.

Eaglstein, W. H., & Falanga, V. (1997). Tissue engineering and the development of Apligraf: a human skin equivalent. *Clinical Therapeutics, 19*, 894–905. doi:10.1016/S0149-2918(97)80043-4

Ealick, S. E. (2000). Advances in multiple wavelength anomalous diffraction crystallography. *Current Opinion in Chemical Biology, 4*(5), 495–499. doi:10.1016/S1367-5931(00)00122-8

Eckmann, J.P., Shimshon, Jacobi, S., Marom, S., Moses, E., Zbinden, C. (2008). Leader neurons in population bursts of 2D living neural networks. *New Journal of Physics, 10*, 19. doi:10.1088/1367-2630/10/1/015011

Edwards, A. M., Arrowsmith, C. H., Christendat, D., Dharamsi, A., Friesen, J. D., Greenblatt, J. F., & Vedadi, M. (2000), *Protein production: feeding the crystallographers and NMR spectroscopists*, Nature Structural Biology, structural genomics supplement, Nov., pp. 970 - 972.

Eggermont, J. J., & Smith, G. M. (1996). Burst-firing sharpens frequency-tuning in primary auditory cortex. *Neuroreport, 7*, 753–757. doi:10.1097/00001756-199602290-00018

Eichelberg, M., Aden, T., & Riesmeier, J. (2005). A survey and Analysis of Electronic Healthcare Record Standards. *ACM Computing Surveys, 5*(14), 1–47.

Eisenhaber, F., Imperiale, F., Argos, P., & Froemmel, C. (1996). Prediction of Secondary Structural Content of Proteins from Their Amino Acid Composition Alone. *New Analytic Vector Decomposition Methods, Proteins, Struct., Funct. Design, 25*(2), 157–168.

El-Bakry, H. M. (2006). New Fast Time Delay Neural Networks Using Cross Correlation Performed in the Frequency Domain. *Neurocomputing Journal, 69*, 2360–2363. doi:10.1016/j.neucom.2006.03.005

El-Bakry, H. M., & Zhao, Q. (2005). Fast Time Delay Neural Networks. *International Journal of Neural Systems, 15*(6), 445–455. doi:10.1142/S0129065705000414

Eliopoulos, E., & Mavridis, I. M. (1996). Molecular graphics approaches in structure prediction and determination. In G. Tsoucaris et al. (eds.), *Crystallography of Supramolecular Compounds* (pp. 491-498), NATO ASI Series: Mathematical and Physical Sciences.

Elmore, J. G., Armstrong, K., Lehman, C. D., & Fletcher, S. W. (2005). Screening for breast cancer. *Journal of the American Medical Association, 293*(10), 1245–1256. doi:10.1001/jama.293.10.1245

Elmore, J. G., Nakano, C. Y., Koepsell, T. D., Desnick, L. M., D'Orsi, C. J., & Ransohoff, D. F. (2003). International variation in screening mammography interpretations in community-based programs. *Journal of the National Cancer Institute, 95*(18), 1384–1393.

Emsley, P., Cowtan, K. (2004), *Coot: model-building tools for molecular graphics* Acta Cryst. D**60**, Part 12 Sp. Iss. 1, pp. 2126-2132

Endres, M., Hutmacher, D. W., Salgado, A. J., Kaps, C., Ringe, J., & Reis, R. L. (2003). Osteogenic induction of human bone marrow derived mesenchymal progenitor cells in novel synthetic polymerhydrogel matrices. *Tissue Engineering, 9*, 689–702. doi:10.1089/107632703768247386

Eng, T. R. (2001). *The eHealth Landscape: A Terrain Map of Emerging Information and Communication Technologies in Health and Health Care*. Princeton, NJ: The Robert Wood Johnson Foundation.

Esserman, L., Cowley, H., Eberle, C., Kirkpatrick, A., Chang, S., Berbaum, K., & Gale, A. (2002). Improving the accuracy of mammography: volume and outcome relationships. *Journal of the National Cancer Institute, 94*(5), 369–375.

Evans, P. R. (1993), *Data reduction*. In *Proceedings of the CCP4 study weekend. Data collection and processing*, edited by L. Sawyer, N. W. Isaacs & S. Bailey, pp. 114–122.

Evans, P. R. (1997), *Scaling of MAD data*. In *Proceedings of the CCP4 study weekend. Recent advances in phasing*, edited by M. Winn, Vol. **33**, pp. 22–24.

Ewens, W., & Grant, G. (2005). Statisical Methods in Bioinformatics: An Introduction. Berlin: Springer Science+Business Media, Inc.

Familydoctor.org Editorial Staff. (2006). *Memory loss with aging: What's normal, what's not*. Retrieved from http://familydoctor.org/online/famdocen/home/seniors/common-older/124.html

Fan, R., Ceded, L., & Toser, O. (2005). Java plus XML: a powerful new combination for SCADA systems. *Computing & Control Engineering Journal, 16*(5), 27–30. doi:10.1049/cce:20050505

Fang, Z., Starly, B., & Sun, W. (2005). Computer aided characterization for effective mechanical properties of porous tissue scaffolds. *Computer Aided Design, 37*, 65–72. doi:10.1016/j.cad.2004.04.002

Fano, A., & Gershman, A. (2002). The future of business services in the age of ubiquitous computing. *Communications of the ACM, 45*(12), 83–87. doi:10.1145/585597.585620

FDA. (2009). *For consumers. U.S. food and drug administration*. Retrieved from http://www.fda.gov/consumer/features/memoryloss0507.html

Fedyukin, I., Reviakin, Y. G., Orlov, O. I., Doarn, C. R., Harnett, D. M., & Merrell, R. C. (2002). Experience in the application of Java Technologies in telemedicine. *Ehealth International, 1*(3), 1–6.

Feinberg, D. A., & Oshio, K. (1991). GRASE (gradient and spin echo) MR imaging: A new fast clinical imaging technique. *Radiology, 181*, 597–602.

Fell, D. (1996). *Understanding the Control of Metabolism, (Frontiers in Metabolism series)*. Aldershot, UK: Ashgate Publishing.

Fiore, S., Mirto, M., Cafaro, M., Vadacca, S., Negro, A., & Aloisio, G. (2008), "A GRelC based Data Grid Management Environment," 21st IEEE International Symposium on Computer-Based Medical Systems.

Floerkemeier, C., & Siegemund, F. (2003). Improving the effectiveness of medical treatment with pervasive computing technologies. In *Proceedings of the UbiHealth 2003: The 2nd International Workshop on Ubiquitous Computing for Pervasive Healthcare Applications. Gloucestor smart house*. (n.d.). Bath Institute of Biomedical Engineering. Retrieved from http://www.bath.ac.uk/bime/projects/smart/index.htm

Flower, D. R. (1995). Multiple molecular recognition properties of the lipocalin protein family. *Journal of Molecular Recognition, 8*, 185–195. doi:10.1002/jmr.300080304

Flower, D. R. (1996). The lipocalin protein family: structure and function. *The Biochemical Journal, 318*, 1–14.

Fondrinier, E., Lorimier, G., Guerin-Boblet, V., Bertrand, A.F., Mayras, C., Dauver, N. (2002). Breast microcalcifications: multivariate analysis of radiologic and clinical factors for carcinoma. *World J Surg. Mar., 26*(3), 290-6.

Ford, M. D., Stuhne, G. R., Nikolov, H. N., Habets, D. F., Lownie, S. P., & Holdsworth, D. W. (2005). Virtual angiography for visualization and validation of computational models of aneurysm hemodynamics. *IEEE Transactions on Medical Imaging, 24*(12), 1586–1592. doi:10.1109/TMI.2005.859204

Ford, M. D., Xie, Y. J., Wasserman, B. A., & Steinman, D. A. (2008). Is flow in the common carotid artery fully developed? *Physiological Measurement, 29*(11), 1335–1349. doi:10.1088/0967-3334/29/11/008

Foster, I., & Kesselman, C. (1999), "The Grid: Blueprint for a Future Computing Infrastructure", Morgan Kaufmann Publishers, USA.

Foster, I., Kesselman, C., & Tuecke, S. (2001). The Anatomy of the Grid: Enabling Scalable Virtual Organizations. *The International Journal of Supercomputer Applications, 15*(3).

Freed, L. E., & Vunjak-Novakovic, G. (1998). Culture of organized cell communities. *Advanced Drug Delivery Reviews, 33*, 15–30. doi:10.1016/S0169-409X(98)00017-9

French, G. S., & Wilson, K. S. (1978). On the treatment of negative intensity observations. *Acta Crystallographica. Section A, Crystal Physics, Diffraction, Theoretical and General Crystallography, 34*, 517–525. doi:10.1107/S0567739478001114

Friedman-Hill, E. (2005). *Jess, the rule engine for the java platform*. Livermore, CA: Sandia National Laboratories.

Frith, M. (n.d.). *Worldwide Bioinformatics Centres.* Retrieved from http://zlab.bu.edu/~mfrith/BioinfoCenters.html

Fuchs, J. R., Nasseri, B. A., & Vacanti, J. P. (2001). Tissue engineering: a 21st century solution to surgical reconstruction. *The Annals of Thoracic Surgery, 72,* 577–591. doi:10.1016/S0003-4975(01)02820-X

Furey, W., & Swaminathan, S. (1997). PHASES-95: a program package for the processing and analysis of diffraction data from macromolecules. *Methods in Enzymology, 277,* 590–620. doi:10.1016/S0076-6879(97)77033-2

Furuie, S. S., Rebelo, M. S., Moreno, R. A., Santos, M., Bertozzo, N., & Mota, G. (2007). Managing Medical Images and Clinical Information: InCor's Experience. *IEEE Transactions on Information Technology in Biomedicine, 11*(1), 17–24. doi:10.1109/TITB.2006.879588

G. Aloisio, M. Cafaro, S. Fiore, M. Mirto, (2005), "ProGenGrid: A Workflow Service Infrastructure for Composing and Executing Bioinformatics Grid Services"

Gail, M. H., & Costantino, J. P. (2001). Validating and improving models for projecting the absolute risk of breast cancer. *Journal of the National Cancer Institute, 93*(5), 334–335. doi:10.1093/jnci/93.5.334

Gail, M. H., Brinton, L. A., Byar, D. P., Corle, D. K., Green, S. B., Schairer, C., & Mulvihill, J. J. (1989). Projecting individualized probabilities of developing breast cancer for white females who are being examined annually. *Journal of the National Cancer Institute, 81*(24), 1879–1886. doi:10.1093/jnci/81.24.1879

Gans, D., Kralewski, J., Hammons, T., & Dowd, B. (2006). Medical groups' adoption of electronic health records and information systems. *Health Affairs (Project Hope), 24*(5), 1323–1333. doi:10.1377/hlthaff.24.5.1323

Gardner, G. G., Keating, D., Williamson, T. H., & Elliott, A. T. (1996). Automatic detection of diabetic retinopathy using an artificial neural network: a screening tool. *The British Journal of Ophthalmology, 80*(11), 940–944. doi:10.1136/bjo.80.11.940

Gardner, M. (1970). Mathematical games: The fantastic combinations of John Conway's new solitaire game Life. *Scientific American, 223,* 120–123.

Gasteiger, E., Hoogland, C., Gattiker, A., Duvaud, S., Wilkins, M. R., & Appel, R. D. Bairoch. (2005). Protein Identification and Analysis Tools on the ExPASy Server. In J. M. Walker (Ed.), *The Proteomics Protocols Handbook* (pp. 571-607). New York: Humana Press.

Gavrielides, M. A., Lo, J. Y., & Floyd, C. E. Jr. (2002). Parameter optimization of a computer-aided diagnosis scheme for the segmentation of microcalcification clusters in mammograms. *Medical Physics, 29*(4), 475–483. doi:10.1118/1.1460874

Geman, S., & Geman, D. (1984). Stochastic Relaxation, Gibbs Distribution and the Bayesian Restoration of Images. *IEEE Transactions on Pattern Analysis and Machine Intelligence, 6,* 721–741. doi:10.1109/TPAMI.1984.4767596

Georgiou, G. M., & Koutsougeras, C. (1992). Complex domain backpropagation. *IEEE Transactions on Circuits and Systems. 2, Analog and Digital Signal Processing, 39*(5), 330–334. doi:10.1109/82.142037

Geourjon, C., & Deleage, G. (1994). SOPM: a self-optimized method for protein secondary structure prediction. *Protein Engineering, 7*(2), 157–164. doi:10.1093/protein/7.2.157

Giacovazzo, C. (1977). A general approach to phase relationships: the method of representations. *Acta Crystallographica. Section A, Crystal Physics, Diffraction, Theoretical and General Crystallography, 33,* 933–944. doi:10.1107/S0567739477002253

Giacovazzo, C. (1980). The method of representations of structure seminvariants. II. New theoretical and practical aspects. *Acta Crystallographica. Section A, Crystal Physics, Diffraction, Theoretical and General Crystallography, 36,* 362–372. doi:10.1107/S0567739480000836

Gill, R., Zayats, M., & Willner, I. (2008). Semiconductor quantum dots for bioanalysis. *Angewandte Chemie International Edition in English, 47*(40), 7602–7625. doi:10.1002/anie.200800169

Gill, S. C., & Von Hippel, P. H. (1989). Calculation of protein extinction coefficients from amino acid sequence data. *Analytical Biochemistry, 182*(2), 319–326. doi:10.1016/0003-2697(89)90602-7

Ginsburg, M. (2007). Pediatric Electronic Health Record Interface Design: The PedOne System. In *Proceedings of the 40th Hawaii International Conference on System Sciences*, (pp. 1-10).

Glassner, O. (1992). *Wilhelm Conrad Roentgen and the early history of the Roentgen rays*. San Francisco: Norman Publishing.

Glatard, T., Montagnat, J., & Pennec, X. (2005), "Grid-enabled workflows for data intensive medical applications", Computer Based Medical Systems, Special Track on Grids for Biomedicine and Bioinformatics.

Goatman, K. A., Whitwam, A. D., Manivannan, A., Olson, J. A., & Sharp, P. F. (2003). Colour normalisation of retinal images. In *Medical Image Understanding and Analysis*.

Goertzel, B. (1992). Self-organizing evolution. *Journal of Social and Evolutionary Systems, 15*(1), 7–53. doi:10.1016/1061-7361(92)90035-C

Gold, P. E. (2006). The many faces of amnesia. *Learning & Memory (Cold Spring Harbor, N.Y.), 13*(5), 506–514. doi:10.1101/lm.277406

Gonda, D. K., Bachmair, A., Wunning, I., Tobias, J. W., Lane, W. S., & Varshavsky, A. (1989). Universality and structure of the N-end rule. *The Journal of Biological Chemistry, 264*(28), 16700–16712.

Gooch, K. J., & Tennant, C. J. (1997). *Mechanical forces: their effects on cells and tissues*. Berlin: Springer Verlag.

Goodman, D., & Brette, R. (2008). Brian: a simulator for spiking neural networks in Python. *Front Neuroinform, 2*, 5. doi:10.3389/neuro.11.005.2008

Graham, I. S. (1995), *The HTML sourcebook*. John Wiley and Sons.

Greenberg, S., & Rounding, M. (2001). The notification collage: Posting information to public and personal displays. *CHI Letters, 3*(1), 515–521.

GRIA. Grid Resources for Industrial Applications, (2008), http://www.gria.org

Gritzalis, S., Iliadis, J., Gritzalis, D., Spinellis, D., & Katsikas, S. (1999). Developing Secure Web-based Medical Applications. *Medical Informatics journal, 24*(1), 75-90.

Gropp, W., Lusk, E., & Skjellum, A. (1999a). *Using MPI-2: Advanced Features of the Message Passing Interface*. Cambridge, MA: MIT Press.

Gropp, W., Lusk, E., & Skjellum, A. (1999b). *Using MPI: Portable Parallel Programming With the Message-Passing Interface*, (2nd Ed.). Cambridge, MA: MIT Press.

Grosjean, M., & Lacoste, M. (1999). *Communication et intelligence collective: le travail a l'hopital*. Paris: Presses universitaires de France.

Grossman, Z., Feinberg, M. B., & Paul, W. E. (1998). Multiple modes of cellular activation and virus transmission in HIV infection: a role for chronically and latently infected cells in sustaining viral replication. *Proceedings of the National Academy of Sciences of the United States of America, 95*(11), 6314–6319. doi:10.1073/pnas.95.11.6314

Gu, T., Pung, H. K., & Zhang, D. Q. (2004). Toward an OSGi-Based Infrastructure for Context-Aware Applications. *IEEE Pervasive Computing / IEEE Computer Society [and] IEEE Communications Society, 3*(4), 66–74. doi:10.1109/MPRV.2004.19

Guadalupe, M., Reay, E., Sankaran, S., Prindiville, T., Flamm, J., McNeil, A., & Dandekar, S. (2003). Severe CD4+ T-cell depletion in gut lymphoid tissue during primary human immunodeciency virus type 1 infection and substantial delay in restoration following highly active antiretroviral therapy. *Journal of Virology, 77*(21), 11708–11717. doi:10.1128/JVI.77.21.11708-11717.2003

Guex, N., & Peitsch, M. C. (1997). SWISS-MODEL and the Swiss-PdbViewer: An environment for comparative protein modeling. *Electrophoresis, 18*, 2714–2723. doi:10.1002/elps.1150181505

Gulsun, M., Demirkazik, F. B., & Ariyurek, M. (2003). Evaluation of breast microcalcifications according to Breast Imaging Reporting and Data System criteria and Le Gal's classification. *European Journal of Radiology, 47*(3), 227–231. doi:10.1016/S0720-048X(02)00181-X

Gurcan, M. N., Chan, H. P., Sahiner, B., Hadjiiski, L., Petrick, N., & Helvie, M. A. (2002). Optimal neural network architecture selection: improvement in computerized detection of microcalcifications. *Academic Radiology*, 9(4), 420–429. doi:10.1016/S1076-6332(03)80187-3

Gurcan, M. N., Sahiner, B., Chan, H. P., Hadjiiski, L., & Petrick, N. (2001). Selection of an optimal neural network architecture for computer-aided detection of microcalcifications--comparison of automated optimization techniques. *Medical Physics*, 28(9), 1937–1948. doi:10.1118/1.1395036

Gurley, L. (2004). *Advantages and disadvantages of the Electronic Medical Record*. Des Plaines, IL: American Academy of Medical Administrators.

Guruprasad, K., Reddy, B. V. B., & Pandit, M. W. (1990). Correlation between stability of a protein and its dipeptide composition: a novel approach for predicting in vivo stability of a protein from its primary sequence. *Protein Engineering*, 4(2), 155–161. doi:10.1093/protein/4.2.155

Haghpanahi, M., & Miramini, S. (2008). Extraction of Morphological Parameters of Tissue Engineering Scaffolds using Two-Point Correlation Function. In *Proceedings of the 6th IASTED International Conference on Biomedical Engineering*, Austria

Haghpanahi, M., Nikkhoo, M., & Peirovi, H. (2008). Mechanobiological models for Intervertebral Disc Tissue Engineering. In *Proceedings of the Biomedical Electronics and Biomedical Informatics*, Greece.

Haghpanahi, M., Nikkhoo, M., Peirovi, H., & Ghanavi, J. (2007). A Poroviscoelastic Finite Element Formulation Including Transport and Swelling for Tissue Engineered Intervertebral Disc. In *Proceedings of the European Society of Biomechanics Workshop 2007, Ireland*, (pp. 46-47).

Haghpanahi, M., Nikkhoo, M., Peirovi, H., & Ghanavi, J. (2007). Mathematical Modeling of the Intervertebral Disc as an Infrastructure for Studying the Mechanobiology of the Tissue Engineering Procedure. *Transactions on Applied and Theoretical Mechanics*, 2(12), 263–275.

Halamka, J. D., Osterland, C., & Safran, C. (1999). CareWeb™, a web-based medical record for an integrated health care delivery system. *International Journal of Medical Informatics*, 54, 1–8. doi:10.1016/S1386-5056(98)00095-1

Hall, T. A. (1999). BioEdit: a user-friendly biological sequence alignment editor and analysis program for

Windows 95/98/NT. *Nucleic Acids Symposium Series*, 41, 95–98.

Han, L., & Biswas, S. K. (1997). Neural networks for sinusoidal frequency estimation. *Journal of the Franklin Institute*, 334B(1), 1–18. doi:10.1016/S0016-0032(96)00079-8

Hansen, I. B. (2006). CipherMe: personal Electronic Health Records in the hands of patients-owners. *Proceedings of the 1st Distributed Diagnosis and Home Healthcare (D2H2) Conference,* (pp. 148-151).

Haralick, R. (1979, May). Statistical and structural approaches to texture. *Proceedings of the IEEE*, 67(5), 786–804. doi:10.1109/PROC.1979.11328

Hardiman, G., (2006). Microarrays Technologies. an overview. *Pharmacogenomics,* 7(8, December), 1153-8.

Haselgrove, J. C., Subramanian, V. H., Christen, R., & Leigh, J. S. (1988). Analysis of in-vivo NMR spectra. *Reviews of Magnetic Resonance in Medicine*, 2, 167–222.

Hauptman, H. A. (1991). The phase problem of x-ray crystallography. *Reports on Progress in Physics*, 1427–1454. doi:10.1088/0034-4885/54/11/002

Hauptman, H. A. (1997). Phasing methods for protein crystallography. *Current Opinion in Structural Biology*, 7(5), 672–680. doi:10.1016/S0959-440X(97)80077-2

Hayes-Roth, B., Hewett, M., Washington, R., Hewett, R., & Seiver, A. (1989). Distributing intelligence within an individual. In L. Gasser & M. Huhns (Ed.), *Distributed Artificial Intelligence*, (Vol. 2, pp. 385-412). San Francisco: Pitman Publishing and Morgan Kaufmann.

Head, J., Wang, F., Lipari, C., & Elliott, R. (2000). The important role of infrared imaging in breast cancer. *IEEE Engineering in Medicine and Biology Magazine*, 19, 52–57. doi:10.1109/51.844380

Health Level 7, HL7. (2000). *ANSI Standard HL7 V 2.4-2000.*

Hecquet, D., Ruskin, H. J., & Crane, M. (2007). Optimisation and parallelisation strategies for Monte Carlo simulation of HIV infection. *Computers in Biology and Medicine*, 37(5), 691–699. doi:10.1016/j.compbiomed.2006.06.010

Helal, A., Mann, W., Giraldo, C., Kaddoura, Y., Lee, C., & Zabbadani, H. (2003). Smart phone based cognitive assistant. In P*roceedings of the 2nd International Workshop on Ubiquitous Computing for Pervasive Healthcare Applications*, Seattle, WA.

Held, A., Buchholz, S., & Schill, A. (2002). Modeling of context information for pervasive computing applications. In the *6th World Multiconference on Systemics, Cybernetics and Informatics* (SCI), Orland, FL.

Helsinger, A., Kleinmann, K., & Brinn, M. (2004). A framework to control emergent survivability of multi agent systems. In *Proceedings of Third International Joint Conference on Autonomous Agents and Multiagent Systems, 1*, 28-35.

Hendrickson, W. A. (1991). Determination of macromolecular structures from anomalous diffraction of synchrotron radiation. *Science, 254*, 51–58. doi:10.1126/science.1925561

Henning, J., Nauerth, A., & Friedburg, H. (1986). RARE imaging: A first imaging method for clinical MR. *Magnetic Resonance in Medicine, 3*(6), 823–833. doi:10.1002/mrm.1910030602

HIMSS Electronic Health Record Committee. (2003). *EHR Definition, Attributes and Essential Requirement.*

Hinton, G. E., & Sejnowski, T. J. (1986). Learning and Relearning in Boltzmann Machine. *Parallel distributed processing: explorations in the microstructure of cognition, vol. 1: foundations* (pp. 282-317). Cambridge, MA: MIT press.

Hirokawa, T., Boon-Chieng, S., & Mitaku, S. (1998). SOSUI: classification and secondary structure prediction system for membrane proteins. *Bioinformatics Applications Note, 14*(4), 378–379.

Hirose, A. (1992a). Dynamics of fully complex-valued neural networks. *Electronics Letters, 28*(16), 1492–1494. doi:10.1049/el:19920948

Hirose, A. (1992b). Proposal of fully complex-valued neural networks. [), Baltimore, MD.]. *Proceedings of International Joint Conference on Neural Networks, 4*, 152–157.

Hobohm, U., & Sander, C. (1994). Enlarged representative set of protein structures. *Protein Science, 3*, 522–525.

Holle, R., & Zahlmann, G. (1999). Evaluation of Telemedical Services. *IEEE Transactions on Information Technology in Biomedicine, 3*(2), 84–91. doi:10.1109/4233.767083

Hollister, S. J. (2005). Porous scaffold design for tissue engineering. *Nature Materials, 4*, 518–524. doi:10.1038/nmat1421

Hollister, S. J., Levy, R. A., Chu, T. M., Halloran, J. W., & Feinberg, S. E. (2002). An image based approach for designing and manufacturing craniofacial scaffolds. *International Journal of Oral and Maxillofacial Surgery, 29*, 67–71. doi:10.1034/j.1399-0020.2000.290115.x

Holm, L., & Sander, C. (1994). Searching protein structure databases has come of age. *Proteins, 19*, 165–173. doi:10.1002/prot.340190302

Holm, L., & Sander, C. (1997). Dali/FSSP classification of three-dimensional protein folds. *Nucleic Acids Research, 25*, 231–234. doi:10.1093/nar/25.1.231

Holm, R. H., Kennepohl, P., & Solomon, E. I. (1996). Structural and Functional Aspects of Metal Sites in Biology. *Chemical Reviews, 96*, 2239–2314. doi:10.1021/cr9500390

Hopfield, J. J. (1982). Neural networks and physical systems with emergent collective computational abilities. *Proceedings of the National Academy of Sciences of the United States of America, 79*, 2554–2558. doi:10.1073/pnas.79.8.2554

Hopfield, J. J. (1984). Neurons with graded response have collective computational properties like those of two-state neurons. *Proceedings of the National Academy of Sciences of the United States of America, 81*, 3088–3092. doi:10.1073/pnas.81.10.3088

Horsch, A., & Balbach, T. (1999). Telemedical Information Systems. *IEEE Transactions on Information Technology in Biomedicine, 3*(3), 166–175. doi:10.1109/4233.788578

Horwitz, P., & Wilcox, B. A. (2005). Parasites, ecosystems and sustainability: an ecological and complex systems perspective. *International Journal for Parasitology, 35*(7), 725–732. doi:10.1016/j.ijpara.2005.03.002

Hountas, A., & Tsoucaris, G. (1995). Twin Variables and Determinants in Direct Methods. *Acta Crystallographica. Section A, Foundations of Crystallography, 51*, 754–763. doi:10.1107/S0108767395004661

Howell, K., Visentin, M., Lavorato, A., Jones, C., Martini, G., & Smith, R. (2004). Thermography, photography, laser doppler flowmetry and 20 Mhz b-scan ultrasound for the assessment of localised scleroderma activity: A pilot protocol. *Thermology International, 14*(4), 144–145.

Hsin, C., & Liu, M. (2006). Self-monitoring of wireless sensor networks. *Computer Communications, 29*(4), 462–476. doi:10.1016/j.comcom.2004.12.031

http://mrwebservice.files.wordpress.com/2008/10/soa-detailed-diagram.png

Huber, R., Schneider, M., Mayr, I., Müller, R., Deutzmann, R., & Suter, F. (1987). Molecular structure of the bilin binding protein (BBP) from Pieris brassicae after refinement at 2.0 A resolution. *Journal of Molecular Biology, 198*, 499–513. doi:10.1016/0022-2836(87)90296-8

Hutchinson, D. (1998). Medline for health professionals: how to search PubMed on the Internet. Sacramento, CA: New Wind.

Hutmacher, D. W. (2000). Scaffolds in tissue engineering bone and cartilage. *Biomaterials, 21*, 2529–2543. doi:10.1016/S0142-9612(00)00121-6

Hutmacher, D. W., Hurzeler, M., & Schliephake, H. (1996). A review of material properties of biodegradable and bioresorbable polymers and devices for GTR and GBR applications. *The International Journal of Oral & Maxillofacial Implants, 11*, 667–678.

Hutmacher, D. W., Sittinger, M., & Risbud, M. V. (2004). Scaffold-based tissue engineering: rationale for computer-aided design and solid free-form fabrication systems. *Trends in Biotechnology, 22*, 354–362. doi:10.1016/j.tibtech.2004.05.005

IEEE Computer Society (1993). *IEEE Recommended Practice for Software Requirements Specifications, Std. 830.*

Indulska, J., Robinsona, R., Rakotonirainy, A., & Henricksen, K. (2003). Experiences in using cc/pp in context-aware systems. In M.-S. Chen, P. K. Chrysanthis, M.

Sloman, & A. Zaslavsky (Eds.), *MDM2003*, (LNCS Vol. 2574, pp. 247–261). Melbourne, Australia: Springer.

Intille, S. S. (2002). Designing a home of the future. *IEEE Pervasive Computing / IEEE Computer Society [and] IEEE Communications Society, 1*(2), 76–82. doi:10.1109/MPRV.2002.1012340

Intille, S., Larson, S. K., & Kukla, C. (2002). Just-in-time context-sensitive questioning for preventative health care. In *Proceedings of the AAAI' 2002 Workshop on Automation as Caregiver: The Role of Intelligent Technology in Elder Care.*

Isaev, A. (2006). *Introduction to Mathematical Methods in Bioinformatics.* Berlin: Springer-Verlag.

Ishaug-Riley, S. L., Crane, G. M., Gurlek, A., Miller, M. J., Yasko, A. W., Yaszemski, M. J., & Mikos, A. G. (1997). Ectopic bone formation by marrow stromal osteoblast transplantation using poly(DL-lactic-co-glycolic acid) foams implanted into the rat mesentery. *Journal of Biomedical Materials Research, 36*, 1–8. doi:10.1002/(SICI)1097-4636(199707)36:1<1::AID-JBM1>3.0.CO;2-P

Ivanov, D. (2006). BioMEMS sensor systems for bacterial infection detection: progress and potential. *BioDrugs, 20*(6), 351–356. doi:10.2165/00063030-200620060-00005

Jacobs, C., Finkelstein, A., & Salesin, D. (1995). Fast multiresolution image querying. *Siggraph, 95*, 277–286.

Jafari, R., Dabiri, F., Brisk, P., & Sarrafzadeh, M. (2005). Adaptive and fault tolerant medical vest for life-critical medical monitoring. In *Proceedings of the ACM symposium on Applied Computing*, Santa Fe, New Mexico. New York: ACM Press.

James, A., Wilcox, Y., & Naguib, R. N. G. (2001). A Telematic System for Oncology Based on Electronic Health and Patient Records. *IEEE Transactions on Information Technology in Biomedicine, 5*(1), 16–17. doi:10.1109/4233.908366

Jankowski, S., Lozowski, A., & Zurada, J. M. (1996). Complex-valued multistate neural associative memory. *Proceedings of IEEE Transactions on Neural Networks, 7*(6), 1491–1496. doi:10.1109/72.548176

Janssen, M. A., & Ostrom, E. (2006). Governing social-ecological systems. In L. Tesfatsion and K. Judd, (Ed.), *Handbook of Computational Economics*, (pp. 1465-1509).

Jayatilaka, D., & Grimwood, D. J. (2001). Wavefunctions derived from experiment. I. Motivation and theory. *Acta Crystallographica. Section A, Foundations of Crystallography*, *57*, 76–86. doi:10.1107/S0108767300013155

Jena, R. K., Aqel, M. M., Srivastava, P., & Mahanti, P. K. (2009). Soft Computing Methodologies in Bioinformatics. *European Journal of Scientific Research*, *26*(2), 189–203.

Jennings, N., Sycara, K., Wooldridge, M. (1998). A roadmap of agent research and development. *Autonomous agents and multi-agents systems,1*(1), 7-38.

Johnson, R. A., & Wichern, D. W. (2002). *Applied Multivariate Statistical Analysis* (5th Ed.). Upper Saddle River, NJ: Prentice-Hall.

Jones, B. (1998). A reappraisal of infrared thermal image analysis for medicine. *IEEE Transactions on Medical Imaging*, *17*(6), 1019–1027. doi:10.1109/42.746635

Jones, E., Oliphant, T., Peterson, P., et al. (2001). *SciPy: Open source scientific tools for Python*. Retrieved from http://www.scipy.org/

Jones, T. A. (1978). A graphics model building and refinement system for macromolecules. *Journal of Applied Crystallography*, *11*, 268–272. doi:10.1107/S0021889878013308

Jones, T. A., Zou, J.-Y., Cowan, S. W., & Kjeldgaard, M. (1991). Improved methods for building protein models in electron density maps and the location of errors in these models. *Acta Crystallographica. Section A, Foundations of Crystallography*, *47*, 110–119. doi:10.1107/S0108767390010224

Joung, H. A., Lee, N. R., Lee, S. K., Ahn, J., Shin, Y. B., & Choi, H. S. (2008). High sensitivity detection of 16s rRNA using peptide nucleic acid probes and a surface plasmon resonance biosensor. *Analytica Chimica Acta*, *630*(2), 168–173. doi:10.1016/j.aca.2008.10.001

Kaartinen, J., Mierisova, S., Oja, J. M. E., Usenius, J. P., Kauppinen, R. A., & Hiltunen, Y. (1998). Automated quantification of human brain metabolites by artificial neural network analysis from in vivo single-voxel 1H NMR spectra. *Journal of Magnetic Resonance (San Diego, Calif.)*, *134*, 176–179. doi:10.1006/jmre.1998.1477

Kabsch, W. (1988a). Automatic indexing of rotation diffraction patterns. *Journal of Applied Crystallography*, *21*, 67–72. doi:10.1107/S0021889887009737

Kabsch, W. (1988b). Evaluation of single-crystal X-ray diffraction data from a position-sensitive detector. *Journal of Applied Crystallography*, *21*, 916–924. doi:10.1107/S0021889888007903

Kabsch, W. (1993). Automatic processing of rotation diffraction data from crystals of initially unknown symmetry and cell constants. *Journal of Applied Crystallography*, *26*, 795–800. doi:10.1107/S0021889893005588

Kang, D., Lee, H., Ko, E., Kang, K., & LEE, J. (2006). A Wearable Context Aware System for Ubiquitous Healthcare. In *the 28th IEEE EMBS Annual International Conference*. New York: IEEE.

Karagiannis, G. E., Stamatopoulus, V. G., Rigby, M., Kotis, T., Negroni, E., Munoz, A., & Mathes, I. (2007). Web-based personal health records: the personal electronic health record (pEHR) multicentred trial. *Journal of Telemedicine and Telecare*, *13*, 32–34. doi:10.1258/135763307781645086

Karkkainen, M. (2003). Increasing efficiency in the supply chain for short life goods using RFID tagging. *International Journal of Retail & Distribution Management*, *31*(10), 529–536. doi:10.1108/09590550310497058

Karssemeijer, N. (1993). Adaptive noise equalization and recognition of microcalcification clusters in mammograms. *Int. J. Patt. Rec. & Im. Analysis.*, *7*.

Katsaloulis, E., Floros, A., Provata, Y., & Cotronis, T. (2006), "Gridification of the SHMap Biocomputational Algorithm", International Special Topic Conference on Information Technology in Biomedicine.

Kentta, O., Merilahti, J., Petakoski-Kult, T., Ikonen, V., & Korhonen, I. (2007). Evaluation of Technology-Based Service Scenarios for Supporting Independent Living. In *The 29th Annual International Conference of the IEEE EMBS*. San Diego, CA: IEEE.

Kidd, C. D., Orr, R., Abowd, G. D., Atkeson, C., Essa, I., MacIntyre, B., et al. (1999). The aware home: A living laboratory for ubiquitous computing research. In *Proceedings of the Second International Workshop on Cooperative Buildings.* Berlin, Germany; Springer-Verlag.

Kim, E. H. (2006). Web-based Personal-Centered Electronic Health Record for Elderly Population. In *Proceedings of the 1st Distributed Diagnosis and Home Healthcare (D2H2) Conference,* (pp. 144-147).

Kim, S. S., Utsunomiya, H., Koski, J. A., Wu, B. M., Cima, M. J., & Sohn, J. (1998). Survival and function of hepatocytes on a novel three-dimensional synthetic biodegradable polymeric scaffold with an intrinsic network of channels. *Annals of Surgery, 228,* 8–13. doi:10.1097/00000658-199807000-00002

Kjeldskov, J., & Graham, C. (2003). A review of mobile HCI research methods. In L. Chittaro (Ed.) *5th International Symposium on Human-Computer Interaction with Mobile Devices and Services, LNCS* (2795: 317-335). Udine, Italy: Springer.

Kjeldskov, J., & Skov, M. (2004). Supporting work activities in healthcare by mobile electronic patient records. In M. Masoodian, S. Jones, & B. Rogers (Eds.) *6th Asia–Pacific Conference on Human–Computer Interaction,* (LNCS Vol. 3101, pp. 191-200). Rotorva, New Zealand: Elsevier Science.

Kjeldskov, J., & Skov, M. B. (2007). Exploring context-awareness for ubiquitous computing in the healthcare domain. *Personal and Ubiquitous Computing, 11*(7), 549–562. doi:10.1007/s00779-006-0112-5

Klein, J. (2005). Integrating Electronic Health Records Using HL7 Clinical Document Architecture.

Kleywegt, G. J., & Jones, T. A. (1996a). Phi/psi-chology: Ramachandran revisited. *Structure (London, England), 4,* 1395–1400. doi:10.1016/S0969-2126(96)00147-5

Kleywegt, G. J., & Jones, T. A. (1996b). Efficient rebuilding of protein structures. *Acta Crystallographica. Section D, Biological Crystallography, 52,* 829–832. doi:10.1107/S0907444996001783

Knies, R. (2007). *Memorable support for SenseCam memory-retention research.* Retrieved from http://research.microsoft.com/en-us/news/features/sensecam.aspx

Ko, E. J., Lee, H. J., & Lee, J. W. (2007). Ontology-Based Context Modeling and Reasoning for U-HealthCare. *IEICE Transactions on Information and Systems. E (Norwalk, Conn.), 90-D*(8), 1262–1270.

Kohda, Y., & Endo, S. (1996). Ubiquitous advertising on the WWW: Merging advertisement on the browser. *Computer Networks and ISDN Systems, 28*(7-11), 1493-1499.

Kohn, L. T., Corrigan, J. M., & Donaldson, M. S. (Eds.). (2000). *To Err is Human: Building a Safer Health System.* Washington, DC: National Academy Press.

Korhonen, I., & Bardram, J. E. (2004). Guest editorial: Introduction to the special section on pervasive healthcare. *IEEE Transactions on Information Technology in Biomedicine, 8*(3), 229–234. doi:10.1109/TITB.2004.835337

Korhonen, I., Paavilainen, P., & Särela, A. (2003). Application of ubiquitous computing technologies for support of independent living of the elderly in real life settings. In *The 2nd International Workshop on Ubiquitous Computing for Pervasive Healthcare Applications* (Ubihealth 2003 Proceedings), Seattle, WA.

Korndorfer, I. P., Schlehuber, S., & Skerra, A. (2003). Structural mechanism of specific ligand recognition by a lipocalin tailored for the complexation of digoxigenin. *Journal of Molecular Biology, 330,* 385–396. doi:10.1016/S0022-2836(03)00573-4

Kourouthanassis, P. (2004). Can technology make shopping fun? *ECR Journal, 3*(2), 37–44.

Kourouthanassis, P., & Giaglis, G. M. (Eds.). (2007). *Pervasive information systems.* New York: M. E. Sharpe Inc.

Krahe, R., & Gabbiani, F. (2004). Burst firing in sensory systems. *Nature Reviews. Neuroscience, 5,* 13–23. doi:10.1038/nrn1296

Kraulis, P. J. (1991). MOLSCRIPT: a program to produce both detailed and schematic plots of protein structures. *Journal of Applied Crystallography, 24,* 946–950. doi:10.1107/S0021889891004399

Kruskal, J., & Wish, M. (1978). *Multidimensional scaling.* London: Sage Publications.

Ku, J., & Schultz, P. G. (1995). Alternate protein frameworks for molecular recognition. *Proceedings of the National Academy of Sciences of the United States of America, 92,* 6552–6556. doi:10.1073/pnas.92.14.6552

Kumar, M., Shirazi, B. A., Das, S. K., Sung, B. Y., Levine, D., & Singhal, M. (2003). PICO: A Middleware Framework for Pervasive Computing. *Pervasive computing, 2*(3), 72-79.

Kunkel, E. J., & Butcher, E. C. (2002). Chemokines and the tissue-specific migration of lymphocytes. *Immunity, 16,* 1–4. doi:10.1016/S1074-7613(01)00261-8

Kunze, C., Grobmann, U., Stork, W., & Muller-Glaser, K. D. (2002). Application of ubiquitous computing in personal health monitoring systems. *Biomedizinische Technik. Biomedical Engineering, 47*(1), 360–362. doi:10.1515/bmte.2002.47.s1a.360

Kuroe, Y. Hashimoto. N. & Mori, T. (2002). On energy function for complex-valued neural networks and its applications, Neural information proceeding. *Proceedings of the 9th International Conference on Neural Information Processing Computational Intelligence for the E-Age* (Vol.3, pp. 1079-1083).

Kuzmak, P. M., & Dayhoff, R. E. (2000). The use of digital imaging and communications in medicine (DICOM) in the integration of imaging into the electronic patient record at the Department of Veterans Affairs. *Digital Imaging, 13*(2), 133–137.

Kuzmak, P. M., & Dayhoff, R. E. (2003). Experience with DICOM for the clinical specialities in the healthcare enterprise. *Proceedings of the Society for Photo-Instrumentation Engineers, 5033,* 18–29. doi:10.1117/12.480668

Kwong, K., Belliveau, J. W., Chesler, D. A., Goldberg, I. E., Weisskoff, R. M., & Poncelet, B. P. (1992). Dynamic magnetic resonance imaging of human brain activity during primary sensory stimulation. *Proceedings of the National Academy of Sciences of the United States of America, 89*(12), 5675–5679. doi:10.1073/pnas.89.12.5675

Kyte, J., & Doolittle, R. F. (1982). A simple method for displaying the hydropathic character of a protein. *Journal of Molecular Biology, 157*(1), 105–132. doi:10.1016/0022-2836(82)90515-0

Lackner, T. E. (2008). Pharmacogenomic dosing of warfarin: ready or not? *The Consultant Pharmacist, 23*(8), 614–619.

Lado, M., Tahoces, P. G., Mendez, A. J., Souto, M., & Vidal, J. J. (2001). Evaluation of an automated wavelet-based system dedicated to the detection of clustered microcalcifications in digital mammograms. *Medical Informatics and the Internet in Medicine, 26*(3), 149–163. doi:10.1080/14639230110062480

Lam, C. X. F., Mo, X. M., Teoh, S. H., & Hutmache, D. W. (2002). Scaffold development using 3D printing with a starch based polymer. *Materials Science and Engineering, 20,* 49–56. doi:10.1016/S0928-4931(02)00012-7

Lamminen, H., Voipio, V., Ruohonen, K., & Uusitalo, H. (2003). Telemedicine in ophthalmology. *Acta Ophthalmologica Scandinavica, 81,* 105–109. doi:10.1034/j.1600-0420.2003.00045.x

Lamzin, V. S., & Wilson, K. S. (1993). Automated refinement of protein models. *Acta Crystallographica. Section D, Biological Crystallography, 49,* 129–147. doi:10.1107/S0907444992008886

Lamzin, V. S., & Wilson, K. S. (1997). Automated refinement for protein crystallography. *Methods in Enzymology, 277,* 269–305. doi:10.1016/S0076-6879(97)77016-2

Landers, R., Pfister, A., Hübner, U., John, H., Schmelzeisen, R., & Mülhaupt, R. (2002). Fabrication of soft tissue engineering scaffolds by means of rapid prototyping techniques. *Journal of Materials Science, 37,* 3107–3116. doi:10.1023/A:1016189724389

Langer, R., & Vacanti, J. P. (1993). Tissue engineering. *Science, 260,* 920–926. doi:10.1126/science.8493529

Lanyi, M. (1977). Differential diagnosis of microcalcifications, X-ray film analysis of 60 intraductal carcinoma, the triangle principle. *Der Radiologe, 17*(5), 213–216.

Lanyi, M. (1985). Microcalcifications in the breast--a blessing or a curse? A critical review. *Diagnostic Imaging in Clinical Medicine, 54*(3-4), 126–145.

Laskowski, R. A., MacArthur, M. W., & Thornton, J. M. (1998). Validation of protein models derived from experiment. *Current Opinion in Structural Biology, 8,* 631–639. doi:10.1016/S0959-440X(98)80156-5

Laskowski, R. A., MacArthur, M. W., Moss, D. S., & Thornton, J. M. (1993). PROCHECK: a program to check the stereochemical quality of protein structures. *Journal of Applied Crystallography, 26,* 283–291. doi:10.1107/S0021889892009944

Laskowski, R. A., MacArthur, M. W., Moss, D. S., & Thornton, J. M. (1993). PROCHECK: a program to check the stereochemical quality of protein structures. *Journal of Applied Crystallography, 26,* 283–291. doi:10.1107/S0021889892009944

Latha Srinivasan and Jem Treadwell (2005), "An Overview of Service-oriented Architecture, Web Services and Grid Computing" HP Software Global Business Unit, 2005

Latre, B., Braem, B., Moerman, I., Blondia, C., Reusens, E., Joseph, W., & Demeester, P. (2007b). A low-delay protocol for multihop wireless body area networks. In *The 4th Annual International Conference on Mobile and Ubiquitous Systems: Networking & Services* (pp. 1-8). Philadelphia: IEEE.

Latre, B., Poorter, E., Moerman, I., & Demeester, P. (2007a). MOFBAN: A Lightweight Modular Framework for Body Area Networks. In T.-W. Kuo et al. (Eds.) 2007 IFIP Int. Conf. on *Embedded and Ubiquitous Computing - EUC 2007* (4808, pp. 610-622). Taipei, Taiwan: Springer.

Le Gal, M., Chavanne, G., & Pellier, D. (1984). Diagnostic value of clustered microcalcifications discovered by mammography (apropos of 227 cases with histological verification and without a palpable breast tumor). *Bulletin du Cancer, 71*(1), 57–64.

Le Gal, M., Durand, J. C., Laurent, M., & Pellier, D. (1976). Management following mammography revealing grouped microcalcifications without palpable tumor. *La Nouvelle Presse Medicale, 5*(26), 1623–1627.

Lee, B., & Richards, F. M. (1971). The interpretation of protein structures: estimation of static accessibility. *Journal of Molecular Biology, 55,* 379–400. doi:10.1016/0022-2836(71)90324-X

Lee, S. W., & Steinman, D. A. (2008). Influence of inlet secondary curvature on image-based CFD models of the carotid bifurcation. *Proceedings of the ASME Summer Bioengineering Conference.* New York: ASME Press.

Leinberger, W., & Kumar, V. (1999). Information Power Grid: The new frontier in parallel computing? *IEEE Concurrency, 7*(4), 75–84. doi:10.1109/MCC.1999.806982

Leslie, A. G. W. (1988), *Profile fitting.* In: J.R. Helliwell, P.A. Machin and M.Z. Papiz, Editors, Proceedings of the CCP4 Study Weekend, Daresbury Laboratory.

Leslie, A.G.W. (1992), *Recent changes to the MOSFLM package for processing film and image plate data,* Joint CCP4 + ESF-EAMCB Newsletter on Protein Crystallography, No. 26.

Lewicki, M. S. (1998). A review of methods for spike sorting: the detection and classification of neural action potentials. *Network (Bristol, England), 9,* R53–R78. doi:10.1088/0954-898X/9/4/001

Ley, D. (2007). Ubiquitous computing. []. Coventry, UK: Becta.]. *Emerging Technologies for Learning, 2,* 64–79.

Li, H., Liu, K. J., & Lo, S. C. (1997). Fractal modeling and segmentation for the enhancement of microcalcifications in digital mammograms. *IEEE Transactions on Medical Imaging, 16*(6), 785–798. doi:10.1109/42.650875

Liedberg, B., Nylander, C., & Lundstrom, I. (1995). Biosensing with surface plasmon resonance, how it all started. *Biosensors & Bioelectronics, 10,* i–ix. doi:10.1016/0956-5663(95)96965-2

Lin, C. Y., Kikuchi, N., & Hollister, S. J. (2004). A novel method for biomaterial scaffold internal architecture design to match bone elastic properties with desired porosity. *Journal of Biomechanics, 37,* 623–636. doi:10.1016/j.jbiomech.2003.09.029

Lipman, D. J., & Pearson, W. R. (1985). Rapid and sensitive protein similarity searches. *Science, 227*(4693), 1435–1441. doi:10.1126/science.2983426

Lisman, J. E. (1997). Bursts as a unit of neural information: making unreliable synapses reliable. *Trends in Neurosciences, 20,* 38–43. doi:10.1016/S0166-2236(96)10070-9

Liszka, K. J., Mackin, M. A., Lichter, M. J., York, D. W., Pillai, D., & Rosenbaum, D. S. (2004). Keeping a Beat on the Heart. *Pervasive Computing, 3*(12), 43–49.

Lo, H. G., Newmark, L. P., Yoon, C., Volk, L. A., Carlson, V. L., & Kittler, A. F. (2007). Electronic Health Records in Specialty Care: A Time-Motion Study. *Journal of the American Medical Informatics Association, 14*(5), 609–615. doi:10.1197/jamia.M2318

Lorenz, H., Richter, G. M., Capaccioli, M., & Longo, G. (1993). Adaptive filtering in astronomical image processing. I. Basic considerations and examples. *Astronomy & Astrophysics, 277*, 321.

Lorincz, K. M., & Fulford-Jones, D. J. (2004). Sensor networks for emergency response: Challenges and opportunities. *IEEE Pervasive Computing / IEEE Computer Society [and] IEEE Communications Society, 3*, 16–23. doi:10.1109/MPRV.2004.18

Los, R., van Ginneken, A. M., de Wilde, M., & van der Lei, J. (2004). OpenSDE: Row Modeling Applied to Generic Structured Data Entry. *Journal of the American Medical Informatics Association, 11*, 162–165. doi:10.1197/jamia.M1375

Lotka, A. J. (1925). *Elements of physical biology*. Baltimore: Williams and Wilkins.

Lyandres, O., Yuen, J. M., Shah, N. C., VanDuyne, R. P., Walsh, J. T., & Glucksberg, M. R. (2008). Progress toward an in vivo surface-enhanced Raman spectroscopy glucose sensor. *Diabetes Technology & Therapeutics, 10*(4), 257–265. doi:10.1089/dia.2007.0288

Lyytinen, K., & Yoo, Y. (2002). Issues and Challenges in Ubiquitous Computing. *Communications of the ACM, 45*(12), 62–96. doi:10.1145/585597.585616

MacArthur, M. W., Laskowski, R. A., & Thornton, J. M. (1994). Knowledge-based validation of protein structure coordinates derived by X-ray crystallography and NMR spectroscopy. *Current Opinion in Structural Biology, 4*, 731–737. doi:10.1016/S0959-440X(94)90172-4

Mackay, C. R., Marston, W. L., & Dudler, L. (1990). Naive and memory T cells show distinct pathways of lymphocyte recirculation. *The Journal of Experimental Medicine, 171*, 801–817. doi:10.1084/jem.171.3.801

Maddams, W. F. (1980). The scope and Limitations of Curve Fitting. *Applied Spectroscopy, 34*(3), 245–267. doi:10.1366/0003702804730312

Maffezzoli, A., Gullo, F., & Wanke, E. (2008). *Detection of intrinsically different clusters of firing neurons in long-term mice cortical networks*. 6th Int. Meeting on Substrate-Integrated Microelectrodes (MEA Meeting 2008), July 8–11, Reutlingen, Germany.

Magenis, R. E., Maslen, C. L., Smith, L., Allen, L., & Sakai, L. Y. (1991). Localization of the fibrillin (FBN) gene to chromosome 15, band q21.1. *Genomics, 11*, 346–351. doi:10.1016/0888-7543(91)90142-2

Maher, M. P., Pine, J., Wright, J., & Tai, Y.-C. (1999). The Neurochip: a new multielectrode device for stimulating and recording from cultured neurons. *Journal of Neuroscience Methods, 87*(1), 45–56. doi:10.1016/S0165-0270(98)00156-3

Main, P., Fiske, S. J., Hull, S. E., Lessinger, L., Germain, G., Declercq, J.-P., & Woolfson, M. M. (1980), *MULTAN80. A system of computer programs for the automatic solution of crystal structures from X-ray diffraction data*. Universities of York, England, and Louvain, Belgium.

Maitra, S. (1979). Moment invariants. *Proceedings of the IEEE, 67*, 697–699. doi:10.1109/PROC.1979.11309

Makalowski, Wojciech (n.d.). *web site notes*. Retrieved from http://www.compgen.uni-muenster.de/

Mannion, R., Ruskin, H. J., & Pandey, R. B. (2000). Effect of mutation on helper T-cells and viral population: a computer simulation model for HIV. *Theory in Biosciences, 119*, 145–155.

Mannion, R., Ruskin, H. J., & Pandey, R. B. (2002). A Monte-Carlo approach to population dynamics of cells in a HIV immune response model. *Theory in Biosciences, 121*, 237–245.

Mansfield, P. (1977). Multi-planar image formation using NMR spin echoes. *Journal of Physics. C. Solid State Physics, 10*, 55–58. doi:10.1088/0022-3719/10/3/004

Marcheschi, P., Mazzarisi, A., Dalmiani, S., & Benassi, A. (2004). HL7 clinical document architecture to share cardiological images and structured data in next generation. *Computers in Cardiology*, 617–620. doi:10.1109/CIC.2004.1443014

Marcheschi, P., Positano, V., Ferdegnini, E. M., Mazzarisi, A., & Benassi, A. (2003). A open source based Application for integration and sharing of multi-modal cardiac image data in a heterogeneous environment. *Computers in Cardiology*, 367–370. doi:10.1109/CIC.2003.1291168

Markatatos G., Atun R., Barthakur N., Jollie C., Kotis T., Stamatopoulos V.G., et al (2004). *e-OpenDay Project Market Analysis.*

Martin, I., Padera, R. F., Vunjak-Novakovic, G., & Freed, L. E. (1998). In vitro differentiation of chick embryo bone marrow stromal cells into cartilaginous and bone-like tissues. *Journal of Orthopaedic Research, 16*, 181–189. doi:10.1002/jor.1100160205

Mata Campos, R., Vidal, E. M., Nava, E., Martinez-Morillo, M., & Sendra, F. (2000). Detection of microcalcifications by means of multiscale methods and statistical techniques. *Journal of Digital Imaging, 13*(2Suppl 1), 221–225.

Mattapallil, J. J., Douek, D. C., Hill, B., Nishimura, Y., Martin, M., & Roederer, M. (2005). Massive infection and loss of memory CD4+ T cells in multiple tissues during acute SIV infection. *Nature, 434*, 1093–1097. doi:10.1038/nature03501

McCarthy, I. P., & Tan, Y. K. (2000). Manufacturing competitiveness and fitness landscape theory. *Journal of Materials Processing Technology, 107*(1-3), 347–352. doi:10.1016/S0924-0136(00)00687-7

McCoy, A. J., Grosse-Kunstleve, R. W., Adams, P. D., Winn, M. D., Storoni, L. C., & Read, R. J. (2007). Phaser crystallographic software. *Journal of Applied Crystallography, 40*, 658–674. doi:10.1107/S0021889807021206

McKinley, B. A. (2008). ISFET and fiber optic sensor technologies: in vivo experience for critical care monitoring. *Chemical Reviews, 108*(2), 826–844. doi:10.1021/cr068120y

Meister, M., Pine, J., & Baylor, D. A. (1994). Multi-neuronal signals from the retina: acquisition and analysis. *Journal of Neuroscience Methods, 51*, 95–106. doi:10.1016/0165-0270(94)90030-2

Melki, P. S., Mulkern, R. V., Panych, L. S., & Jolesz, F. A. (1991). Comparing the FAISE method with conventional dual-echo sequences. *Journal of Magnetic Resonance Imaging, 1*, 319–326. doi:10.1002/jmri.1880010310

Memory loss. (2006). *Encyclopedia of alternative medicine.* Gale Cengage. Retrieved from http://www.enotes.com/alternative-medicine-encyclopedia/memory-loss

Merla, A., Romano, V., Zulli, F., Saggini, R., Di Donato, L., & Romani, G. (2002). Total body infrared imaging and postural disorders. *24th IEEE Int. Conference on Engineering in Medicine and Biology.*

Merritt, E. A., & Bacon, D. J. (1997). Raster3D: photorealistic molecular graphics. *Methods in Enzymology, 277*, 505–524. doi:10.1016/S0076-6879(97)77028-9

Merritt, E. A., & Murphy, M. E. P. (1994). Raster3D version 2.0. A program for photorealistic molecular graphics. *Acta Crystallographica. Section D, Biological Crystallography, 50*, 869–873. doi:10.1107/S0907444994006396

Meyer, C. H., Hu, B. S., Nishimura, D. G., & Macovski, A. (1992). Fast Spiral Coronary Artery Imaging. *Magnetic Resonance in Medicine, 28*(2), 202–213. doi:10.1002/mrm.1910280204

Middlebrooks, J. C., Clock, A. E., Xu, L., & Green, D. M. (1994). A panoramic code for sound location by cortical neurons. *Science, 264*, 842–844. doi:10.1126/science.8171339

Miersová, S., & Ala-Korpela, M. (2001). MR spectroscopy quantification: a review of frequency domain methods. *NMR in Biomedicine, 14*, 247–259. doi:10.1002/nbm.697

Mikos, A. G., Sarakinos, G., Lyman, M. D., Ingber, D. E., Vacanti, J. P., & Langer, R. (1993). Prevascularization of porous biodegradable polymers. *Biotechnology and Bioengineering, 42*, 716–723. doi:10.1002/bit.260420606

Milenkovic, A., Otto, C., & Jovanov, E., (2006). Wireless sensor networks for personal health monitoring: Issues and an implementation. *Computer Communications, Wireless Sensor Networks and Wired/Wireless Internet Communications, 29*(13-14), 2521–2533.

Milojicic, D., Messer, A., Bernadat, P., Greenberg, I., Fu, G., Spinczyk, O., et al. (2001). *Pervasive Services Infrastructure* (Tech. Report). HP Labs.

Minar, N., Burkhart, R., Langton, C., & Askenazi, M. (1996). *The Swarm simulation system: A toolkit for building multi-agent simulations.* (Working Paper 96-06-042, Santa Fe Institute, Santa Fe, NM).

Mirto, M., Fiore, S., Cafaro, M., Passante, M., & Aloisio, G. (2008), "A Grid-Based Bioinformatics Wrapper for Biological Databases,", 21st IEEE International Symposium on Computer-Based Medical Systems.

Mitchell, S., Spiteri, M., Bates, J., & Coulouris, G. (2000). Context aware multimedia computing in the intelligent hospital. In the *9th ACM SIGOPS European Workshop*. Denmark: ACM.

Mizoguchi, F., Nishiyama, H., Ohwada, H., & Hiraishi, H. (1999). Smart office robot collaboration based on multi-agent programming. *Artificial Intelligence, 114*, 57–94. doi:10.1016/S0004-3702(99)00068-5

Mood, A. M., Graybill, F. A., & Boes, D. C. (1974). *Introduction to the Theory of Statistics*. Columbus, OH: McGraw-Hill, Inc.

Morita, N., & Konishi, O. (2004). A Method of Estimation of Magnetic Resonance Spectroscopy Using Complex-Valued Neural Networks. *Systems and Computers in Japan, 35*(10), 14–22. doi:10.1002/scj.10705

Morris, A. L., MacArthur, M. W., Hutchinson, E. G., & Thornton, J. M. (1992). Stereochemical quality of protein structure coordinates. *Proteins, 12*, 345–364. doi:10.1002/prot.340120407

Morris, G. M., Goodsell, D. S., Halliday, R. S., Huey, R., Hart, W. E., Belew, R. K., & Olson, A. J. (1998). Automated docking using a Lamarckian genetic algorithm and an empirical binding free energy function. *Journal of Computational Chemistry, 19*, 1639–1662. doi:10.1002/(SICI)1096-987X(19981115)19:14<1639::AID-JCC10>3.0.CO;2-B

Morris, R. G. (1986). Short-term forgetting in senile dementia of the Alzheimer's type. *Cognitive Neuropsychology, 3*(1), 77–97. doi:10.1080/02643298608252670

Mount, D. W. (2001). Bioinformatics: Sequence and genome Analysis. Cold Spring Harbor, NY: CSHL Press.

Mulder, N. J., Apweiler, R., Attwood, T. K., Bairoch, A., Bateman, A., & Binns, D. (2007). New developments in the InterPro database. *Nucleic Acids Research, 35*, D224–D228. doi:10.1093/nar/gkl841

Muller, H. N., & Skerra, A. (1993). Functional expression of the uncomplexed serum retinol binding protein in Escherichia coli. Ligand binding and reversible unfolding characteristics. *Journal of Molecular Biology, 230*, 725–732. doi:10.1006/jmbi.1993.1194

Munoz, M., Rodriguez, M., Favela, J., Martinez-Garcia, A., & Gonzalez, V. (2003). Context-aware mobile communication in Hospitals. *IEEE Computer, 36*(9), 38–46.

Murdoch, I. (1999). Telemedicine. *The British Journal of Ophthalmology, 83*, 1254–1256. doi:10.1136/bjo.83.11.1254

Murphy, W. L., Dennis, R. G., Kileny, J. L., & Mooney, D. J. (2002). Salt Fusion: An approach to improve pore interconnectivity within tissue engineering scaffolds. *Tissue Engineering, 8*, 43–52. doi:10.1089/107632702753503045

Murshudov, G. N., Vagin, A. A., & Dodson, E. J. (1997). Refinement of macromolecular structures by the maximum-likelihood method. *Acta Crystallographica. Section D, Biological Crystallography, 53*, 240–255. doi:10.1107/S0907444996012255

Murshudov, G. N., Vagin, A. A., Lebedev, A., Wilson, K. S., & Dodson, E. J. (1999). Efficient anisotropic refinement of macromolecular structures using FFT. *Acta Crystallographica. Section D, Biological Crystallography, 55*, 247–255. doi:10.1107/S090744499801405X

Najjar, L., Thompson, J. C., & Ockerman, J. J. (1997). A wearable computer for quality assurance in a food-processing plant. In *Proceedings of the 1st International Symposium on Wearable Computers*. Los Alamitos, CA: IEEE Press.

Narasimhan, V. L., & Croll, P. R. (2008). Communicating Security Policies to Trusted e-Health Information Systems: A Specification Process based Approach. In R. Merrell, R.A. Cooper (Ed.) *Proceedings of Telehealth and Assistive Technologies (TeleHealth/AT 2008)*.

Naresh, R., Tripathi, A., & Omar, S. (2006). Modelling the spread of AIDS epidemic with vertical transmission. *Applied Mathematics and Computation, 178*(2), 262–272.

Naressi, A., Couturier, C., Castang, I., de. Beer, R., & Graveron-Demilly, D. (2001). Java-based graphical user interface for MRUI, a software package for quantitation of in vivo medical magnetic resonance spectroscopy signals. *Computers in Biology and Medicine, 31*, 269–286. doi:10.1016/S0010-4825(01)00006-3

National Eye Institute [Online]. (2007). Retrieved from http://www.nei.nih.gov.

Navaza, J. (1994). AMoRe: an automated package for molecular replacement. *Acta Crystallographica. Sec-*

tion A, Foundations of Crystallography, 50, 157–163. doi:10.1107/S0108767393007597

Ndifon, W. (2005). A complex adaptive systems approach to the kinetic folding of RNA. *Bio Systems, 82*(3), 257–265. doi:10.1016/j.biosystems.2005.08.004

Nebes, R. D., & Butters, M. A. (2000). Decreased working memory and processing speed mediate cognitive impairment in geriatric depression. *Psychological Medicine, 30,* 679–691. doi:10.1017/S0033291799001968

NEMA. (2000). *Digital Imaging and Communications in Medicine (DICOM): Version 3.0.* Hornero, R., López, M.I., Acebes, M. and Calonge, T. (2003). Teleophthalmology for diabetic retinopathy screening in a rural area of Spain, *Eighth Annual Meeting of the American Telemedicine Association (ATA'2003).*

Newman, K., & Mason, R. S. (2006). Organic mass spectrometry and control of fragmentation using a fast flow glow discharge ion source. *Rapid Communications in Mass Spectrometry, 20*(14), 2067–2073.

Nikkhoo, M., Haghpanahi, M., Peirovi, H., & Ghanavi, J. (2007). Mathematical model for tissue engineered intervertebral disc as a saturated porous media. *Proceedings of the 3rd WSEAS International Conference on Applied and Theoretical Mechanics, Spain,* (pp. 197-201).

Niknejad, H., Peirovi, H., Jorjani, M., Ahmadiani, A., Ghanavi, J., & Seifalian, A. M. (2008). Properties of the amniotic membrane for potential use in tissue engineering. *European Cells & Materials, 15,* 88–99.

Nitta, T. (1997). An Extension of the Back-Propagation Algorithm to Complex Numbers. *Neural Networks, 10*(8), 1392–1415. doi:10.1016/S0893-6080(97)00036-1

Nitta, T. (2000). An Analysis of the Fundamental Structure of Complex-Valued Neurons. *Neural Processing Letters, 12*(3), 239–246. doi:10.1023/A:1026582217675

Nitta, T. (2002). Redundancy of the Parameters of the Complex-valued Neural Networks. *Neurocomputing, 49*(1-4), 423–428. doi:10.1016/S0925-2312(02)00669-0

Nitta, T. (2003). On the Inherent Property of the Decision Boundary in Complex-valued Neural Networks. *Neurocomputing, 50*(c), 291–303. doi:10.1016/S0925-2312(02)00568-4

Nitta, T. (2003). Orthogonality of Decision Boundaries in Complex-Valued Neural Networks. *Neural Computation, 16*(1), 73–97. doi:10.1162/08997660460734001

Nitta, T. (2003). Solving the XOR Problem and the Detection of Symmetry Using a Single Complex-valued Neuron. *Neural Networks, 16*(8), 1101–1105. doi:10.1016/S0893-6080(03)00168-0

Nitta, T. (2003). The Uniqueness Theorem for Complex-valued Neural Networks and the Redundancy of the Parameters. *Systems and Computers in Japan, 34*(14), 54–62. doi:10.1002/scj.10363

Nitta, T. (2004). Reducibility of the Complex-valued Neural Network. *Neural Information Processing - Letters and Reviews, 2*(3), 53-56.

Nyborg, J. and A.J. Wonacott A. J. (1977), *The Rotation Method in Crystallography,* U.W. Arndt & A. J. Wonacott, eds, North Holland Publishing Co.

O' Sullivan, T., O'Donoghue, J., Herbert, J., & Studdert, R. (2006). CAMMD: Context-Aware Mobile Medical Devices. *Journal of Universal Computer Science, 12*(1), 45–58.

O'Brien, F. J., Harley, B. A., Yannas, I. V., & Gibson, L. (2004). Influence of freezing rate on pore structure in freeze-dried collagen-GAG scaffolds. *Biomaterials, 25,* 1077–1086. doi:10.1016/S0142-9612(03)00630-6

Ogawa, S., Lee, T. M., Nayak, A. S., & Glynn, P. (1990). Oxygenation-sensitive contrast in magnetic resonance image of rodent brain at high magnetic fields. *Magnetic Resonance in Medicine, 14*(1), 68–78. doi:10.1002/mrm.1910140108

Oldfield, T. J. (2002). Data Mining the Protein Data Bank: Residue Interactions. *Proteins, 49,* 510–528. doi:10.1002/prot.10221

Online Mendelian Inheritance in Man, Johns Hopkins University, (2000). *Marfan syndrome, Type I; MIM Number 154700: 2/25/00.* Retrieved October 10th, 2007 from http://www.ncbi.nlm.nih.gov/omim/

Online Mendelian Inheritance in Man, Johns Hopkins University, (2008). *Fibrillin 1; MIM Number 134797: 12/11/08.* Retrieved October 11th, 2008 from http://www.ncbi.nlm.nih.gov/sites/entrez

Open Grid Services Architecture (OGSA). (2008), http://www.ggf.org/documents/GFD.30.pdf

Oracle Database Online Documentation 10g Release 2 [Online]. (2007). Retrieved from http://youngcow.net/doc/oracle10g/index.htm.

Osareh, A., Mirmehdi, M., Thomas, B., & Markham, R. (2003). Automated identification of diabetic retinal exudates in digital colour images. *The British Journal of Ophthalmology*, *87*(10), 1220–1223. doi:10.1136/bjo.87.10.1220

Osman, K. A., Ashford, R. L., & Oldacres, A. (2007). Homecare Hub - A Pervasive Computing Approach to Integrating Data for Remote Delivery of Personal and Social Care. In *Proceedings of the 2nd International Conference on Pervasive Computing and Applications* (pp. 348-353).

Otsu, N. (1979). A threshold selection method from grey-level histograms. *IEEE Transactions on Systems, Man, and Cybernetics*, *9*(1), 62–66. doi:10.1109/TSMC.1979.4310076

Otwinowski, Z., & Minor, W. (1997), *Processing of X-ray Diffraction Data Collected in Oscillation Mode*, Methods in Enzymology, Volume **276**: Macromolecular Crystallography, part A, pp.307-326.

Ouzounis, C. A., & Valencia, A. (2003). Early bioinformatics: the birth of a discipline— a personal view. *Bioinformatics (Oxford, England)*, *19*(17), 2176–2190.

Padlan, E. A. (1994). Anatomy of the antibody molecule. *Molecular Immunology*, *31*, 169–217. doi:10.1016/0161-5890(94)90001-9

Paganelli, F., Spinicci, E., & Giuli, D. (2008). ERMHAN: A Context-Aware Service Platform to Support Continuous Care Networks for Home-Based Assistance. *International Journal of Telemedicine and Applications*, (Vol. 2008).

Panagiotakopoulos, T. H., Lymberopoulos, D. K., & Manwlessos, G. M. (2008). Monitoring of patients suffering from special phobias exploiting context and profile information. In the *8th IEEE International Conference on Bioinformatics and BioEngineering (BIBE)*. Athens, Greece: IEEE.

Pandev, V., Ghosa, D., & Mukheriee, B. (2002). Exploring user profiles to support differentiated services in next-generation wireless networks. *IEEE Network*, *18*(5), 40–48. doi:10.1109/MNET.2004.1337734

Pandey, R. B., Mannion, R., & Ruskin, H. J. (2000). Effect of cellular mobility on immune response. *Physica A*, *283*, 447–450. doi:10.1016/S0378-4371(00)00206-5

Papoulis, A., & Pillai, S. U. (2002). *Probability, Random Variables and Stochastic Processes*. Columbus, OH: McGraw-Hill, Inc.

Paradiso, R., Belloc, C., Loriga, G., & Taccini, N. (2005). Wearable Healthcare Systems, New Frontiers of e-Textile. *Studies in Health Technology and Informatics*, *117*, 9–16.

Park, S., & Jayaraman, S. (2003). Enhancing the quality of life through wearable technology. *IEEE Engineering in Medicine and Biology Magazine*, *22*(3), 41–48. doi:10.1109/MEMB.2003.1213625

Parliamentary Office of Science and Technology. UK (2006), *Pervasive Computing, 263*.

Paterson, G. I., Shepherd, M., & Wang, X. Watters, C. & Zitner, D. (2002). Using the XML-based Clinical Document Architecture for Exchange of Structured Discharge Summaries. In *Proceedings of the 35th Hawaii International Conference on System Sciences*, (pp. 119-128).

Pawson, D. (2002). *XSL-FO Making XML Look Good in Print*.

PDB. *protein data bank* (n.d.). Retrieved from http://www.rcsb.org/pdb/

Pelosi, L., & Slade, T. (2000). Working memory dysfunction in major depression: An event-related potential study. *Clinical Neurophysiology*, *11*(9), 1531–1543. doi:10.1016/S1388-2457(00)00354-0

Perelson, A. S., & Nelson, P. W. (1999). Mathematical analysis of HIV-1 dynamics in vivo. *SIAM Review*, *41*(1), 3–44. doi:10.1137/S0036144598335107

Perelson, A. S., & Oster, G. F. (1979). Theoretical studies of clonal selection: Minimal antibody repertoire size and reliability of self-non-self discrimination. *Journal of Theoretical Biology*, *81*(4), 645–670. doi:10.1016/0022-5193(79)90275-3

Perrakis, A., Morris, R., & Lamzin, V. S. (1999). Automated protein model building combined with iterative structure refinement. *Nature Structural Biology, 6,* 458–463. doi:10.1038/8263

Perrin, D., Ruskin, H. J., & Crane, M. (2006c). HIV modelling - a parallel implementation of a lymph network. In *Selected proceedings of Third International Conference on Cluster and Grid Computing Systems* (CGCS 2006), Venice, Italy.

Pillow, J., Shlens, J., Paninski, L., Sher, A., Litke, A., Chichilnisky, E. J., & Simoncelli, E. (2008). Spatiotemporal correlations and visual signalling in a complete neuronal population. *Nature, 454,* 995–999. doi:10.1038/nature07140

Plassmann, P., & Ring, E. (1997). An open system for the acquisition and evaluation of medical thermological images. *European Journal of Thermology, 7,* 216–220.

Podichetty, V.K, Biscup, R.S. (2003). E-Health: A New Approach in Healthcare Practice. *The Internet Journal of Allied Health Sciences and Practice, 1*(2).

Pohanka, M., Skládal, P., & Pavlis, O. (2008). Label-free piezoelectric immunosensor for rapid assay of Escherichia coli. *Journal of Immunoassay & Immunochemistry, 29*(1), 70–79. doi:10.1080/15321810701735120

Polanski, A., & Kimmel, M. (2007). *Bioinformatics.* Berlin: Springer-Verlag.

Pollack, M. E. (2003). Autominder: An intelligent cognitive orthotic system for people with memory impairment. *Robotics and Autonomous Systems, 44*(3-4), 273–282. doi:10.1016/S0921-8890(03)00077-0

Poon, C. C. Y., & Zhang, Y. (2006). A Novel Biometrics Method to Secure Wireless Body Area Sensor Networks for Telemedicine and M-Health. *IEEE Communications Magazine, 44*(4), 73–81. doi:10.1109/MCOM.2006.1632652

Popov, A. N., & Bourenkov, G. P. (2003). Choice of data-collection parameters based on statistic modeling. *Acta Crystallogr. D, 59,* 1145–1153. doi:10.1107/S0907444903008163

Potter, S. M. (2001). Distributed processing in cultured neuronal networks. *Progress in Brain Research, 130,* 49–62. doi:10.1016/S0079-6123(01)30005-5

Potterton, L., McNicholas, S., Krissinel, E., Gruber, J., Cowtan, K., & Emsley, P. (2004). Developments in the CCP4 molecular-graphics project. *Acta Crystallographica. Section D, Biological Crystallography, 60,* 2288–2294. doi:10.1107/S0907444904023716

Prasad, P. N. (2003). *Introduction to Biophotonics* (pp. 312-314). New York: John Wiley & Sons.

Prater, E., Frazier, G. V., & Reyes, P. M. (2005). Future impacts of RFID on e-supply chains in grocery retailing. *Supply Chain Management: An International Journal, 10*(2), 134–142. doi:10.1108/13598540510589205

Preuveneers, D., Van Den Bergh, J., Wagelaar, D., Georges, A., Rigole, P., Clerckx, T., & Berbers, Y. Coninx K., Jonckers V., & De Bosschere K. (2004). Towards an extensible context ontology for ambient intelligence. In P. Markopoulos, B. Eggen, E. Aarts, & J. L. Crowley (Eds.), *2nd European Symposium on ambient Intelligence, Ambient Intelligence* (3295, 138-159). Eindhoven: Springer.

Provencher, S., W. (2001). Automatic quantification of localized in vivo ^{1}H spectra with LCModel. *NMR in Biomedicine, 14*(4), 260–264. doi:10.1002/nbm.698

Punie, Y. (2003). *A social and technological view of Ambient Intelligence in Everyday Life: What bends the trend?* European Media, Technology and Everyday Life Research Network, (EMTEL2) Key Deliverable Work Package 2, EC DG-JRC IPTS, Sevilla.

Rakotonirainy, A., Loke, S. W., & Fitzpatrick, G. (2000). Context-Awareness for the Mobile Environment. *Position Paper for CHI2000 Workshop on "The What, Who, Where, When, Why, and How of Context Awareness".* The Hague, The Netherlands.

Ramachandran, G. N., Ramakrishnan, C., & Sasisekharan, V. (1963). Stereochemistry of polypeptide chain configurations. *Journal of Molecular Biology, 7,* 95–99. doi:10.1016/S0022-2836(63)80023-6

Ramakrishnan, C., & Ramachandran, G. N. (1965). Stereochemical criteria for polypeptide and protein chain conformations. II. Allowed conformations for a pair of peptide units. *Biophysical Journal, 5,* 909–933. doi:10.1016/S0006-3495(65)86759-5

Ranganathan, A., & Campbell, R. H. (2003). A Middleware for Context-Aware Agents in Ubiquitous Computing Environments. In M. Endler, & D. Schmidt (Eds.), *ACM/IFIP/USENIX International Middleware Conference, LNCS* (2672: 143-161). Rio de Janeiro, Brazil: Springer.

Ras Mol (n.d.). molecular visualization freeware. Retrieved from http://www.umass.edu/microbio/- rasmol/

Reddy, M., & Dourish, P. (2002). A finger on the pulse: temporal rhythms and information seeking in medical work. In the *ACM Conference on Computer Supported Cooperative Work*, CSCW. New Orleans, Louisiana: ACM.

Reynaud, S., Mathieu, G., Girard, P., & Hernandez, F. (2006), "LAVOISIER: A Data Aggregation and Unification Service", Proceedings of Computing in High Energy and Nuclear Physics (CHEP06), Mumbai, India, February 2006.

Rhodes, G. (2006), *Crystallography Made Crystal Clear*, Elsevier.

Richmond, B., & Wiener, M. (2004). Recruitment order: a powerful neural ensemble code. *Nature Neuroscience*, 7, 97–98. doi:10.1038/nn0204-97

Riecken, D. (2004). Personalized Views of Personalization. *Communications of the ACM*, 43(3), 74–82.

Rieke, F., Warland, D., Van Steveninck, R. de R., & Bialek, W. (1997). *Spikes: Exploring the Neural Code*. Cambridge, MA: MIT Press.

Rivas, G. A., Rubianes, M. D., Rodríguez, M. C., Ferreyra, N. F., Luque, G. L., & Pedano, M. L. (2007). Carbon nanotubes for electrochemical biosensing. *Talanta*, 74(3), 291–307. doi:10.1016/j.talanta.2007.10.013

Rizzoli, A., Rosa, R., Rosso, F., Buckley, A., & Gould, E. (2007). West Nile virus circulation detected in northern Italy in sentinel chickens. *Vector Borne and Zoonotic Diseases (Larchmont, N.Y.)*, 7(3), 411–417. doi:10.1089/vbz.2006.0626

Roberts, R. J. (2001). PubMed Central: The GenBank of the published literature. *Proceedings of the National Academy of Sciences of the United States of America*, 98(2), 381–382.

Roque, A. C., & Andre, T. C. (2002). Mammography and computerized decision systems: a review. *Annals of the New York Academy of Sciences*, 980, 83–94. doi:10.1111/j.1749-6632.2002.tb04890.x

Rossmann, M. G., & Arnold, E. (2001), Editors of International Tables for Crystallography, Volume **F**: Crystallography of biological macromolecules, Part 25: *Macromolecular Crystallography Programs*, pp. 685-743, International Union of Crystallography.

Rossmann, M. G., & Blow, D. M. (1962). The detection of sub-units within the crystallographic asymmetric unit. *Acta Crystallographica*, 15, 24–31. doi:10.1107/S0365110X62000067

Rumelhart, D. E., Hinton, G. E., & Williams, R. J. (1986). Learning internal representations by error propagation. In Rumelhart, D,E and McClelland, J.L.(Eds.), *Parallel Distributed Processing*, (Vol. 1: *Foundations*, pp. 318-362). Cambridge, MA: MIT press.

Ruskin, H. J., & Burns, J. (2005). *Network emergence in immune system shape space*. (. *LNCS*, 3481, 1254–1263.

Ruskin, H. J., & Burns, J. (2006). Weighted networks in immune system shape space. *Physica A*, 365(2), 549–555. doi:10.1016/j.physa.2005.11.006

Ruskin, H. J., Pandey, R. B., & Liu, Y. (2002). Viral load and stochastic mutation in a Monte Carlo simulation of HIV. *Physica A*, 311(1-2), 213–220. doi:10.1016/S0378-4371(02)00832-4

Sachlos, E., & Czernuszka, J. T. (2003). Making tissue engineering scaffolds work Review on the application of solid freeform fabrication technology to the production of tissue engineering scaffolds. *European Cells & Materials*, 5, 29–40.

Sander, C., & Schneider, R. (1991). Database of homology derived protein structures and the structural meaning of sequence alignment. *Proteins*, 9, 56–68. doi:10.1002/prot.340090107

Sarela, A., Korhonen, I., Lotjonen, J., Sola, M., & Myllymaki, M. (2003). IST Vivago - an intelligent social and remote wellness monitoring system for the elderly. In the conference on *Information Technology Applications in Biomedicine* (4th International IEEE EMBS proceedings). Birmingham, UK: IEEE.

Sarvazyan, A. P., Lizzi, F. L., & Wells, P. N. (1991). A new philosophy of medical imaging. *Medical Hypotheses, 36*(4), 327–335. doi:10.1016/0306-9877(91)90005-J

Satyanarayanan, M. (2001). Pervasive Computing: Vision and Challenges. *IEEE Personal Communications, 8*(4), 10–17. doi:10.1109/98.943998

Schaefer, G., & Starosolski, R. (2008). A comparison of two methods for retrieval of medical images in the compressed domain. *30th IEEE Int. Conference Engineering in Medicine and Biology*, (pp. 402–405).

Schaefer, G., Nakashima, T., & Yokota, Y. (2008). Fuzzy Classification for Gene Expression Data Analysis. In A. Kelemen, A. Abraham, & Y. Chen (Eds.), *Computational Intelligence in Bioinformatics*. Berlin: Springer-Verlag.

Schaefer, G., Tait, R., & Zhu, S. (2006). Overlay of thermal and visual medical images using skin detection and image registration. *28th IEEE Int. Conference Engineering in Medicine and Biology*, (pp. 965–967).

Schaefer, G., Zavisek, M., & Nakashima, T. (2009). Thermography based breast cancer analysis using statistical features and fuzzy classification. *Pattern Recognition*.

Schaefer, G., Zhu, S., & Ruszala, S. (2005). Visualisation of medical infrared image databases. *27th IEEE Int. Conference Engineering in Medicine and Biology*, (pp. 1139–1142).

Schilit, B. N., Adams, N. L., & Want, R. (1994). Context-aware computing applications. Paper presented at the *IEEE Workshop on Mobile Computing Systems and Applications*, Santa Cruz, CA, USA.

Schlehuber, S., & Skerra, A. (2001). Duocalins: engineered ligand-binding proteins with dual specificity derived from the lipocalin fold. *Biological Chemistry, 382*, 1335–1342. doi:10.1515/BC.2001.166

Schlehuber, S., & Skerra, A. (2002). Tuning ligand affinity, specificity, and folding stability of an engineered lipocalin variant a so-called anticalin using a molecular random approach. *Biophysical Chemistry, 96*, 213–228. doi:10.1016/S0301-4622(02)00026-1

Schlehuber, S., & Skerra, A. (2005). Lipocalins in drug discovery: from natural ligand-binding proteins to anticalins. *Drug Discovery Today, 10*, 23–33. doi:10.1016/S1359-6446(04)03294-5

Schlehuber, S., Beste, G., & Skerra, A. (2000). A novel type of receptor protein, based on the lipocalin scaffold, with specificity for digoxigenin. *Journal of Molecular Biology, 297*, 1105–1120. doi:10.1006/jmbi.2000.3646

Schmidt, T. G. M., & Skerra, A. (1994). One-step affinity purification of bacterially produced proteins by means of the "Strep tag" and immobilized recombinant core streptavidin. *Journal of Chromatography. A, 676*, 337–345. doi:10.1016/0021-9673(94)80434-6

Schumacher, M., Fernandez, B. E., Hybertson, D., Buschmann, F., & Peter Sommerlad, P. (2006). *Security Patterns: Integrating Security and Systems Engineering*. Hoboken, NJ: Wiley.

Segev R., Baruchi, I., Hulata, E., Ben-Jacob, E. (2004). Hidden Neuronal Correlations in Cultured Networks. *Phys. Rev. Lett. 92*(11): 118102(1)-118102(4).

Seiden, P., & Celada, F. (1992). A model for simulating cognate recognition and response in the immune system. *Journal of Theoretical Biology, 158*, 329–357. doi:10.1016/S0022-5193(05)80737-4

Sen, P. K. (2002). Computational sequence analysis: Genomics and statistical controversies. In Y. P. Chaubey (ed.), *Recent Advances in Statistical Methods* (pp. 274-289). London: World Scien. Publ.

Shadlen, M. N., & Newsome, W. T. (1998). The variable discharge of cortical neurons: implications for connectivity, computation and information coding. *The Journal of Neuroscience, 18*, 3870–3896.

Shah, A. J., Wang, J., Yamada, T., & Fajardo, L. L. (2003). Digital mammography: a review of technical development and clinical applications. *Clinical Breast Cancer, 4*(1), 63–70. doi:10.3816/CBC.2003.n.013

Shahaf, G., Eytan, D., Gal, A., Kermany, E., & Lyakhov, V. (2008). Order-Based Representation in Random Networks of Cortical Neurons. *PLoS Computational Biology, 4*(11), e1000228. doi:10.1371/journal.pcbi.1000228

Sharkasi, A., Crane, M., Ruskin, H. J., & Matos, J. A. O. (2006). The reaction of stock markets to crashes and events: A comparison study between emerging and mature markets using wavelet transforms. *Physica A, 368*(2), 511–521. doi:10.1016/j.physa.2005.12.048

Shein, M., Volman, V., Raichman, N., Hanein, Y., & Ben-Jacob, E. (2008). Management of synchronized network activity by highly active neurons. *Physical Biology*, *5*(3), 36008. doi:10.1088/1478-3975/5/3/036008

Sheldrick, G. M. (2008). A short history of SHELX. *Acta Crystallographica. Section A, Foundations of Crystallography*, *64*, 112–122. doi:10.1107/S0108767307043930

Sheldrick, G. M., & Schneider, T. R. (1997). SHELXL: high resolution refinement. *Methods in Enzymology*, *277*, 319–343. doi:10.1016/S0076-6879(97)77018-6

Shen, G., Liu, M., Cai, X., & Lu, J. (2008). A novel piezoelectric quartz crystal immnuosensor based on hyperbranched polymer films for the detection of alpha-Fetoprotein. *Analytica Chimica Acta*, *630*(1), 75–81.

Shen, L., Rangayyan, R. M., & Desautels, J. E. L. (1994). Application of shape analysis to mammographic calcifications. *IEEE Transactions on Medical Imaging*, *13*, 263–274. doi:10.1109/42.293919

Shepperd, S., Charnock, D., & Gann, B. (1999). Helping patients access high-quality health information. *BMJ (Clinical Research Ed.)*, *319*, 764–766.

Sijens, P. E., Dagnelie, P. C., Halfwrk, S., van Dijk, P., Wicklow, K., & Oudkerk, M. (1998). Understanding the discrepancies between 31P MR spectroscopy assessed liver metabolite concentrations from different institutions. *Magnetic Resonance Imaging*, *16*(2), 205–211. doi:10.1016/S0730-725X(97)00246-4

Sinthanayothin, C., Boyce, J. F., Williamson, T. H., Cook, H. L., Mensah, E., Lal, S., & Usher, D. (2002). Automated detection of diabetic retinopathy on digital fundus images. *Diabetic Medicine*, *19*(2), 105–112. doi:10.1046/j.1464-5491.2002.00613.x

Siponen, M. (2005). Analysis of modern IS security development approaches: towards the next generation of social and adaptable ISS methods. *Information and Organization*, *15*(4), 339–375. doi:10.1016/j.infoandorg.2004.11.001

Skerra, A. (2000). Engineered protein scaffolds for molecular recognition. *Journal of Molecular Recognition*, *13*, 167–187. doi:10.1002/1099-1352(200007/08)13:4<167::AID-JMR502>3.0.CO;2-9

Skerra, A. (2000). Lipocalins as a scaffold. *Biochimica et Biophysica Acta*, *1482*, 337–350.

Skerra, A. (2001). Anticalins: a new class of engineered ligand-binding proteins with antibody-like properties. *Journal of Biotechnology*, *74*, 257–275.

Small, G. W. (2002). What we need to know about age related memory loss. *British Medical Journal*, *324*, 1502–1505. doi:10.1136/bmj.324.7352.1502

Smith, D., & Newell, L. M. (2002). A Physician's Perspective: Deploying the EMR. *Journal of Healthcare Information Management*, *16*(2), 71–79.

Smith, J. R., Fishkin, K., Jiang, B., Mamishev, A., Philipose, M., & Rea, A. D. (2005). RFID-based techniques for human-activity detection. *Communications of the ACM*, *48*(9), 39–44. doi:10.1145/1081992.1082018

Snel, J., Olabarriaga, S., Alkemade, J., Andel, H., Nederveen, A., Majoie, C., et al. (2006), "A Distributed Workflow Management System for Automated Medical Image Analysis and Logistics", 19th IEEE International Symposium on Computer-Based Medical Systems.

Soerensen, R. A., & Nygaard, J. M. (2008). Distributed zero configuration base station. In *Proceedings of the 2nd International Conference on Pervasive Computing Technologies for Healthcare*, Tampere, Finland.

Softky, W. R., & Koch, C. (1993). The highly irregular firing of cortical cells is inconsistent with temporal integration of random EPSPs. *The Journal of Neuroscience*, *13*, 334–350.

Soille, P. (1999). *Morphological image analysis: Principles and applications*. Berlin: Springer-Verlag.

Sousa, S. F., Fernandes, P. A., & Ramos, M. J. (2006). Protein–Ligand Docking: Current Status and Future Challenges. *PROTEINS: Structure, Function, and Bioinformatics*, *65*, 15–26. doi:10.1002/prot.21082

Spyrou, G., Nikolaou, M., Koufopoulos, K., & Ligomenides, P. (2002). A computer based model to assist in improving early diagnosis of breast cancer. *In Proceedings of the 7th World Congress on Advances in Oncology and 5th International Symposium on Molecular Medicine*, October 10-12, 2002, Creta Maris Hotel, Hersonissos, Crete, Greece.

Spyrou, G., Nikolaou, M., Koussaris, M., Tsibanis, A., Vassilaros, S., & Ligomenides, P. (2002). *A System for Computer Aided Early Diagnosis of Breast Cancer based on Microcalcifications Analysis*. 5th European Conference on Systems Science, 16-19 October 2002, Creta Maris Hotel, Crete, Greece.

Spyrou, G., Pavlou, P., Harissis, A., Bellas, I., & Ligomenides, P. (1999). Detection of Microcalcifications for Early Diagnosis of Breast Cancer. In *Proceedings of the 7th Hellenic Conference on Informatics, University of Ioannina Press, August 26-28, 1999*, Ioannina, Greece, (p. V104).

Stamatakos, G., Dionysiou, D., Zacharaki, E., et al. (2002), "In silico radiation oncology: combining novel simulation algorithms with current visualization techniques", Proceedings of the IEEE, Special Issue on "Bioinformatics: Advances and Chalenges", Volume 90, Issue 11, pp. 1764- 1777.

Stanford, V. (2002). Using pervasive computing to deliver elder care. *IEEE Pervasive Computing / IEEE Computer Society [and] IEEE Communications Society, 1*, 10–13. doi:10.1109/MPRV.2002.993139

Steinman, D. A. (2002). Image-based computational fluid dynamics modeling in realistic arterial geometries. *Annals of Biomedical Engineering, 30*(4), 483–497. doi:10.1114/1.1467679

Steinman, D. A., Milner, J. S., Norley, C. J., Lownie, S. P., & Holdsworth, D. W. (2003). Image-based computational simulation of flow dynamics in a giant intracranial aneurysm. *AJNR. American Journal of Neuroradiology, 24*(4), 559–566.

Steipe, B. (1998). Protein Design Concepts. In P. v. R. Schleyer, et al. (Ed.), *The Encyclopedia of Computational Chemistry* (pp. 2168-2185). Chichester, UK: John Wiley & Sons.

Stoltz, J. F., Wang, X., Muller, S., & Labrador, V. (1999). Introduction to mechanobiology of cells. *Applied Mechanics and Engineering, 4*, 177–183.

Strang, T., & Linnoff-Popien, C. (2004). *A Context Modeling Survey*. Presented at the Workshop on Advanced Context Modelling, Reasoning and Management (proceedings of the 6th International Conference on Ubiquitous Computing – UbiComp), Nottingham, England.

Su, J., Zhang, J., Liu, L., Huang, Y., & Mason, R. P. (2008). Exploring feasibility of multicolored CdTe quantum dots for in vitro and in vivo fluorescent imaging. *Journal of Nanoscience and Nanotechnology, 8*(3), 1174–1177.

Sun, J., Zhu, M. Q., Fu, K., Lewinski, N., & Drezek, R. A. (2007). Lead sulfide near-infrared quantum dot bioconjugates for targeted molecular imaging. *International Journal of Nanomedicine, 2*(2), 235–240.

Sun, W., Darling, A., Starly, B., & Nam, J. (2004). Computer-aided tissue engineering: overview, scope and challenges. *Biotechnology and Applied Biochemistry, 39*, 29–47. doi:10.1042/BA20030108

Surridge, M., Taylor, S., De Roure, D., & Zaluska, E. (2005), "Experiences with GRIA-Industrial Applications on a Web Services Grid", in Proceedings of the First International Conference on e-Science and Grid Computing, pp. 98-105. IEEE Press.

Taboas, J. M., Maddox, R. D., Krebsbach, P. H., & Hollister, S. J. (2003). Indirect solid free form fabrication of local and global porous, biomimetic and composite 3D polymerceramic scaffolds. *Biomaterials, 24*, 181–194. doi:10.1016/S0142-9612(02)00276-4

Tait, R., Schaefer, G., & Hopgood, A. (2008). Intensity-based image registration using multiple distributed agents. *Knowledge-Based Systems, 21*(3), 256–264. doi:10.1016/j.knosys.2007.11.013

Tait, R., Schaefer, G., Howell, K., Hopgood, A., Woo, P., & Harper, J. (2006). Automated overlay of visual and thermal medical images. *Int. Biosignal Conference.*

Tan, K. H., Chua, C. K., Leong, K. F., Cheah, C. M., Cheang, P., & Abu Bakar, M. S. (2003). Scaffold development using selective laser sintering of polyetheretherketone-hydroxyapatite biocomposite blends. *Biomaterials, 24*, 3115–3123. doi:10.1016/S0142-9612(03)00131-5

Tancred, D. C., Carr, A. J., & McCormack, B. A. O. (1998). Development of a new synthetic bone graft. *Journal of Materials Science. Materials in Medicine, 9*, 819–823. doi:10.1023/A:1008992011133

Tancred, D. C., McCormack, B. A. O., & Carr, A. J. (1998). A synthetic bone implant macroscopically identical to cancellous bone. *Biomaterials, 19*, 2303–2311. doi:10.1016/S0142-9612(98)00141-0

Tangherlini, A., Merla, A., & Romani, G. (2006). Field-warp registration for biomedical high-resolution thermal infrared images. *28ᵗʰ IEEE Int. Conference on Engineering in Medicine and Biology*, (pp. 961–964).

Taylor A.N. (2000). *Royal Brompton Hospital Research Open Day 2000 Proceedings*.

Taylor, G. (2003). The phase problem. *Acta Crystallogr. D, 59*, 1881–1890. doi:10.1107/S0907444903017815

Terwilliger, T. C. (2000). Maximum likelihood density modification. *Acta Crystallographica. Section D, Biological Crystallography, 56*, 965–972. doi:10.1107/S0907444900005072

Terwilliger, T. C. (2003). Automated main-chain model building by template matching and iterative fragment extension. *Acta Crystallographica. Section D, Biological Crystallography, 59*, 38–44. doi:10.1107/S0907444902018036

Terwilliger, T. C., & Berendzen, J. (1999). Automated MAD and MIR structure solution. *Acta Crystallographica. Section D, Biological Crystallography, 55*, 849–861. doi:10.1107/S0907444999000839

The AKOGRIMO Project. http://www.mobileGrids.org/

The BRIDGES Project. http://www.brc.dcs.gla.ac.uk/projects/bridges/

The EGEE Project. http://www.eu-egee.org/

The gLite middleware, (2008), http://glite.web.cern.ch/glite

The WISDOM Initiative, http://wisdom.eu-egee.fr/

Thireou, T., Altamazoglou, V., Levakis, M., Eliopoulos, E., Hountas, A., Tsoucaris, G., & Bethanis, K. (2007). CrystTwiv: a webserver for automated phase extension and refinement in X-ray crystallography. *Nucleic Acids Research, 35*, W718–W722. doi:10.1093/nar/gkm225

Thompson, J. D., Higgins, D. G., & Gibson, T. J. (1994). ClustalW: improving the sensitivity of progressive multiple sequence alignment through sequence weighting, position specific gap penalties and weight matrix choice. *Nucleic Acids Research, 22*, 4673–4680. doi:10.1093/nar/22.22.4673

Thomson, A. M. (1997). Activity-dependent properties of synaptic transmission at two classes of connections made by rat neocortical pyramidal axons in vitro. *The Journal of Physiology, 502*, 131–147. doi:10.1111/j.1469-7793.1997.131bl.x

Thorpe, S., Delorme, A., & VanRullen, R. (2001). Spike based strategies for rapid processing. *Neural Networks, 14*(6-7), 715–726. doi:10.1016/S0893-6080(01)00083-1

Thurner, S., & Biely, C. (2007). The eigenvalue spectrum of lagged correlation matrices. *Acta Physiologica Polonica, 38*(13), 4111–4122.

Tickle, I. J., Laskowski, R. A., & Moss, D. S. (1998). Error estimates of protein structure coordinates and deviations from standard geometry by full-matrix refinement of B- and B2-crystallin. *Acta Crystallographica. Section D, Biological Crystallography, 54*, 243–252. doi:10.1107/S090744499701041X

Timins, J. K. (2005). Controversies in mammography. *New Jersey Medicine, 102*(1-2), 45–49.

Tobias, J. W., Shrader, T. E., Rocap, G., & Varshavsky, A. (1991). The N-end rule in bacteria. *Science, 254*(5036), 1374–1377. doi:10.1126/science.1962196

Tran, Q., & Mynatt, E. (2002). *Cook's collage: Two exploratory designs*. Paper presented at the New Technologies for Families Workshop, CHI 2002, Minneapolis, MN.

Tran, Q., Calcaterra, G., & Mynatt, E. (2005). Cook's collage: Deja vu display for a home Kkitchen. In. *Proceedings of HOIT, 2005*, 15–32.

Tsang, V. L., & Bhatia, S. N. (2004). Three dimensional tissue fabrication. *Advanced Drug Delivery Reviews, 56*, 1635–1647. doi:10.1016/j.addr.2004.05.001

Tsoucaris, G. (1970a). A new method for phase determination. The 'maximum determinant rule'. *Acta Crystallographica. Section A, Crystal Physics, Diffraction, Theoretical and General Crystallography, 26*, 492–499. doi:10.1107/S0567739470001298

Tsoucaris, G. (1970b). The strengthening of direct methods of crystal structure determination by use of data from isomorphous compounds. *Acta Crystallographica. Section A, Crystal Physics, Diffraction, Theoretical and General Crystallography, 26*, 499–501. doi:10.1107/S0567739470001304

Tzamalis, P., Bethanis, K., Hountas, A., & Tsoucaris, G. (2003). The crystallographic symmetry test for the correctness of a set of phases. *Acta Crystallographica. Section A, Foundations of Crystallography, 59*, 28–33. doi:10.1107/S0108767302018810

Tzung-Shi Chen. Yi-Shiang Chang, Hua-Wen Tsai, Chih-Ping Chu, (2007), "Data Aggregation for Range Query in Wireless Sensor Networks", IEEE Wireless Communications & Networking Conference (WCNC 2007), Hong Kong.

Uematsu, S. (1985). Symmetry of skin temperature comparing one side of the body to the other. *Thermology, 1*(1), 4–7.

Uhrmacher, A. M., Tyschler, P., & Tyschler, D. (2000). Modeling and simulation of mobile agents. *Future Generation Computer Systems, 17*, 107–118. doi:10.1016/S0167-739X(99)00107-7

Usón, I., & Sheldrick, G. M. (1999). Advances in direct methods for protein crystallography. *Current Opinion in Structural Biology, 9*(5), 643–648. doi:10.1016/S0959-440X(99)00020-2

Vagin, A. A., & Teplyakov, A. (1997). MOLREP: an Automated Program for Molecular Replacement J. *Appl. Cryst., 30*, 1022–1025. doi:10.1107/S0021889897006766

Vail, N. K., Swain, L. D., Fox, W. C., Aufdlemorte, T. B., Lee, G., & Barlow, J. W. (1999). Materials for biomedical applications. *Materials & Design, 20*, 123–132. doi:10.1016/S0261-3069(99)00018-7

van den Boogaart, A., Van Hecke, P., Van Hulfel, S., Graveron-Dermilly, D., van Ormondt, D., & de Beer, R. (1996). MRUI: a graphical user interface for accurate routine MRS data analysis. *Proceeding of the European Society for Magnetic Resonance in Medicine and Biology 13th Annual Meeting*, Prague, (p. 318).

van Huffel, S., Chen, H., Decanniere, C., & Hecke, P. V. (1994). Algorithm for time-domain NMR data fitting based on total least squares. *Journal of Magnetic Resonance. Series A., 110*, 228–237. doi:10.1006/jmra.1994.1209

Van Rullen, R., & Thorpe, S. J. (2001). Rate coding versus temporal order coding: what the retinal ganglion cells tell the visual cortex. *Neural Computation, 13*, 1255–1283. doi:10.1162/08997660152002852

Varshavsky, A. (1997). The N-end rule pathway of protein degradation. *Genes to Cells, 2*(1), 13–28. doi:10.1046/j.1365-2443.1997.1020301.x

Varshney, U. (2003). Pervasive Healthcare. *IEEE Computer, 36*(12), 138–140.

Varshney, U. (2007). Pervasive healthcare and wireless health monitoring. *Mobile Networking Applications, 2*(2-3), 113–127. doi:10.1007/s11036-007-0017-1

Vasileios G. Stamatopoulos V.G., Karagiannis, G.E. (2005). *e-OpenDay Project Final Report.*

Vats, A., Tolley, N. S., Polak, J. M., & Gough, J. E. (2003). Scaffolds and biomaterials for tissue engineering: a review of clinical applications. *Clinical Otolaryngology, 28*, 165–172. doi:10.1046/j.1365-2273.2003.00686.x

Vogels, T. P., Rajan, K., & Abbott, L. F. (2005). Neural network dynamics. *Annual Review of Neuroscience, 28*, 357–376. doi:10.1146/annurev.neuro.28.061604.135637

Wagner, S. (2008a). Towards an open and easily extendible home care system infrastructure. In *Proceedings of the 2nd International Conference on Pervasive Computing Technologies for Healthcare*, Tampere, Finland.

Wagner, S. (2008b). Zero-configuration of pervasive healthcare sensor networks. In *Proceedings of the 3rd International Conference on Pervasive Computing and Applications (ICPCA' 2008)*, Alexandria, Egypt.

Waldo, J., Wyant, G., Wollrath, A., & Kendall, S. (1994), "A Note on Distributed Computing", http://research.sun.com/techrep/1994/smli_tr-94-29.pdf

Wallace, A. C., Laskowski, R. A., & Thornton, J. M. (1995). LIGPLOT: a program to generate schematic diagrams of protein-ligand interactions. *Protein Engineering, 8*, 127–134. doi:10.1093/protein/8.2.127

Walter, T. (2003). *Application de la morphologie mathématique au diagnostic de la rétinopathie diabétique à partir d images couleur.* PhD Thesis, Ecole Nationale Superieure des Mines de Paris.

Walter, T., Klein, J.-C., Massin, P., & Erginay, A. (2002). A contribution of image processing to the diagnosis of diabetic retinopathy-detection of exudates in color fundus images of the human retina. *IEEE Transactions on Medical Imaging, 21*(10), 1236–1243. doi:10.1109/TMI.2002.806290

Walters, G. J. (2001). Privacy and security: An ethical analysis. *Computers & Society, 31*(2), 8–23. doi:10.1145/503345.503347

Wang, B. C. (1985). Resolution of phase ambiguity in macromolecular crystallography. *Methods in Enzymology, 115*, 90–112. doi:10.1016/0076-6879(85)15009-3

Wang, X. H., Zhang, D. Q., Gu, T., & Pung, H. K. (2004). Ontology Based Context Modeling and Reasoning using OWL. In the *2nd IEEE Annual Conference on Pervasive Computing and Communications* (pp. 18-22). Orlando, FL: IEEE.

Web Services Description Language (WSDL) 1.1, (2008), http://www.w3.org/TR/wsdl

Weeks, C. M., & Miller, R. (1999). The design and implementation of SnB version 2.0. *Journal of Applied Crystallography, 32*, 120–124. doi:10.1107/S0021889898010504

Weeks, C. M., Blessing, R. H., Miller, R., Mungee, R., Potter, S. A., & Rappleye, J. (2002). Towards automated protein structure determination: BnP, the SnB-PHASES interface. *Zeitschrift fur Kristallographie, 217*, 686–693. doi:10.1524/zkri.217.12.686.20659

Weiser, M. (1989). The computer for the 21st century. *ACM SIGMobile Mobile Computing and Communications Review, 3*, 3–11. doi:10.1145/329124.329126

Weiser, M. (1991). The computer of the 21st century. *Scientific American, 265*(3), 66–75.

Whitaker, M. J., Quirk, R. A., Howdle, S. M., & Shakesheff, K. M. (2001). Growth factor release from tissue engineering scaffolds. *The Journal of Pharmacy and Pharmacology, 53*, 1427–1437. doi:10.1211/0022357011777963

Whitehouse, P. J. (Ed.). (1993). *Dementia*. Philadelphia: F. A. Davis Company.

Wiecek, B., Zwolenik, S., Jung, A., & Zuber, J. (1999). Advanced thermal, visual and radiological image processing for clinical diagnostics. *21st IEEE Int. Conference on Engineering in Medicine and Biology.*

Wikipedia, Service Oriented Infrastructure Definition, (2008), http://en.wikipedia.org/wiki/Service_Oriented_Infrastructure

Wintermantel, E., Mayer, J., Blum, J., Eckert, K. L., Luscher, P., & Mathey, M. (1996). Tissue engineering scaffolds using superstructures. *Biomaterials, 17*, 83–91. doi:10.1016/0142-9612(96)85753-X

Wireless World Research Forum. Working Group 2, (2003). Service Architectures for the Wireless World. Whitepaper on Service Adaptability, version 1.0. Retrieved June 23, 2007, from http://www.wireless-world-research.org.

Witherden, D. A., Kimpton, W. G., Washington, E. A., & Cahill, R. N. P. (1990). Non-random migration of CD4+, CD8+ and γδ+T19+ lymphocytes through peripheral lymph nodes. *Immunology, 70*, 235–240.

Wohlgemuth, S., & Ronacher, B. (2007). Auditory discrimination of amplitude modulations based on metric distances of spike trains. *Journal of Neurophysiology, 97*, 3082–3092. doi:10.1152/jn.01235.2006

Wong, J. Y., Velasco, A., Rajagopalan, P., & Pham, Q. (2003). Directed movement of vascular smooth muscle cells on gradient-compliant hydrogels. *Langmuir Journal, 19*, 1908–1913. doi:10.1021/la026403p

Wooldridge, M., & Jennings, N. (1995). Intelligent agents: Theory and practice. *The Knowledge Engineering Review, 2*(10), 115–152. doi:10.1017/S0269888900008122

Wright, D., Gutwirth, S., Friedewald, M., Vildjiounaite, E., & Punie, Y. (Eds.). (2008). *Safeguards in a World of Ambient Intelligence*. Springer.

Wright, T., & McGechan, A. (2003). Breast cancer: new technologies for risk assessment and diagnosis. *Molecular Diagnosis, 7*(1), 49–55. doi:10.2165/00066982-200307010-00009

Wu, C. H., & McLarty, J. (2000). *Neural Networks and Genome Informatics*. New York: Elsevier Science.

Wu, Y., Doi, K., Giger, M. L., & Nishikawa, R. M. (1992). Computerized detection of clustered microcalcifications in digital mammograms: applications of artificial neural networks. *Medical Physics, 19*, 555–560. doi:10.1118/1.596845

Wu, Y., Giger, M. L., Doi, K., Schmidt, R. A., & Metz, C. E. (1993). Artificial neural networks in mammography: application to decision making in the diagnosis of breast cancer. *Radiology, 187*, 81–87.

Wutoh, R., Boren, S. A., & Balas, E. A. (2004). eLearning: a review of Internet-based continuing medical education. *The Journal of Continuing Education in the Health Professions, 24*, 20–30. doi:10.1002/chp.1340240105

Job Description Language (JDL) Attributes Specification. (2008), https://edms.cern.ch/document/590869/1/

MEA types of Multi Channel Systems (n.d.). Retrieved from http://www.multichannelsystems.com/products-mea/microelectrode-arrays.html

Xiang, Y., Gu, Q., & Li, Z. (2003). A Distributed Framework of Web-based Telemedicine System. In *Proceedings of the 16th IEEE Symposium on Computer-Based Medical Systems,* (pp. 108-113).

Xiao, Y., Lasome, C., Moss, J., Mackenzie, C., & Faraj, S. (2001). Cognitive properties of a whiteboard: A case study in a trauma center. In *Proceedings of the 7th European Conference on Computer Supported Cooperative Work*, Bonn, Germany (pp. 16-20). Amsterdam: Kluwer.

Xiong, Z., Yan, Y., Zhang, R., & Sun, L. (2001). Fabrication of porous poly(L-lactic acid) scaffolds for bone tissue engineering via precise extrusion. *Scripta Materialia, 45*, 773–779. doi:10.1016/S1359-6462(01)01094-6

Xu, D., Keller, J. M., Popescu, M., & Bondugula, R. (2009). Applications of Fuzzy Logic in Bioinformatics. *Series on Advances in Bioinformatics and Computational Biology,* (Vol. 9).

Yamane, S., Iwasaki, N., Kasahara, Y., Harada, K., Majima, T., & Monde, K. (2007). Effect of pore size on in vitro cartilage formation using chitosan-based hyaluronic acid hybrid polymer fibers. *Journal of Biomedical Materials Research, 81*, 586–593.

Yang, S., Leong, K. F., Du, Z., & Chua, C. K. (2001). The design of scaffolds for use in tissue engineering: Part I. Traditional factors. *Tissue Engineering, 7*, 679–689. doi:10.1089/107632701753337645

Yankaskas, B. C., Schell, M. J., Bird, R. E., & Desrochers, D. A. (2001). Reassessment of breast cancers missed during routine screening mammography: a community-based study. *AJR. American Journal of Roentgenology, 177*(3), 535–541.

Yeates, T. O. (1997). Detecting and Overcoming Crystal Twinning. *Methods in Enzymology, 276*, 344–358. doi:10.1016/S0076-6879(97)76068-3

Yogesan, K., Constable, I., Eikelboom, R., & van Saarloos, P. (1998). Tele-ophthalmic screening using digital imaging devices. *Australian and New Zealand Journal of Ophthalmology, 26*(Suppl. 1), S9–S11. doi:10.1111/j.1442-9071.1998.tb01385.x

Yuan, W., Guan, D., Lee, S., & Lee, Y. (2007). The Role of Trust in Ubiquitous Healthcare. In the *9th International Conference on e-Health Networking, Application and Services*. Taipei, Taiwan: IEEE.

YunHee Kang. (2004), "An Extended OGSA Based Service Data Aggregator by Using Notification Mechanism", Grid and Cooperative Computing, Springer.

Zeng, H., Wang, H., Chen, F., Xin, H., Wang, G., & Xiao, L. (2006). Development of quartz-crystal-microbalance-based immunosensor array for clinical immunophenotyping of acute leukemias. *Analytical Biochemistry, 351*(1), 69–76. doi:10.1016/j.ab.2005.12.006

Zhang, S., Yang, J., Wu, Y., & Liu, J. (2005). *An enhanced massively multi-agent system for discovering HIV population dynamics.* (. *LNCS, 3645*, 988–997.

Zhang, W., Doi, K., Giger, M. L., Wu, Y., Nishikawa, R. M., & Schmidt, R. A. (1994). Computerized detection of clustered microcalcifications in digital mammograms using a shift-invariant artificial neural networks. *Medical Physics, 21*, 517–524. doi:10.1118/1.597177

Zhang, W., Doi, K., Giger, M. L., Wu, Y., Nishikawa, R. M., & Schmidt, R. A. (1996). An improved shift-invariant artificial neural networks for computerized detection of clustered microcalcifications in digital mammograms. *Medical Physics, 23*, 595–601. doi:10.1118/1.597891

Zhang, W., Yoshida, H., Nishikawa, R. M., & Doi, K. (1998). Optimally weighted wavelet transform based on supervised training for detection of microcalcifications in digital mammograms. *Medical Physics, 25*(6), 949–956. doi:10.1118/1.598273

Zhang, Y., Wang, H., Yan, B., Zhang, Y., Li, J., Shen, G., & Yu, R. (2008). A reusable piezoelectric immunosensor using antibody-adsorbed magnetic nanocomposite. *Journal of Immunological Methods, 332*(1-2), 103–111. doi:10.1016/j.jim.2007.12.019

Zhou, C., & Liu, L. (1993). Complex Hopfield model. *Optics Communications, 103*(1-2), 29–32. doi:10.1016/0030-4018(93)90637-K

Zitova, B., & Flusser, J. (2003). Image registration methods: a survey. *Image and Vision Computing, 21,* 977–1000. doi:10.1016/S0262-8856(03)00137-9

About the Contributors

Athina Lazakidou currently works at the University of Peloponnese, Department of Nursing in Greece as Lecturer in Health Informatics, and at the Hellenic Naval Academy as a Visiting Lecturer in Informatics. She worked as a Visiting Lecturer at the Department of Computer Science at the University of Cyprus (2000-2002) and at the Department of Nursing at the University of Athens (2002-2007). She did her undergraduate studies at the Athens University of Economics and Business (Greece) and received her BSc in Computer Science in 1996. In 2000, she received her Ph.D. in Medical Informatics from the Department of Medical Informatics, University Hospital Benjamin Franklin at the Free University of Berlin, Germany. She is also an internationally known expert in the field of computer applications in health care and biomedicine, with six books and numerous papers to her credit. She was also Editor of the "Handbook of Research on Informatics in Healthcare and Biomedicine" and "Handbook of Research on Distributed Medical Informatics and E-Health", the best authoritative reference sources for information on the newest trends and breakthroughs in computer applications applied to health care and biomedicine. Her research interests include health informatics, e-Learning in medicine, software engineering, graphical user interfaces, (bio)medical databases, clinical decision support systems, hospital and clinical information systems, electronic medical record systems, telematics, and other web-based applications in health care and biomedicine.

* * *

Theodoros E. Athanaileas received his Dipl.-Ing degree in Electrical and Computer Engineering from the National Technical University of Athens (NTUA), Greece in 2005 and received his PhD degree from the School of Electrical and Computer Engineering, NTUA in 2009. His current research interests focus on parallel algorithms, distributed computing and grid computing and their applications to a number of research fields, with particular emphasis on the simulation and design of wireless systems and biomedicine. Dr. Athanaileas is a member of the Technical Chamber of Greece.

Vassilis Atlamazoglou received his PhD in Biomedical Engineering from National Technical University of Athens (NTUA), Greece, in 2001. He was a postdoctoral fellow in Bioinformatics at the Biomedical Research Foundation of the Academy of Athens from 2004 to 2007. From 2005 to 2008, he was an adjunct lecturer on Bioinformatics at the University of Thessaly. He has also been working as a postdoctoral researcher at the Genetics Laboratory of Agricultural University of Athens and a research assistant at the Laboratory of Biomedical Optics and Applied Biophysics of the NTUA. His research interests include structural bioinformatics, molecular simulations, data mining and biomedical image analysis.

Eduard Babulak is international scholar, researcher, consultant, educator, professional engineer and polyglot with more than twenty five years of teaching experience and industrial experience as a professional engineer and consultant. Invited Speaker at University of Cambridge, UK in March, 2009. Invited Speaker at MIT, USA in September 2005, Expert-Evaluator for the European Commission in Brussels, June, 2007. Professor Babulak is Fellow of the Royal Society for the encouragement of Arts, Manufactures and Commerce (RSA), Fellow of British Computer Society (BCS), Nominated Fellow of the IET, Nominated Distinguished Member & Senior Member of ACM, Mentor and Senior Member of IEEE, served as a Chair of the IEEE Vancouver Ethics, Professional and Conference Committee. He works as Full Professor of Computing Science and Information Systems at the University of the South Pacific in Suva, Fiji. He worked as Full Professor and Head of MIS Department in Cyprus, held five Visiting Professorships in Canada (B.C. and Quebec), Spain, in Czech Republic (Prague and Pardubice). He worked as Associate Professor in California, Senior Lecturer in UK, Lecturer in Pennsylvania, Germany, Austria, Lecture and Teaching Assistant in Canada and College Instructor in Czechoslovakia. His academic and engineering work was recognized internationally by the Engineering Council in UK, European Federation of Engineers and credited by the British Columbia and Ontario Society of Professional Engineers in Canada. Prof. Babulak is Editor-in-Chief, Honorary Editor, Co-Editor and Guest Editor. His research interests are in Future Networks and Ubiquitous Computing and QoS, E-Commerce, E-Health, IT, MIS, Applied Informatics in Transportation, E-Manufacturing, Human Centric Computing, E-Learning, Automation and Applied Mathematics. Professor Babulak speaks 14 languages, a member of the Institution of Engineering Technology (MIET), American Society for Engineering Education (MASEE), American Mathematical Association (MAMA) and Mathematical Society of America (MMSA). Professor Babulak's biography was cited in the Cambridge Blue Book, Cambridge Index of Biographies and number of issues of Who's Who.

Kostas Bethanis studied physics at the University of Athens and attended *Part III* of Mathematical *Tripos* in the Department of Theoretical Physics and Applied Mathematics at the University of Cambridge, U.K. He received his PhD in theory and applications of x-ray crystallography of biological macromolecules from Agricultural University of Athens (AUA), Greece, in 2001. He also conducted post doctorate studies at AUA from 2001 to 2003. From 2003 to 2008, he was a lecturer and from 2008 he is an assistant professor at the Physics Laboratory, Dept. of Science, AUA. His teaching subjects include biophysics, macromolecular crystallography, supramolecular chemistry and bio-nanotechnology. His research interests include the development of new methods and algorithms for phase determination in macromolecular crystallography, the determination of protein structures, their complexes and of supramolecular structures and the study of inclusion phenomena.

Martin Crane received his B.A. B.A.I. (Mech. Eng.) degrees from Trinity College Dublin in 1989 and his Ph.D. the same institution in 1993. He has worked in a variety of areas of Computational Science such as CFD, Combustion Modelling, Financial Data Analysis and, more recently, Systems Biology. He has been a College Lecturer in Dublin City University since 1999.

Hazem Mokhtar El-Bakry (Mansoura, EGYPT 20-9-1970) received B.Sc. degree in Electronics Engineering, and M.Sc. in Electrical Communication Engineering from the Faculty of Engineering, Mansoura University – Egypt, in 1992 and 1995 respectively. Dr. El-Bakry received Ph. D degree from University of Aizu - Japan in 2007. Currently, he is assistant professor at the Faculty of Computer Science and Information Systems – Mansoura University – Egypt. His research interests include neural networks, pattern recognition, image processing, biometrics, cooperative intelligent systems and electronic circuits. In these areas, he has published more than 55 papers in major international journals and 120 papers in refereed international conferences. El-Bakry has the patent No. 2003E 19442 DE HOL / NUR, Magnetic Resonance, SIEMENS Company, Erlangen, Germany, 2003. Furthermore, he is associate editor for journal of computer science and network security (IJCSNS) and journal of convergence in information technology (JCIT). Furthermore, he is a referee for IEEE Transactions on Signal Processing, Journal of Applied Soft Computing, the International Journal of Machine Graphics & Vision, the International Journal of Computer Science and Network Security, Enformatika Journals, WSEAS Journals and many different international conferences organized by IEEE. In addition, he has been awarded the Japanese Computer & Communication prize in April 2006. Moreover, he has been selected in who is who in Asia 2007 and BIC 100 educators in Africa 2008.

Maria-Anna S. Fengou was born in Athens, Greece, in 1986. She obtained her diploma in Electrical and Computer Engineering from the University of Patras, Greece, in July 2008. Since October 2008, she is a Ph.D student at the Wire Communications Laboratory in the Department of Electrical and Computer Engineering of the University of Patras, Greece. Her main research fields are on telemedicine systems and applications, medical communication networks, telemedicine platforms, next generation networks (NGNs), web services, context awareness and biosignal processing. Until now she has participated in a European research project on ICT technology. She is a member of the Technical Chamber of Greece.

Petros Giastas received his PhD in Chemistry from National and Kapodistrian University of Athens, in collaboration with the Institute of Physical Chemistry, NCSR Demokritos, Athens, Greece, in 2007. He is currently a postdoctoral researcher at the Department of Biochemistry of Hellenic Pasteur Institute. His research interests span the whole workflow of structural biology, protein expression and purification, crystallization, and structure determination via X-ray crystallographic methods.

Professor Michael A. Gatzoulis is the academic head of the Adult Congenital Heart Centre and the Centre for Pulmonary Hypertension at Royal Brompton Hospital and a professor of cardiology, congenital heart disease at the National Heart & Lung Institute, Imperial College, London. His research interests include congenital heart disease, pulmonary arterial hypertension and heart disease and pregnancy. He has served as associate editor, guest editor or reviewer for most major journals in the cardiovascular field. He is also the author of over 150 peer-reviewed publications (including papers in Nature, New England Journal of Medicine, The Lancet and Circulation).

Roberto Hornero was born in Plasencia, Spain, in 1972. He received the degree in telecommunication engineering and Ph.D. degree from the University of Valladolid, Spain, in 1995 and 1998, respectively. He is currently "Profesor Titular" in the Department of Signal Theory and Communications at the University of Valladolid. His main research interest is nonlinear analysis of biomedical signals to help physicians in the clinical diagnosis. He founded the Biomedical Engineering Group in 2004. The research interests of this group are connected with the field of nonlinear dynamics, chaotic theory, and wavelet transform with applications in biomedical signal and image processing.

Mohammad Haghpanahi received the B.S. and M.Sc. degree in mechanical engineering from the Shiraz University, the M.Sc. degree in solid mechanics from the ENSAM, Paris, and the Ph.D. in biomechanics from the ENSAM, Paris. He is now an Associate Professor in the Department of Mechanical Engineering in Iran University of Science and Technology. His interests are in biomechanics of orthopedics, stress analysis of systems by finite element analysis, biomechanics of tissue engineering, extraction of spine biomechanics, sport biomechanics and fatigue analysis of welded joints.

Stamatia Ilioudi graduated from the Department of Informatics, University of Piraeus, Greece and she received recently her MSc in Information Systems from the Athens University of Economics and Business. Currently she is a PhD Student at the Department of Nursing of the University of Peloponnese in Greece. Her research interests include electronic communication systems, healthcare information systems, computer applications in health organizations and other areas.

George E. Karagiannis BSc, MSc, MCMI currently works as European R&D Projects Manager at the Royal Brompton & Harefield NHS Trust, London, UK. He has significant experience in managing and conducting research in the framework of Health and ICT related projects of a multinational and multidisciplinary nature. His project portfolio includes the following EC co-funded studies: EVINCI, eurIPFnet, iWebCare, Clinicip, PharmacoV, pEHR, Panaceia-iTV, eOpenDay and Iremma. His main research interests lie in the development, implementation and evaluation of health information systems, measuring the return on investment in health information technologies, market deployment of R&D products and entrepreneurship in healthcare.

Dimitrios C. Karaiskos (dimkar@aueb.gr) is a PhD Student at the Department of Management Science and Technology at the Athens University of Economics and Business, Greece. His main research interests lie in the areas of pervasive and ubiquitous computing, software engineering, information systems design and evaluation. He is also a researcher in the ISTLAB Wireless Research Center (http://istlab.dmst.aueb.gr/wrc/).

Anastasia N. Kastania was born in Athens, Greece. She received her B.Sc. in Mathematics and her Doctor of Philosophy Degree in Medical Informatics from the National Kapodistrian University of Athens. Research productivity is summarized in various articles (monographs or in collaboration with other researchers) in International Journals, International Conference Proceedings, International Book Series and International Book Chapters. She works at the Athens University of Economics and Business since 1987 and currently, she is Visiting Assistant Professor (PD 407/80) of Applied Informatics in the Department of Accounting and Finance, Athens University of Economics and Business. She has participated in many Research Projects in Greece and in European Union. She is the writer of many didactic books in the areas of Informatics and Statistics. She is also a Visiting Researcher at the Bioinformatics and Medical Informatics Team, Centre of Basic Research, Biotechnology Division, Foundation for Biomedical Research of the Academy of Athens.

D. P. Klemer is an Associate Professor in the College of Engineering and Applied Science at the University of Wisconsin-Milwaukee. In 1982 he received his Ph.D. degree in Electrical Engineering from the University of Michigan, Ann Arbor, and subsequently received an M.D. degree in 1999 from the College of Physicians and Surgeons at Columbia University in New York City. He completed his

Residency in Internal Medicine at NYU/Bellevue Hospital Center in New York City, and is a Diplomate of the American Board of Internal Medicine. He currently conducts research in microelectronic solid-state biosensors at the University of Wisconsin-Milwaukee; affiliations also include the Medical College of Wisconsin and the Clement J. Zablocki Veterans Affairs Medical Center in Milwaukee, Wisconsin.

Sophia Kossida received her BSc degree in Biology from the University of Crete. She was then awarded a DPhil in 1998 from Oxford University. She carried out a post-doc at Harvard University, USA at the Molecular & Cellular Biology Department. She was employed as Senior Scientist within the Target Discovery Group of Lion Bioscience Research Inc.(LBRI) in Cambridge, MA, US. She moved over to Toulouse, in France where she was appointed Director of Bioinformatics of Endocube. In parallel, she was appointed Associate Professor of Bioinformatics at the University of Paul Sebatier in Toulouse, France. She joined Novartis in Switzerland in 2002 as Lab head within the Functional Genomics Group. She joined Biomedical Research Foundation in July 2004 as tenure track research Bioinformatician. Sophia lectures at the University of Helsinki and she has joint projects with Dr Baumann's group (Biomedicum in Helsinki).

Dimosthenis Kyriazis received the diploma from the Dept. of Electrical and Computer Engineering of the National Technical University of Athens, Athens, Greece in 2001, the MS degree in Techno-Economic Systems (MBA) co-organized by the Electrical and Computer Engineering Dept - NTUA, Economic Sciences Dept - National Kapodistrian University of Athens, Industrial Management Dept - University of Piraeus and his Ph.D. from the Electrical and Computer Engineering Department of the National Technical University of Athens in 2007. He is currently a Researcher in the Telecommunication Laboratory of the Institute of Communication and Computer Systems (ICCS). Before joining the ICCS he has worked in the private sector as Telecom Software Engineer. He has participated in numerous EU / National funded projects (such as IRMOS, NextGRID, Akogrimo, BEinGRID, HPC-Europa, GRIA, Memphis, CHALLENGERS, HellasGRID, etc). His research interests include Grid computing, scheduling, Quality of Service provision and workflow management in heterogeneous systems and service oriented architectures.

Panos A. Ligomenides is Professor Emeritus of the University of Maryland, USA, and a Life (Regular) Member of the Academy of Athens, since 1993. He has been elected to be the President of the Academy of Athens for 2009. He is the President of the Center for the Greek Language (Ministry of Education), and of the Hellenic Language Heritage. Professor Ligomenides has worked in the USA as researcher of IBM Corporation, as Professor of the Universities Stanford, UCLA, Maryland, and of the Polytechnic University of Madrid, Spain. He has been the Vice-President (Research) of Caelum Research Corporation, Technical Consultant of many Corporations, and of Public Organizations. He has been credited with many honorary distinctions, like Distinguished Professor of Electrical Engineering, Outstanding Educator of America, and Ford Foundation Fellow. Professor Ligomenides has a remarkable research and an extended published scientific record.

Maria Isabel López was born in Valladolid, Spain, in 1960. She received his degree in Medicine and Ph.D. from the University of Valladolid in 1985 and 1991, respectively. She is currently "Associate Profesor" in Ophthalmology in the Department of Surgery of the Medicine Faculty at the University of Valladolid and also works in the University Hospital of this town as a clinician. Her main research interest is teleophtalmology from a clinical point of view and ocular diabetes. She is the director of the ocular diabetes unit at the "Instituto de Oftalmobiología Aplicada" in the University of Valladolid. She is a member of the American Academy of ophthalmology and the Spanish Society of Retina and Vitreous.

Miguel López is a telecommunications professor in the University of Valladolid (Spain). He was born in Barcelona, Spain; in 1950. He has a PhD in Telecommunications Engineering from the Polytechnic University of Madrid, in 1982. Since 1991 he has been devoted to the promotion of Information Society in Castille and Leon region from several positions: Director of the Technical School of Telecommunications, R&D General Manager of a Telecommunications Technological Centre and also CEO of a cable telecommunications operator. Now, his research interests are biomedical signal, Telemedicine, Information Society, and to contribute to the promotion of the entrepreneurial character of University.

Dimitrios K. Lymberopoulos was born in Tripolis, Greece, in 1956. He received the Electrical Engineering diploma and the Ph.D. degree from the University of Patras, Patras, Greece, in 1980 and 1988, respectively. He is currently a Professor in the Department of Electrical and Computer Engineering, University of Patras, where he lectures on communication systems, multimedia communications, and telemedicine services. Since 1982, he has been involved as a Technical Supervisor in various research projects funded by the Greek Government, the European Union, the Greek Telecommunication Organization, and the major Greek Telecommunication industries. He has authored or co-authored over 130 papers in international journals, conferences and technical reports. His research interests include medical communication protocols, telemedicine, context awareness, ontologies in the medical information domain, next generation networks, web multimedia services, data management in medical applications, teleworking (telemedicine) development platforms, and medical communication networks. Prof. Lymberopoulos is a member of the Technical Chamber of Greece and the Greek Society of Electrical and Mechanical Engineers.

Nikos E. Mastorakis received his B.Sc. and M.Sc. (Diploma) in Electrical Engineering from the National Technical University of Athens and the Ph.D. in Electrical Engineering and Computer Science from the same university. He also received the B.Sc. (Ptychion) in Pure Mathematics from the National University of Athens. He have served as special scientist on Computers and Electronics in the Hellenic (Greek) Army General Staff (1993-1994) and taught several courses in the Electrical and Computer Engineering Department of the National Technical University of Athens (1998-1994). He has also served as Visiting Professor at the University of Exeter, School of Engineering (UK, 1998), Visiting Professor in the Technical University of Sofia (Bulgaria, 2003-2004) while he is Full Professor and Head of the Department of Computer Science at the Military Institutions of University Education (MIUE) -Hellenic Naval Academy, Greece since 1994.

Andreas Menychtas graduated from the School of Electrical and Computer Engineering, National Technical University of Athens in 2004. In 2009, he received his Ph.D. in area of Distributed Computing from the School of Electrical and Computer Engineering of the National Technical University of Athens. Currently, he works as researcher in the Institute of Communication and Computer Systems of National Technical University of Athens. He has been involved in several EU and National funded projects such as GRIA, NextGRID, EGEE, IRMOS, HellasGRID and GRID-APP. His research interests include Distributed Systems, Web Services, Service Oriented Architectures, Security and Information Engineering.

Arcangelo Merla holds a Ph.D. in advanced biomedical technologies and bioimaging from the University of Chieti (Italy) and a M.S. (laurea) in physics from the University of Bologna (Italy). His expertise is in the area of biomedical imaging and modeling, with special reference to biomedical and clinical applications of thermal and infrared imaging. Dr. Merla published extensively in these areas in clinical and biomedical journals and refereed conference proceedings over the past years. Dr. Merla had joined the faculty of the Clinical Sciences and Bioimaging Department - School of Medicine, at the University of Chieti in August 2002. He is the Director of the functional Infrared Imaging Lab at the Institute of Advanced Biomedical Technologies at the Foundation University G.'Annunzio of Chieti-Pescara, and Visiting Assistant Professor at the Computational Physiology Lab, at the University of Houston.

Mohammad Nikkhoo received the B.S. degree in mechanical engineering from the IAUCTB, the M.Sc. degree in biomechanical engineering from the Iran University of Science and technology and is currently a Ph.D. student at the Mechanical Engineering Department, Iran University of Science and Technology. He is now a Research Assistant in the Biomechanics Laboratory in IUST. His interests are in biomechanics of tissue engineering, biomechanics of soft tissues, biomechanics of orthopedics and intelligent rehabilitation systems.

Theodor C Panagiotakopoulos was born in Patras in 1981. He took his diploma in Electrical and Computer Engineering from the University of Patras, Greece in 2006. Since 2007 he is a Ph.D researcher in the field of bioinformatics and bioengineering in the Department of Electrical and Computer Engineering of the University of Patras. His fields of interest comprise at amongst others: context-awareness, user modeling, emotion recognition, telemedicine systems and applications and biosignals processing. Until now, he has published more than 6 articles in conferences and book chapters, he has participated in 2 European research projects on ICT technologies and he is a member of the Technical Chamber of Greece.

Habib Allah Peirovi received the M.D. degree in Shaheed Beheshti University of Medical Sciences, Specialty in Surgery, Fellowship, and Sub-specialty on Transplantation and Vascular Surgery from Faculty of Medicine, Shaheed Beheshti University of Medical Sciences. He is now Professor of Surgery in the Faculty of Medicine, Shaheed Beheshti University of Medical Sciences. His interests are in nanomedicine, tissue engineering, bio-artificial Liver, differentiation of cells, transplantation and vascular surgery.

Dimitri Perrin received his M.Sc. in Computing (2005) from Institut Supérieur d'Informatique de Modélisation et de leurs Applications (ISIMA, Clermont-Ferrand, France), his M.Sc. in Models, Systems, Intelligence (2005) from Université Blaise Pascal (Clermont-Ferrand, France), and his Ph.D. in Computing (2008) from Dublin City University. His research interest and expertise include Complex Systems modelling, (with particular focus on infection progression models and epigenetic mechanisms), Computational Biology and Bioinformatics, Parallel Computing and applications, and novel techniques for large dataset analysis, (e.g. DNA microarray, SNP).

Heather J. Ruskin received her B.Sc. degree in Physics and M.Sc. in Medical Statistics from London University (Kings/London School of Hygiene and Tropical Medicine) and her Ph.D. in Statistical and Computational Physics from Trinity College Dublin. She is currently a Professor in the School of Computing and Associate Dean of Research in Engineering and Computing in Dublin City University. Her research interests include Computational Models for Complex Systems; spatiotemporal processes and many-body problems in biosystems (biomimetics) and in socioeconomic systems (traffic and finance).

Gerald Schaefer gained his BSc. in Computing from the University of Derby and his PhD in Computer Vision from the University of East Anglia. He previously worked at the Colour & Imaging Institute, University of Derby (1997-1999), the School of Information Systems, University of East Anglia (2000-2001), the School of Computing and Informatics, Nottingham Trent University (2001-2006) and the School of Engineering and Applied Science at Aston University (2006-2009). In May 2009 he joined the Department of Computer Science at Loughborough University. His research interests include colour image analysis, physics-based vision, image retrieval, and medical imaging. He is the author of more than 150 publications in these areas.

Konstantinos Siassiakos holds a diploma (1995) of Electrical and Computer Engineer from the Department of Electrical and Computer Engineering Studies, University of Patras, Greece, and a Ph.D. (2001) diploma from the Department of Electrical and Computer Engineering, National Technical University of Athens, Greece. Dr. K. Siassiakos currently works as visiting lecturer at the Technological Educational Institute of Chalkida, at the University of Piraeus in Greece and Hellenic Naval Academy. Also He currently works as Special Scientific Staff at the Civil Service Staffing Council (ASEP). He has worked as an IT consultant at Ministry of Development (General Secretariat of Industry), Ministry of National Education and Religious Affairs (General Secretariat for Adult Education), Ministry of Health and Social Solidarity and as a researcher at the Department of Technology Education & Digital Systems, University of Piraeus. His research interests include logistics management, electronic customer relationship management systems, applied cryptography, management information systems, web-based learning systems, educational technologies, human computer interaction, quality assurance, business process reengineering, and e-government technologies.

K. Sivakumar [M.Sc., M.Phil., Annamalai University; Ph.D, Sri Chandrasekharendra Saraswathi Viswa Mahavidyalaya University, India]. Currently working as a Lecturer in Chemistry at Sri Chandrasekharendra Saraswathi Viswa Maha Vidyalaya University in the Department of Science and Humanities, Tamilnadu, India. His research areas are computational analysis and characterization of various human proteins using In silico and computational methods. He has published many research papers, book chapters and articles in International and National journals, books and magazines. He is a recipient of Summer Research Fellowship for the year 2008 offered jointly by the Indian Academy of Sciences, Indian National Science Academy and The National Academy of Sciences, India.

George M. Spyrou received his BSc on Physics from National and Kapodistrian University of Athens, Greece. He holds Masters of Science on Medical Physics and on Bioinformatics as well. During his PhD on Medical Physics he worked on algorithms and simulations applied on medical issues, especially on breast cancer imaging. Currently, he is working as a Senior Research Scientist (Professor Level) in the Biomedical Informatics Unit of the Biomedical Research Foundation of the Academy of Athens

(BRFAA) leading the Modeling and Computational Intelligence in Biomedical Informatics (MCIBI) Group. Also, he has been assigned as the Head of the Department of Informatics and New Technologies in BRFAA. The main aim of Spyrou is the application of his knowledge and skills in Physics, Mathematics and Informatics to medically and biologically relevant issues, mainly on the 'mining' of the information either in direct medical data (e.g. medical images) or in genomes and proteomes relevant to medical problems.

Vasileios G. Stamatopoulos graduated from National Technical University of Athens, receiving a degree in Electrical and Computer Engineering. Flowingly he was awarded an MSc and a PhD in Bioengineering from Strathclyde University in Glasgow. He has worked at the same university as a Research Assistant. He was then employed by the Royal Brompton & Harefield NHS Trust as a Project Manager.. For the past year he is a Visiting Lecturer at the University of Central Greece, Department of Biomedical Informatics. He is currently employed by the "Praxi help-forward Network" as a Technology Transfer Consultant while he is academically affiliated with the Biomedical Research Foundation of the Academy of Athens (BRFAA).

David Steinman, a pioneer of image-based CFD modelling, has spent more than a decade working to integrate the fields of computer modelling and medical imaging. He is a Professor of Mechanical and Biomedical Engineering at the University of Toronto, where he heads the Biomedical Simulation Laboratory. Dr. Steinman has authored more than 130 peer-reviewed journal articles and conference papers, and serves on the Editorial Board of the ASME Journal of Biomechanical Engineering and the Advisory Board of the journal Physiological Measurement. He currently holds a Career Investigator Award from the Heart & Stroke Foundation of Canada.

Dolores Steinman trained as a paediatrician, before finishing doctoral and postdoctoral research in cancer cell biology. An accomplished photographer, she is interested in the meaning attached to the various visual medical/ scientific representations. While staying involved in the research work she is also passionate about the relationship/connection between humanities and science. Dr. Steinman is currently a Guest Researcher in the Department of Mechanical Engineering at the University of Toronto.

Trias Thireou received her PhD in Biomedical Engineering from National Technical University of Athens, Greece, in collaboration with the German Cancer Research Center, in 2002. She was a postdoctoral fellow in Bioinformatics at the Foundation for Research and Technology - Hellas (FORTH) until 2006. She is currently an adjunct lecturer on Bioinformatics at the University of Thessaly and at National Technical University of Athens, and a postdoctoral researcher at the Genetics Laboratory of Agricultural University of Athens. Her research interests include structural bioinformatics, molecular simulations, data mining, machine learning and biomedical image analysis.

Isabel de la Torre was born in Zamora, Spain, in 1979. She received the Engineer degree in telecommunications engineering from the University of Valladolid, Valladolid, Spain, in 2003. Currently, she is an assistant professor in the Department of Signal Theory and Communications at the University of Valladolid, where she is working towards the Ph.D. degree. Her research has been mainly focused in development of telemedicine applications and EHR (Electronic Health Record) standards in ophthalmology.

Konstantinos Tserpes is a Senior Research Engineer in the Distributed, Knowledge and Media Systems Lab (DKMS) of the Institute of Communication and Computer Systems (ICCS) and the National Technical University of Athens (NTUA). He graduated from the Computer Engineering and Informatics department, University of Patras, Greece in 2003. In 2005 he received his master's degree in Information Systems Management from National Technical University of Athens (NTUA). In 2008, he acquired his PhD in the area of Service Oriented Architectures with a focus on quality aspects from the school of Electrical and Computer Engineers of ICCS-NTUA. He has been involved in several EU and National funded projects (e.g. NextGRID, AkoGRIMO, HPC-Europa, EchoGRID, CHALLENGERS, HellasGRID, USNES) and his research interests are revolving around Service Oriented Computing and its application and business extensions.

Theodora A. Varvarigou received the B. Tech degree from the National Technical University of Athens, Athens, Greece in 1988, the MS degrees in Electrical Engineering (1989) and in Computer Science (1991) from Stanford University, Stanford, California in 1989 and the Ph.D. degree from Stanford University as well in 1991. She worked at AT&T Bell Labs, Holmdel, New Jersey between 1991 and 1995. Between 1995 and 1997 she worked as an Assistant Professor at the Technical University of Crete, Chania, Greece. Since 1997 she was elected as an Assistant Professor while since 2007 she is a Professor at the National Technical University of Athens, and Director of the Postgraduate Course "Engineering Economics Systems". Prof. Varvarigou has great experience in the area of semantic web technologies, scheduling over distributed platforms, embedded systems and grid computing. In this area, she has published more than 150 papers in leading journals and conferences. She has participated and coordinated several EU funded projects such as IRMOS, SCOVIS, POLYMNIA, Akogrimo, NextGRID, BEinGRID, Memphis, MKBEEM, MARIDES, CHALLENGERS, FIDIS, and other.

Stelios Zimeras was born in Piraeus, Greece. He received his B.Sc. in Statistics from University of Piraeus, Depart. of Statistics and Insurance Sciences, and his her Doctor of Philosophy Degree in Statistics from University of Leeds, U.K. Research productivity is summarized in various articles (monographs or in collaboration with other researchers) in International Journals, International Conference Proceedings, International Book Series and International Book Chapters. Since 2008 he is Assistant Professor at the University of the Aegean, Depart. of Statistics and Actuarial – Financial Mathematics. He has participated in many Research Projects in Greece and in European Union. He is the writer of many papers in the areas of Informatics and Statistics.

Index

Symbols

2D tissue 81
3D Bioplotter 82
3D lattice 58
3D reconstruction 78
3D structures 43, 45
3D tissues 81
3D virtual model 79
(TwiV) algorithm 20
β-lactoglobulin 46, 47, 48, 49, 50
β-tricalcium phosphate (β-TCP) 77

A

ab initio 4, 15, 17, 35, 58
Ab initio methods 4
ab intio 4
ABNR method 45
action manager 274
action reasoner 274
activation-dependent 63
Adopted Basis Newton-Raphson (ABNR) 45
agent-based approach 58, 59
agent-based design 59
agent-based models 58
agent-based paradigm 55, 56, 58, 59, 69
algorithmic development 129
aliphatic index 143, 147, 148, 150, 151, 155
Allochromatium vinosum 24, 25, 35
Alzheimer Disease 123
ambient intelligence 255, 279
American National Standards Institute (ANSI) 90
AmI 255
amino acid 38, 40, 41, 46, 49, 54

Analyte 185
aneurysm 228, 229, 239
angiography 229, 231, 233, 235, 239, 240
antibody 38, 39, 51, 52, 53
antibody-antigen bindings 57
anticalins 38, 41, 42, 53
antigen 39, 42
application architecture 92
arterio-venous fistulae (AVF) 191
artificial intelligence 131
artificial neural networks (ANN) 205
artificial vascular system present 81
asymmetric unit 5, 11, 15, 17, 25, 29, 34
atherosclerosis 228, 229, 234, 239
atomic coordinate entries 7
atomicity 4
auto-build options 20
availability 286

B

backbone plasticity 42
BAN 264, 266
Bayesian inference 19
Bayesian methods 27
beta-barrel architecture 42
Beta-Lactoglobulin (Blg) 46
bilin-binidng protein (BBP) 42, 46
binding modes 45, 47
bioanalytical purposes 42
biochemical functions 39
biochemical techniques 46
Biochimica Biophysica Acta 41
biocompatible 76, 77, 83
bioinformatics 128, 129, 130, 131, 134, 139, 140, 141